Lecture Notes in Control and Information Sciences

Edited by M. Thoma and A. Wyner

For information about Vols. 1–61 please contact your bookseller or Springer-Verlag.

Lecture Notes in Control and Information Sciences

Edited by M. Thoma and A. Wyner

125

J. Simon (Editor)

Control of Boundaries and Stabilization

Proceedings of the IFIP WG 7.2 Conference
Clermont Ferrand, France, June 20–23, 1988

Springer-Verlag
Berlin Heidelberg New York
London Paris Tokyo Hong Kong

Editor of Conference Proceedings of the series:
Computational Techniques in Distributed Systems IFIP-WG 7.2
Irena Lasieka
Dept. of Applied Mathematics
Thornton Hall
University of Virginia
Charlottesville, VA 22903
USA

Editor
Jacques Simon
Université Blaise Pascal
epartement de Mathématiques Appliquées
63177 Aubière Cedex
France

ISBN 978-3-540-51239-4 Springer-Verlag Berlin Heidelberg New York
ISBN 978-0-387-51239-6 Springer-Verlag New York Berlin Heidelberg

Offsetprinting: Mercedes-Druck, Berlin

2161/3020-543210

PREFACE

This book contains the text of lectures which have been presented during the conference *Control of Boundaries and Stabilization* held at Clermont-Ferrand, France, June 20-23, 1988.

The Conference was devoted to two subjects
- **Stabilization** with emphasis on exact controllability : considering a physical system, such as a vibrating plate, can one reach a steady state in a finite time by acting on the boundary.
- **Control of boundaries** : given a physical system find the geometry of the domain (optimal shape) which minimizes a cost related to the solution of a boundary value problem in this domain, for example find a minimum drag profile. Many lectures included mathematical analysis as well as engineering applications and numerical simulation.

Participants included three main speakers, J.L. Lions, J. Rauch and D.L. Russell, 17 invited lecturers, and 60 other mathematicians, from 12 different countries.

I wish to express my acknowledgement to the lecturers for their high level contributions, to the sessions chairpersons who directed very interesting discussions, and to all the participants who enliven the discussions and who contributed to create a friendly atmosphere which reinforced the scientific interest.

The Conference was held under the auspices both of
- *Année spéciale sur les phénomènes non linéaires* - supported by *CNRS* and *MRES* with the assistance of *SMAI*, *SMF*, *DRET* and *CEA*
- *IFIP* (TC7-WG7.2).

The organizing institutions were
Département de Mathématiques Appliquées de l'Université Blaise Pascal (Clermont 2)
Laboratoire d'Analyse Numérique de l'Université Pierre et Marie Curie (Paris 6).

Let the following organizations find here the thanks of the organizers for financial support :
Direction des Recherches Etudes et Techniques du Ministère des Armées (DRET)
Centre National de la Recherche Scientifique (CNRS)
Ministère de la Recherche et de l'Enseignement Supérieur (MRES)
Centre National d'Etudes Spatiales (CNES)
Conseil Régional d'Auvergne
Université Blaise-Pascal (Clermont 2) et son *UFR Recherches Scientifiques et Techniques*
Société de Mathématiques Appliquées et Industrielles (SMAI)
Ville de Clermont-Ferrand
International Federation for Information Processing (IFIP)
Commissariat à l'Energie Atomique (CEA)
Société Mathématique de France (SMF).

A special gratitude is expressed, for giving us the run of it's magnificent premises, to the director of the

Ecole Nationale des Impôts.

Sincere gratitude is also expressed to F.X. Le Dimet who co-organized the conference, and to Misses D. Courageot, I. Fontaine-Gilmour and M.F. Rebord and Mr. M. Ouberdous who contributed the organization.

Jacques SIMON

PROGRAM AND ORGANIZATION COMMITTEES

ANNEE SPECIALE SUR LES PHENOMENES NON LINEAIRES

Organized under the auspices of CNRS and MRES

Honorary Committee

J. Leray (Président), R. Dautray, P. Germain, J.L. Lions

Scientific Committee

J. Ball, H. Brézis, F. Brezzi, C. Cercignani, P. Clavin, C. Dafermos
A. Douady, G. Geymonat, J.P. Guiraud, P. Lallemand, I.D. Landau
F. Mignot, J. Ockendon, G. Papanicolaou, J. Rauch, P.A. Raviart
Y. Yamaguti

Organizing committee

C. Bardos, O. Pironneau, J.P. Puel

INTERNATIONAL FEDERATION FOR INFORMATION PROCESSING

International Program Committee (Technical Committee 7, Working Group 7.2)

A. Bermudez, A. Butkowski, R. Curtain, G. Da Prato, R. Glowinski
K. Hoffman, W. Krabs, I. Lasiecka, J.L. Lions, U. Mosco
O. Pironneau, J.P. Yvon, J.P. Zolesio

CONTROL OF BOUNDARIES AND STABILIZATION

International Conference at Clermont-Ferrand, June 20-23, 1988

Organizing Committee

F.X. Le Dimet, J. Simon

LIST OF CONTRIBUTORS

TABLE OF CONTENTS

Part 1. PLENARY LECTURES

Part 2. INVITED LECTURES

PART 1

PLENARY LECTURES

REMARKS ON THE EXACT CONTROLLABILITY

FOR LARGE TIMES

Jacques Louis LIONS
Collège de France and C.N.E.S.

1. Setting of the problem.

Let Ω be a bounded open set of $\mathbb{R}^n$, with smooth boundary Γ. We consider in Ω the *wave equation*

(1.1) $$\frac{\partial^2 y}{\partial t^2} - \Delta y = 0 \quad t > 0 \quad ,$$

subject to the *initial conditions*

(1.2) $$y(0) = y^0 \quad , \quad \frac{\partial y}{\partial t}(0) = y^1 \quad \text{in } \Omega$$

(where we write $y(0)$ for $x \to y(x, 0)$), and subject to the *boundary condition*

(1.3) $$y = v \quad \text{on } \Sigma = \Gamma \times (0,T).$$

In (1.3) v is at our disposal in $L^2(\Sigma)$; it is the *control function* .

The (classical) problem of *Exact Controllability* (E.C.) is as follows (cf. D.L. RUSSELL [1]) : given $T > 0$, one says that one has E.C. (at T) if for any couple $\{y^0 , y^1\}$ in a suitable function space one can find v such that the solution y of (1.1) (1.2) (1.3) satisfies

(1.4) $$y(T) = \frac{\partial y}{\partial t}(T) = 0 \ .$$

In short : one can drive the system to rest at time T .

It follows immediately from the finite speed of propagation for the solutions of (1.1) that E.C. may be possible ***only if*** T ***is large enough.***

If T is large enough, one can prove that for any couple such that

(1.5) $$y^0 \in L^2(\Omega) \ , \ y^1 \in H^{-1}(\Omega) = \text{dual of } H_0^1(\Omega)$$

where $H_0^1(\Omega) = \{\varphi \mid \varphi \ , \nabla\varphi \in L^2(\Omega) \ , \varphi = 0 \text{ on } \Gamma\}$, there exists v such that one has (1.4).

In fact there exist an ***infinite number*** of such v's .

In the (affine) space spanned by all v's giving (1.4) ***one can choose*** a particular element, a natural candidate (but not the only one ! And indeed we shall see later on that it may be useful to make other choices) is to choose ***the unique element*** v ***such that***

(1.6) $$\| v \|_{L^2(\Sigma)} = \min.$$

This element $v = v_T$ ***actually depends on*** T (provided T is large enough). The goal of the present paper is ***to study the behaviour of*** v_T ***as*** $T \to +\infty$.

Let us make first more precise the construction of v_T .

2. Construction of v_T .

We extend a little the situation of Section 1 . We consider ***a weight function*** q

(2.1) $$q \in L^\infty(T) \ , \ q \geq q_0 > 0 \quad \text{on } \Gamma$$

and we define

(2.2) $$\| v \|_{L^2(\Sigma \, ; \, q)} = \left(\int_\Sigma q\, v^2 \, d\Gamma \, dt \right)^{1/2} .$$

We have in this way a new norm on $L^2(\Sigma)$, which is equivalent to the usual one $(q = 1)$.

We want to construct $v_{T,q} = v_T$ such that

(i) one has (1.4)

(ii) $\| v_T \|_{L^2(\Sigma;q)} = \min$. ■

We apply the method H.U.M. (Hilbert Uniqueness Method) introduced in J.L. LIONS [1] and studied in [2] [3].

Let $\{\varphi^0, \varphi^1\}$ be given in $H_0^1(\Omega) \in L^2(\Omega)$. Let φ be the solution of

(2.3) $$\varphi'' - \Delta\varphi = 0 \quad \text{in } \Omega \times (0,T)$$

(where we write φ', φ'' for $\frac{\partial \varphi}{\partial t}, \frac{\partial^2 \varphi}{\partial t^2}$), subject to

(2.4) $$\varphi(0) = \varphi^0 \; , \; \varphi'(0) = \varphi' \; ,$$

$$\varphi = 0 \quad \text{on } \Sigma \; .$$

Let ψ be the solution of

(2.5) $$\psi'' - \Delta\psi = 0 \quad \text{in } \Omega \times (0,T)$$

subject to

(2.6) $$\psi(T) = \psi'(T) = 0$$

(2.7) $$\psi = q^{-1} \; \partial\varphi/\partial\nu \quad \text{on } \Sigma \; .$$

We then define

(2.8) $$\Lambda \{\varphi^0, \varphi^1\} = \{\psi'(0), -\psi(0)\} .$$

One verifies that

(2.9) $$< \Lambda \{\varphi^0, \varphi^1\}, \{\varphi^0, \varphi^1\} > = \int_\Sigma q^{-1} \left(\frac{\partial\varphi}{\partial\nu}\right)^2 d\Gamma \, dt \; .$$

[Indeed multiply (2.5) by φ. Integrations by parts lead to (2.9)].

One has always (i.e. for an arbitrary $T > 0$)

(2.10) $$\int_\Sigma q^{-1} \left(\frac{\partial \varphi}{\partial \nu}\right)^2 d\Gamma\, dt \leqslant c_1 \| \{\varphi^0 , \varphi^1\} \|^2_{H_0^1(\Omega) \times L^2(\Omega)}$$

(cf. J.L. LIONS [4] and [3]). In (2.10) c_1 ***dépends on*** T - But we return to that below - One has ***the reverse inequality*** due to L.F. HO [1] : there exists c_2 (also depending on T) such that

(2.11) $$\int_\Sigma q^{-1} \left(\frac{\partial \varphi}{\partial \nu}\right)^2 d\Gamma\, dt \geqslant c_2 \| \{\varphi^0 , \varphi^1\} \|^2_{H_0^1(\Omega) \times L^2(\Omega)} \cdot$$

Inequalities (2.10) (2.11), and equality (2.9), imply that Λ is ***an isomorphism*** from $H_0^1(\Omega) \times L^2(\Omega)$. Therefore the equation

(2.12) $$\Lambda \{\varphi^0 , \varphi^1\} = \{y^1 , -y^0\}$$

admits a unique solution. Of course Λ ***depends on*** Γ so that ***the solution*** of (2.12) ***depends on*** Γ . We shall denote by

(2.13) $$\{\varphi_T^0 , \varphi_T^1\} = \text{solution of } (2.12).$$

We then compute

(2.14) $$\varphi = \varphi_T = \text{solution of } (2.3)\ (2.4) \text{ with } \varphi^0 = \varphi_T^0 \ , \ \varphi^1 = \varphi_T^1 \ .$$

Looking at (2.5) (2.6) (2.7) (2.8) (2.12) we see that

(2.15) $$v_* = q^{-1} \frac{\partial \varphi}{\partial \nu} \quad (\varphi = \varphi_T)$$

gives E.C.

Let us verify now that

(2.16) $$\| v_* \|_{L^2(\Sigma ; q)} = \min .$$

If we set

(2.17) $z = y - \psi$

we have

$$z'' - \Delta z = 0 \ ,$$

(2.18) $$z(0) = z'(0) = z(T) = z'(T) = 0 \ ,$$

$$z = v - q^{-1}\,\partial\varphi / \partial\nu \quad \text{on } \Sigma \ (\varphi = \varphi_T)\ .$$

If we multiply the first equation in (2.18) by $\varphi = \varphi_T$, we obtain, after integrations by parts,

(2.19) $$\int_\Sigma (v - q^{-1} \frac{\partial\varphi}{\partial\nu}) \frac{\partial\varphi}{\partial\nu}\, d\Gamma\, dt = 0$$

or, equivalently

(2.20) $$\int_\Sigma q\, v_\bullet^2\, d\Gamma\, dt = \int_\Sigma \sqrt{q}\, v\, \sqrt{q}\, v_\bullet\, d\Gamma\, dt$$

hence

(2.21) $$\| v_\bullet \|_{L^2(\Sigma\,;\,q)} \leqslant \| v \|_{L^2(\Sigma\,;\,q)}$$

and (2.16) follows. ∎

Remark 2.1.

The above proof of (2.16) is more a verification than a proof ! In fact it is by solving directly (2.16) that one is lead to (2.3)... (2.7) which is ***the optimality system*** of (2.16). For the case $q = 1$ cf. J.L. LIONS [3], Chapter VIII . ∎

Remark 2.2.

Solving (2.12) is equivalent to finding

$$(2.22)\qquad \inf_{\varphi^0,\varphi^1} \frac{1}{2} < \Lambda\{\varphi^0,\varphi^1\},\{\varphi^0,\varphi^1\}> - \int_\Omega (y^1 y^0 - y^0 \varphi^1)\,dx$$

i.e. , using (2.9) :

φ_T^0, φ_T^1 *is the element* $H_0^1(\Omega) \times L^2(\Omega)$ *which minimizes*

$$(2.23)\qquad \frac{1}{2}\int_{\Gamma\times(0,T)} q^{-1}\left(\frac{\partial\varphi}{\partial\nu}\right)^2 d\Gamma\,dt - \int_\Omega (y^1 y^0 - y^0 \varphi^1)\,dx\ .$$

Remark 2.4.

It is now this formulation (2.23) that we use for letting $T \to +\infty$.

Before doing in the next section ***a renormalization*** , we observe that the minimization of (2.23) is the ***dual problem*** of the minimization of $\|v\|_{L^2(\Sigma\,;\,q)}$ among all v's giving (1.4). ■

3. A renormalization

We introduce

$$(3.1)\qquad \begin{aligned} &T\varphi_T^0 = \varrho_T^0\ ,\quad T\varphi_T^1 = \varrho_T^1 \\ &T\varphi_T = \varrho_T \end{aligned}$$

and we observe, after multiplying (2.23) by T , that we want to minimize

$$\frac{T}{2}\int_{\Gamma\times(0,T)} q^{-1}\left(\frac{\partial\varphi}{\partial\nu}\right)^2 d\Gamma\,dt - \int_\Omega (y^1\, T\varphi^0 - y^0\, T\varphi^1)\,dx\ .$$

If we set $T\varphi = \varrho$, $T\varphi^0 = \varrho^0$, $T\varphi^1 = \varrho^1$

the problem is equivalent to minimizing

$$(3.2)\qquad \frac{1}{2T}\int_{\Gamma\times(0,T)} q^{-1}\left(\frac{\partial\varrho}{\partial\nu}\right)^2 d\Gamma\,dt - \int_\Omega (y^1\varrho^0 - y^0\varrho^1)\,dx\ ,$$

where

$$\varrho'' - \Delta\varrho = 0 \quad \text{in } \Omega \times (0,T)\,,$$

(3.3) $$\varrho(0) = \varrho^0\,,\ \varrho'(0) = \varrho^1$$

$$\varrho = 0 \qquad \text{on } \Sigma\,.$$

We now study the solution $\{\varrho_T^0, \varrho_T^1\}$ *of* (3.2) *as* $T \to +\infty$.

Up to now q is ***arbitrary,*** provided $q \in L^\infty(\Gamma)$ and $q \geq q_0 > 0$.

We are going to show that ***in some cases*** for ***a suitable choice of*** q everything becomes simple.

In general we shall have

(3.4) $$\frac{1}{2T}\int_{\Gamma\times(0,T)} q^{-1}\left(\frac{\partial\varrho}{\partial\nu}\right)^2 d\Gamma\, dt \to Q(\varrho^0, \varrho^1) \text{ as } T \to \infty\,,$$

where

(3.5) $Q(\varrho^0, \varrho^1)$ = positive definite quadratic form on $H_0^1(\Omega) \times L^2(\Omega)$.

Therefore, as $T \to +\infty$,

(3.6) $$\{\varrho_T^0, \varrho_T^1\} \to \{\varrho_\infty^0, \varrho_\infty^1\} \text{ in } H_0^1(\Omega) \times L^2(\Omega)$$

where $\{\varrho_\infty^0, \varrho_\infty^1\}$ ***is the couple solution of***

(3.7) $$\inf_{\varrho^0, \varrho^1}.\ Q(\varrho^0, \varrho^1) - \int_\Omega (y^1\varrho^0 - y^0\varrho^1)\, dx\,.$$

The problem is now to compute $Q(\varrho^0, \varrho^1)$.

4. A formula.

We use for (3.3) *a classical multiplier,* introduced by several authors in particular by L.F. HO, loc. cit. We define

(4.1) $m(x) = x - x^0$, x^0 chosen arbitrarely in $\mathbb{R}^n$

and we multiply the first equation (3.3) by $m \nabla \rho$. After some integrations by parts (cf. details in J.L. LIONS [3]) one finds

(4.2) $$(\rho', m\nabla\rho + \frac{n-1}{2}\rho)_0^T + TE_0(\rho^0, \rho^1) = \int_{\Gamma\times(0,T)} \frac{m\nu}{2}\left(\frac{\partial\rho}{\partial\nu}\right)^2 d\Gamma\, dt$$

where $(f, g)_0^T = (f(T), g(T)) - (f(0), g(0))$,

$(f, g) = \int_\Omega f g\, dx$, $|f|^2 = (f, f)$,

ν = unitary normal to Γ directed toward the exterior of Ω ,

(4.3) $$E_0(\rho^0, \rho^1) = \text{Energy} = \frac{1}{2}\left[|\nabla \rho^0|^2 + |\rho^1|^2\right],$$

$$|\nabla \rho^0|^2 = \sum_{i=1}^{n} \left|\frac{\partial \rho^0}{\partial x_i}\right|^2 .$$

It follows immediately from (4.2) that, as $T \to +\infty$,

(4.4) $$\lim. \frac{1}{2T}\int_{\Gamma\times(0,T)} (m\nu)\left(\frac{\partial\rho}{\partial\nu}\right)^2 d\Gamma\, dt \to E_0(\rho^0, \rho^1).$$

5. Application of (4.4) (I).

Let us make the ***geometrical hypothesis :***

(5.1) Ω is ***strictly star shaped*** with respect to a point (that we can choose as 0).

Choosing then $x^0 = 0$ in (4.1), (5.1) means that

(5.2) $m\nu \geqslant \gamma_0 > 0$ on Γ .

We can therefore choose

(5.3) $q = (m\nu)^{-1}$.

For this choice of q , (3.7) *becomes*

(5.4) $$\inf_{\rho^0,\rho^1} \; E_0(\rho^0, \rho^1) - \int_\Omega (y^1 \rho^0 - y^0 \rho^1)\, dx$$

The solution of (5.4) is straightfoward. One has :

ρ^0_∞ is the solution of

(5.5) $-\Delta \rho^0_\infty = y^1 \quad , \quad \rho^0_\infty = 0$ on Γ

and

(5.6) $\rho^1_\infty = -y^0$.

Remark 5.1.

The method HUM (as briefly introduced in Section 2) can be used from ***a numerical view point*** (with some care and with some non trivial regularization procedures). cf. R. GLOWINSKI, C. LI, J.L. LIONS [1] and a series of other papers.

One computes φ^0_T , φ^1_T with ***simple*** q's . In R. GLOWINSKI et al., loc. cit, we chose $q = 1$, Ω = square. One can then compute $T\varphi^0_T$, , $T\varphi^1_T$ for a series of larger and larger T . After a proper scaling, one can compare with ρ^0_∞ , ρ^1_∞ as given by (5.5) (5.6). ***This gives a very good method to test the computations !*** (Test is passed succesfully in GLOWINSKI et al forth coming publications...). ■

6. Application of (4.4) (II).

One can apply (4.4) ***even without the hypothesis*** (5.1) (or with (5.1) but choosing x^0 such that Ω is ***not*** star shaped with respect to x^0). In other words $m\nu$ ***can change sign on*** Γ ***, or can degenerate on part of*** Γ .

We can construct *a particular control* v_T^* giving E.C. as follows :

We start from $\{\varphi^0, \varphi^1\} \in H_0^1(\Omega) \times L^2(\Omega)$ and we define φ by (2.3) (2.4). We then define ψ as the solution of (2.5)(2.6) subject to the boundary condition.

(6.1) $$\psi = m\nu \frac{\partial \varphi}{\partial \nu} .$$

We then define Λ as in (2.8). We have

(6.2) $$< \Lambda \{\varphi^0, \varphi^1\}, \{\varphi^0, \varphi^1\} > = \int_{\Gamma \times (0,T)} m\nu \left(\frac{\partial \varphi}{\partial \nu}\right)^2 d\Gamma \, dt .$$

The expression $\int_{\Gamma \times (0,T)} m\nu \left(\frac{\partial \varphi}{\partial \nu}\right)^2 d\Gamma \, dt$ is a quadratic form on $H_0^1(\Omega) \times L^2(\Omega)$. By virtue of (4.4), this form is *for* T *large enough* positive definite on $H_0^1(\Omega) \times L^2(\Omega)$.

Therefore, for T large enough, Λ defines an isomorphism from $H_0^1(\Omega) \times L^2(\Omega)$ into its dual. Therefore the equation

(6.3) $$\Lambda \{\varphi^0, \varphi^1\} = y^1, -y^0 \}$$

admits a unique solution. Let φ_T^{0*}, φ_T^{1*} be this solution, and let φ_T^* be the corresponding solution of (2.3) (2.4).

Then

(6.4) $$v_T^* = m\nu \frac{\partial \varphi_T^*}{\partial \nu}$$

defines a particular control, which gives E.C., i.e. (1.4).

If we renormalize, as in Section 3, *we obtain,* using (4.4), *that,* as $T \to +\infty$

(6.5) $$\{T\varphi_T^{0*}, T\varphi_T^{1*}\} \to \{\rho_\infty^0, \rho_\infty^1\} \text{ in } H_0^1(\Omega) \times L^2(\Omega)$$

where $\rho_\infty^0, \rho_\infty^1$ *are given by* (5.5) (5.6).

7. Various Remarks.

Using HUM and the type of normalization introduced above, one can study in a somewhat simular fashion several other situations where one can let $T \to +\infty$.

Several examples will be given in subsequent publications with R. GLOWINSKI et al.

BIBLIOGRAPHY

L.F. HO [1] Observabilité frontière de l'équation des ondes. C.R.A.S. Paris, 302, (1986), p. 443-446.

R. GLOWINSKI, C. LI, J.L. LIONS [1] Japanese J. of Applied Math. 1988.

J.L. LIONS [1] Contrôlabilité exacte des systèmes distribués. C.R.A.S. Paris 302 (1986), p. 471-475.

[2] Exact Controllability, Stabilization and perturbations for distributed systems. SIAM Review. 30 (1), (1988), p. 1-68.

[3] ***Contrôlabilité exacte, perturbations et stabilisation de systèmes distribués.*** t. 1. ***Contrôlabilité exacte.*** t. 2. ***Perturbations.*** Masson. R.M.A. 1988.

[4] ***Contrôle des systèmes distribués singuliers.*** Gauthier Villars, Collection MMI, t. 13, 1983.

D.L. RUSSELL [1] Controllability and stabilization theory for linear partial differential equations. Recent progress and open questions. SIAM Rev. 20 (1978), p. 639-739.

MICROLOCAL IDEAS IN CONTROL AND STABILIZATION

Claude Bardos and Gilles Lebeau
Centre de Mathematiques Appliquees
Ecole Normale Superieure
Paris, 75230 France

Jeffrey Rauch
Department of Mathematics
University of Michigan
Ann Arbor, MI 48109, USA

Problems of observability, controlability, and stabilization lead to the search for the same type of inequalities for solutions of Partial Differential Equations. The problems discussed below concern the observation, control or stabilization of waves governed by hyperbolic partial differential equations. A neccessary condition for any of the three processes is that the intervention must be on a large enough set that it encounters every ray of geometric optics. Otherwise a wave localized near a ray can escape observation, control and stabilization. To capture these ideas of propagation along rays we use methods of microlocal analysis. Recent progress has rendered these tools sufficiently powerful to study control and stabilization at the boundary. In the present lecture I will describe some of the basic ideas and results of microlocal analysis and give an application to interior stabilization where the method is clearest and the technicality minimized. Then we describe some of our recent results for the case of stabilization from the boundary.

§1. A simple example serves to show the difference between

global and local energy estimates. Generalization to only slightly more complicated hyperbolic equations force one toward microlocal analysis.

Consider the simplest hyperbolic equation, the tranport equation,

$$(1.1)\qquad (\partial_t + \Sigma\, a_j(t,x)\partial_j + b(t,x))\, u=0, \qquad \partial_j \equiv \partial/\partial x_j .$$

We suppose that u and the a_j are real, and that the coefficients have bounded derivatives of all orders. Multiplication by u yields the "energy" identity

$$\partial_t(u^2/2) + \Sigma\, \partial_j(a_j u^2/2) + (b-\Sigma\partial_j a_j/2)u^2 = 0.$$

Integrating over $x \in \mathbb{R}^d$ yields the differential inequality for the $L^2(\mathbb{R}^d)$ norm,

$$\partial_t(\|u\|^2) \le c\, \|u\|^2, \qquad c \equiv \inf\, (b-\Sigma\partial_j a_j/2).$$

Thus, the L^2 norm grows at most exponentially..

The equation 1.1 is explicitely solvable by integrating along the integral curves of the vector field $\partial_t+\Sigma a_j\partial_j$. These curves are the <u>rays</u> for this problem. The values of u in a set $A\subset\mathbb{R}_t\times\mathbb{R}^d_x$ are influenced only by the Cauchy data in the set of points at t=0 which are connected to points of A by rays.

For this equation, it is easy to modify the energy method to reveal this more refined structure. It suffices to multiply the equation by qu with a suitable cutoff function, q(t,x). Multiplying as indicated yields the identity,

$$(1.2)\quad \partial_t(qu^2/2)+\Sigma\partial_j(qa_ju^2/2) + (b-\Sigma\partial_j a_j/2)qu^2 = ((\partial_t+\Sigma a_j\partial_j)q)u^2.$$

If we choose q so that it is constant along rays, the right hand side vanishes and we obtain an estmate,

$$\|q^{1/2}u(t)\| \le e^{\text{const.}|t|}\, \|q^{1/2}u(0)\|.$$

Choosing q supported close to a single ray, this captures the propagation along rays for solutions of 1.1.

§2. The next simplest example reveals that the usual method of multipliers, though fine for deriving global energy estimates, is a tool which is not refined enough to derive localized energy estimates.

Consider the speed one wave equation in d-dimensional space,

$$(2.1)\qquad \Box u = 0, \qquad \Box \equiv \partial_t^2 - \Sigma\, \partial_j^2.$$

Multiplication by $\partial_t u$ yields the energy conservation law,

$$(2.2)\qquad \partial_t e - \Sigma\, \partial_j (u_t u_{x_j}) = 0, \qquad 2e \equiv (\partial_t u)^2 + \Sigma\, (\partial_j u)^2.$$

Integration over a truncated backward light cone shows that for $t>0$,

$$(2.3)\qquad \int_{B_r(y)} e(t,x)\, dx \le \int_{B_{r+t}(y)} e(0,x)\, dx,$$

Thus, energy cannot travel faster than speed 1.

However, much more is true for this equation. Sharp signals propagate with speed nearly exactly equal to 1. The group velocities associated with this equation all have magnitude exactly equal to 1. Moreover, writing the explicit formula for the solution one finds that the data in the interior of $B_t(y)$ at time zero influence the values of $u(t)$ in $B_r(y)$ via an integral operator with smooth kernel. Thus for any $\varepsilon>0$ there is a $c=c(\varepsilon)$ so that

$$(2.4) \qquad \int_{B_r(y)} e(t,x)\,dx \le \int_{B_{r+t}(y)\setminus B_{t-\varepsilon}} e(0,x)\,dx + c\|u(0)\|^2_{L^2(B_{t-\varepsilon})} .$$

This estimate is an L^2 estimate which captures the essence of the fact that signals propagate essentially at speed 1 and not slower.

One can try to prove such an estimate by using the mutiplier $q(t,x)u_t$ motivated by the success in the previous section. Recall that the key was to have the cutoff function q be constant along rays. Unfortunately, for $\square$, rays travel in all directions at unit speed. It is not possible to have $(\partial_t+\omega\cdot\partial_x)q=0$ for all vectors ω of unit length except when $q\equiv$constant.

An astute observer will remark that it suffices for q to decrease along rays in order that the right hand side of 1.2 be nonpositive. A further weakening is discussed at the end of §3.

§3. Faced with the dilemma that signals travel in all directions we ask how is it possible to recover localized estimates which say roughly that the energy found over here at time t=0 progagates over there at a later time. There are two classical ideas in mathematical physics which point the way.

The first is to take the Fourier transform with respect to x. The "components" with frequency ξ travel at speed one in the directions $\pm\xi/|\xi|$. Thus, if we localize in Fourier we recover a nearly unique direction of motion. In a crude sense, if we localize in "frequency space" the wave equation resembles the transport equation of the first section.

The second is Huyghen's construction of "secondary wavelets" with its link to Hamilton-Jacobi Theory. We will not present that motivation for lack of time.

To carry out the frequency space localizations hinted at above, a key is the idea of wavefront set introduced by Sato in the analytic category and molded by Hormander into one of the central themes of the analysis of partial differential equations. The idea is simple. We know that the regularity of a function is linked to the rate of decay of its Fourier transform. The more regular, the faster is the decay. Where a function is not infinitely smooth there must be directions where the Fourier transform is not rapidly decreasing. For a function u(y) the wavefront set is a set of pairs y,η consisting of points y and directions η such that the Fourier transform of localized functions do not decay. Precisely,

__Definition__. A point y,η is not in the wavefront set of a distribution $u\in\mathcal{D}'(\Omega)$, denoted WFu, if and only if there is a cutoff function $\varphi \in \mathcal{D}(\Omega)$ with $\varphi(y)\neq 0$ so that $\mathcal{F}(\varphi u)$ is rapidly decreasing for directions near η. That is, there is a conic neighborhood Γ of η in $\mathbb{R}^{\dim y}\backslash 0$ such that for all m there is a $c=c(m)$ so that

$$|\mathcal{F}(\varphi u)| \le c(1+|\xi|^2)^{-m} \quad \text{for all } \xi \text{ in } \Gamma.$$

Similarly we say that u is is the <u>Sobolev space H^s microlocally at</u> y,η denoted $u \in H^s(y,\eta)$ if there is a cutoff and conic neighborhood Γ as above with

$$(3.1) \qquad (1+|\xi|^2)^{-s/2}\mathcal{F}(\varphi u) \in L^2(\Gamma)$$

Let $\chi(\xi)$ be a function which is smooth on $\xi\neq 0$, homogeneous of degree zero, supported in Γ and nonzero on a neighborhood of η. then 3.1 yields

3.2 $$\chi(D)\varphi(x)u \in L^2(\mathbb{R}^{\dim y}).$$

The operator on the left is a pseudodifferential operator with essential support near y,η. It achieves a localization in space and frequency direction. Note that the region in frequency space is infinite in extent though localized in direction. This is consistent with Heisenberg's uncertainty principle.

The above discussion suggests a new way to derive estimates for solutions of partial differential equations. One can multiply by $q(y,D_y)u$ where q is pseudodifferential. One can find such multipliers whose symbol is constant along bicharacteristics. Typically $q(y,\eta)$ will not be polynomial in η, that is, q will not be differential.

Let $y=(t,x)$, $\eta=(\tau,\xi)$ and consider solutions of the second order equation

(3.3) $$\partial_t^2 u - \Sigma\, a_{ij}(x)\partial_i\partial_j u + \Sigma\, a_j(x)\partial_j u + a(x)\partial_t u + b(x)u = 0.$$

Let $p(t,x,\tau,\xi)$ be the principal symbol of the differential operator,

$$p(y,\eta) = p(t,x;\tau,\xi) \equiv \tau^2 - a_{ij}\xi_i\xi_j.$$

Where p is not zero it is not unreasonable to think that one can, in some sense, divide by p and that u will be microlocally two derivatives smoother that Pu. This is the content of the <u>Microlocal Elliptic Regularity Theorem</u>.

<u>Definition</u>. The <u>characteristic variety</u> of P, denoted char P is the set of t,x,τ,ξ such that $p(t,x,\tau,\xi)=0$.

Theorem 1. If t,x,τ,ξ is not in the characteristic variety of P and $Pu \in H^s(t,x,\tau,\xi)$ then $u \in H^{s+2}(t,x,\tau,\xi)$. If $Pu \in C^\infty$ then the wavefront set of u is contained in char P.

Let H_p denote the Hamiltonian vector field of p in y,η space,

$$H_p \equiv \sum ((\partial p/\partial \eta_j)\partial/\partial y_j - (\partial p/\partial y_j)\partial/\partial \eta_j).$$

Performing a calculation similar to the one for the transport equation in §1, but based on the elementary calculus of pseudodifferential operators, one finds energy estimates when $H_p q=0$. That is, if q is constant on the integral curves of H_p. Those integral curves along which $p=0$ are called bicharacteristics. Their projection on t,x space are called the rays of geometrical optics. This yields the fundamental theorem of Hormander,

Theorem 2. (Hormander) If $Pu \in C^\infty$ then the wavefront set of u is invariant under the flow of H_p. Moreover if u is H^s at one point y_1,η_1 of a bicharacteristic γ, then, u is H^s at all points of γ.

Each of these results comes in a quantitative form asserting that suitable functions expected to lie in L^2 have L^2 norms dominated by expressions known to be finite by hypothesis. Excellent, and brief, introductions to the basic methods of microlocal analysis are presented in [H] and [N].

Mulitipliers q give good information whenever

(3.4) $\quad H_p q \le 0$ modulo multiples of p.

The interested reader can check the validity of this relation for the multiplier ∂_t and the multiplier of Morawetz. There are many solutions of 3.4, but, few are polynomial in η. This is the main reason that pseudodifferential methods are more penetrating.

§4. We show how these ideas are applied to a problem of stabilization. Consider the dissipative wave equation on a compact manifold M without boundary,

$$u_{tt} - \Sigma\, \partial_i (a_{ij}(x) \partial_j u) + a(x) u_t + b(x) u = 0. \tag{4.1}$$

The difficulties introduced by the presence of boundaries will be briefly discussed below. The coefficients are smooth, a_{ij} is a positive definite real symmetric matrix valued function, $a \ge 0$, and $b \ge 0$. We suppose that b is not identically equal to zero to avoid having to pay special attention to constant solutions. Let $\omega \equiv \{x \in M : a > 0\}$ be the region on which dissipation takes place. The energy at time t, e(t) is defined by

$$2e(t) \equiv \int_M u_t^2(t,x) + a_{ij}(x) \partial_i u(t,x) \partial_j u(t,x) + b(x) u^2 \, dx. \tag{4.2}$$

The energy norm turns the space of Cauchy data, $(u_t, u) \in L^2(M) \oplus H^1(M)$, into a Hilbert space. Denote by S(t) the operator carrying Cauchy data at time zero to Cauchy data at time t.

Multiplying the equation by u_t and integrating over $[0,t] \times M$ yields the law of energy decay,

$$e(t) = e(0) - \int_0^t \int a u_t^2 \, dx \, dt. \tag{4.3}$$

Thus, S(t), is a contraction for $t \ge 0$. The techniques of Iwasaki [I] show that if $a(x)$ is not identically zero, then all solutions tend to zero in energy, that is $e(t) \longrightarrow 0$ as t tends to infinity. In fact his proof yields the following simple but useful criterion.

Proposition . **If $S(t)$ is a stongly continuous contraction semigroup on a Hilbert space H then as t tends to $+\infty$, s-lim $S(t) = 0$ provided 1. the domain of some power of the generator (with the graph topology) is compactly imbedded in H, and, 2. the generator has no purely imaginary eigenvalues.**

The second property is often verified using a unique continuation theorem for elliptic equations as in 4.7 below.

The decay to zero shows that all oscillations are eventually damped out by the dissipation. The problem of stabilization is to determine under what conditions this damping is effective in the sense that for any $\varepsilon>0$ there is a time $T>0$ so that $e(T)\le\varepsilon^2 e(0)$ for all solutions of finite energy. This is the same thing as saying that $\|S(T)\|\le\varepsilon$. Thanks to the semigroup property the following conditions are equivalent:

i. $\|S(t)\|$ tends to zero as t tends to infinity.

ii. There is a $T>0$ such that $\|S(T)\|<1$.

iii. There is a $T>0$ such that the spectral radius of $S(T)$ is strictly less than one.

iv. There are constants M and $c>0$ so that for all $t>0$, $\|S(t)\|\le Me^{-ct}$.

If ρ is a ray of geometric optics, one can construct

solutions which have their energy concentrated as close to ρ as one likes [Ralston]. Thus, if there is a ray which is disjoint from the set $[0,T]\times\omega$ then for such a solution, the change in energy during the time interval $[0,T]$ can be made arbitrarily small so $\|S(T)\|=1$. Considering nearby rays and solutions supported so close to the corresponding rays that the solutions are essentially orthogonal we obtain the stronger conclusion that the essential spectrum of $S(T)$ must intersect $\{z:|z|=1\}$.

Theorem 3. If there is a ray of geometric optics which is disjoint from $[0,T]\times\omega$ then ess spec $S(T)\cap\{|z|=1\}$ is nonempty.

A construction of localised solutions following ideas of Taylor is sketched in [BLR1].

This Theorem shows that a necessary condition for stabilization is that there must be a time $T>0$ such that every ray of geometric optics meets the set $[0,T]\times\omega$. The beauty of this simple observation is that the condition is sufficient.

Theorem 4. If $T>0$ is such that every ray of geometric optics intersects the set $]0,T[\times\omega$ then the map $S(T)$ sending Cauchy data at time zero to Cauchy data at time T has norm less than one.

Exponential decay of $S(t)$ is proved in [RT]. The result on the norm of $S(T)$ for the sharp value of T uses a trick from [BLR1].

outline of proof. Theorem 2 implies that if u is in $H^1(]0,T[\times\omega)$ then $u\in H^1(]0,T[\times M)$, in particular $e<\infty$. Note that for $0\le t\le T$ the norms $e(t)^{1/2}$ are uniformly equivalent so the energy norm of u makes sense. The quantitative version asserts that for ε small positive and $\omega_1 \subset\subset \omega$ sufficiently large, there are constants $c_1 > 0$ and c_2 such that

$$(4.4) \qquad \int\int_{[\varepsilon, T-\varepsilon] \times \omega_1} u_t^2 + (\partial_j u)^2 \, dxdt \geq c_1 e(T) - c_2 \int\int_{[0,T] \times M} u^2 \, dxdt,$$

Next we use the fact that the wave front set of u is contained in the characteristic variety of P. Roughly speaking this says that near (t,x), u can oscillate only like $e^{i(\tau t + x\xi)}$ with $p(t,x,\tau,\xi)=0$. At such points $\tau \neq 0$ so the time derivative is as large as any other first derivative. A version of the Microlocal Elliptic Regularity Theorem yields the estimate,

$$\int\int_{[0,T] \times \omega_1} u_t^2 \, dxdt \geq c_3 \int\int_{[\varepsilon, T-\varepsilon] \times \omega_1} u_t^2 + (\partial_j u)^2 \, dxdt - c_4 \int\int_{[0,T] \times M} u^2 \, dxdt.$$

Together with 4.4 this yields

$$(4.5) \qquad \int\int_{[0,T] \times \omega_1} u_t^2 \, dxdt \geq c_5 e(T) - c_6 \int\int_{[0,T] \times M} u^2 \, dxdt.$$

To show that S(T) is a strict contraction it suffices to show that there is a constant c, independent of u, with

$$(4.6) \qquad \int_0^T \int a u_t^2 \, dx \, dt \geq ce(T).$$

Inequality 4.5 yields 4.6 modulo a lower order, L^2 norm of u, term. To prove 4.6 it suffices to show that the L^2 norm of u_t dominates the L^2 norm of u. If not, we could choose finite energy solutions u_n such that

i. The initial energy of u_n is equal to one, and,

ii. $\int\int_{]0,T[\times\omega} a\partial_t u_n^2 \, dxdt \longrightarrow 0$ as t tends to ∞.

As the u_n are bounded in $H^1(]0,T[\times M)$, we may pass to a subsequence which converges strongly in $L^2(]0,T[\times M)$. According to 4.5, the L^2 norms are strictly positive so the limit, u, is nonzero. Thus u is a finite energy solution such that u_t vanishes on $]0.T[\times\omega$. To complete the proof of 4.6 it suffices to show that no such u exists.

Let N denote the space of distributions on $]0,T[\times M$ which satisfy Pu=0, and, $\partial_t u=0$ on $]0,T[\times\omega$. It suffices to show that $N=\{0\}$.

On $]0,T[\times\omega$ the Microlocal Elliptic Regularity Theorem applied to the two equations Pu=0 and $\partial_t u=0$ implies that

$$WFu \subset \text{char } P \cap \text{char } \partial_t = \phi.$$

thus $N \subset C^\infty(]0,T[\times\omega)$.

Propagating this regularity using Hormander's Theorem yields $u \in C^\infty(]0,T[\times M)$ since each bicharacteristic passing over $]0,T[\times M$ passes over $]0,T[\times\omega$ where it is not in the wavefront set of u. Thus $N\subset C^\infty(]0,T[\times M)$. In particular, the solutions in N lie in $H^s(]0,T\times M)$ for any s. The quantitative version of this argument is that any of these H^s norms is equivalent to the energy norm on N.

Since for s>1, the H^s norm is compact with respect to energy Riesz' Theorem implies that N is finite dimensional.

For u∈N, let $w\equiv u_t$. Then, w is a solution of Pw=0 and w vanishes on $]0,T[\times\omega$. Thus w_t also vanishes on $]0,T[\times\omega$ so w∈N. Thus, ∂_t is a linear map of the finite dimensional space N into itself. If N were nontrivial it would follow that there is a nonzero $u \in N$ with $\partial_t u = \lambda u$ for some $\lambda \in \mathbb{C}$. Thus for $0<t<T$, $u = e^{\lambda t} v(x)$ with $v \in C^\infty$. Since $u_t \equiv 0$ on ω, we have $v=0$ on ω. On the other hand, the function v satisfies the reduced wave equation

$$(4.7) \qquad (\lambda^2 - \partial_i a_{ij} \partial_j + \lambda + b(x))v = 0.$$

The unique continuation principle for this second order elliptic equation implies that $v=0$. We conclude that $N=\{0\}$ and the proof is complete. /////

§5 Next we consider a problem where the stabilization is a result of influence at the boundary. Here M is a nice bounded open set in $\mathbb{R}^d$ and the friction term is absent,

$$(5.1) \qquad u_{tt} - \Sigma\, \partial_i (a_{ij}(x) \partial_j u) + b(x) u = 0, \qquad x \in M \subset \mathbb{R}^d, \quad b \geq 0.$$

At the boundary, we consider dissipative boundary conditions related to the Neumann boundary condition, $\partial_\nu u = 0$, where

$$\partial_\nu \equiv \sum_{i,j} a_{ij}(x) n_i(x) \partial_j$$

is the natural normal derivative associated with the elliptic operator $\Sigma\, \partial_i (a_{ij}(x) \partial_j u)$. Here $(n_1, \ldots, n_d)$ is the unit outward pointing normal to ∂M. Our dissipative condition is

$$(5.2) \qquad \partial_\nu u + a(x) u_t + a_0(x) u = 0, \qquad x \in \partial M, \quad \text{with} \quad a \geq 0, \ a_0 \geq 0.$$

The energy is given by one half of

$$\int_M u_t^2(t,x) + \Sigma a_{ij}(x) \partial_i u(t,x) \partial_j u(t,x) + b(x) u^2 \, dx \quad + \int_{\partial M} a_0(x) u^2 \, d\sigma$$

The energy law is

$$(5.3) \qquad e(t) = e(0) - \int_0^t \int_{\partial M} a(x) u_t^2 \, d\sigma \, dt.$$

We suppose that not both b and a_0 are identically zero. Then the energy defines a norm on the space of Cauchy data, $L^2 \oplus H^1$, and this norm is a nonincreasing function of time. It is worth noting that if $d>1$ and $a(x)$ is not identically zero, then the mixed initial boundary value problem (5.1)-(5.2) is only well posed in $t \geq 0$.

Iwasaki's Proposition shows that for all solutions of finite energy, the energy tends to zero as t tends to plus infinity. Let $S(t)$ denote the evolution operator on Cauchy data. We seek conditions which guarantee that the four equivalent conditions from §4 hold.

For boundary value problems, the set of rays must be enlarged to include the possibility of (i) reflection at the boundary following the classical law, (ii) limits of such rays which can glide along the boundary, and, (iii) rays which just kiss the boundary at diffractive points.

If there are bicharacteristics which have infinite order of contact with $T^*(\mathbb{R}\times\partial\Omega)$ there may be more than one way to continue rays for the mixed problem and there may be gliding rays whose approximablility as above is an open problem. To lighten the exposition, we suppose that such infinite order contact does not occur. In that case, solutions of localized energy are known to exist for all the above genralized rays. Let $\beta = \{x \in M : a > 0\}$. Then a necessary condition for stabilization is that every such ray must intersect $[0,T]\times\beta$.

To prove that $S(T)$ is a strict contraction it suffices to find $\beta_1 \subset\subset \beta$, $T>0$, and $c_1>0$, such that

5.4 $$\int\int_{[0,T]\times\beta_1} u_t^2 \, d\sigma dt \geq c_1 e(T).$$

We think of 5.4 as asserting that the wave leaves a strong imprint on the boundary. It is not surprising that a ray which just kisses the boundary at a point of tangency might not leave a strong imprint. Thus, it is not wise to observe, control, or stabilize rays at such diffractive points.

Theorem 5. If $T>0$ is such that every ray of geometric optics meets the set $]0,T[\times\beta$ in at least one nondiffractive point, then, the operator $S(T)$ carrying Cauchy data at time zero to Cauchy data at time T has norm less than one.

The proof is similar in spirit to that outlined in §4. Microlocal techniques yield an analogue of the fundamental inequality 4.7,

5.5 $$\int\int_{[0,T]\times\beta_1} u_t^2 \, d\sigma dt \geq c_2 e(T) - c_3 \int\int_{[0,T]\times M} u^2 \, dxdt.$$

This is weaker than 5.4 by a compact term which is handled as in §4. The technically difficult part in the proof of 5.5 is to show that rays which glide along the boundary leave a strong enough imprint. Our methods, reported in [BLR-1,3], are inspired by the work of Melrose-Sjostrand on the propagation of singularities.

In case $b\equiv 0$ and $a_0\equiv 0$ constants are solutions which do not decay to zero and the energy is not a norm on $L^2\oplus H^1$. However, the quantity

$$\int_M u_t \, dx + \int_{\partial M} a(x)u \, d\sigma$$

is independent of time. One then works in the Hilbert space of Cauchy data for which this functional is zero. Energy is then a norm and the above argument proves exponential decay. For arbitrary data one gets exponential decay to the constant state c such that

$$c\int_{\partial M} a(x)\, d\sigma = \int_M u_t(0,x)\, dx + \int_{\partial M} a(x)u(0,x)\, d\sigma$$

Our methods are quite robust. Mass terms, $\rho(x)u_{tt}$ cause no trouble. Nonlocal boundary conditions, $\partial_\nu u+K(\partial_t u)=0$, where K satisfies a lower bound $\int u_t K(u_t)dx \geq c\int_\beta u_t^2 dx$ also yield to the same arguments.

References

[BLR1] C. Bardos, G. Lebeau, and J. Rauch, Controle et stabilisation dans les problemes hyperboliques, Appendice II in Vol 1 of the recent book of J.L. Lions, Masson Publ.

[BLR2] C. Bardos, G. Lebeau, and, J. Rauch, Un exemple d'utilisation des notions de propagation pour le controle et la stabilisation des problemes hyperboliques, Proceedings of the workshop on nonlinear hyperbolic equations in applied science, Turin, June 1987. ed. N. Bellomo.

[BLR3] C. Bardos, G. Lebeau, and J. Rauch, article in preparation.

[H] L. Hormander, *The Analysis of Linear Partial Differential Operators,* vol.1, Springer-Verlag, 1983.

[I] N. Iwasaki, Local decay of solutions for symmetric hyperbolic systems with dissipative and coercive boundary conditions in exterior domains, Publ. RIMS Kyoto 5(1969) 193-218.

[N] L. Nirenberg. *Lectures on Linear Partial Differential Equations,* Regional Conference Series in Math. #17,

Amer.Math.Soc., Providence, 1967.

[RT] J. Rauch and M.Taylor, Exponential decay of solutions to hyperbolic equations in bounded domains, Indiana University Math. Journal 24(1974)

[R] J. Ralston, Solutions of the wave equation with localised energy, Comm. Pure and Appl. Math 22(1969) 807-823.

SPECTRAL AND ASYMPTOTIC PROPERTIES OF LINEAR ELASTIC SYSTEMS WITH INTERNAL ENERGY DISSIPATION*

by

David L. Russell
Department of Mathematics
Virginia Tech., Blacksburg, VA 24061, U.S.A.

Abstract

In this article we consider the problems inherent in developing mathematical models replicating observed damping phenomena in simple linear elastic systems. The Euler - Bernoulli and Timoshenko beam equations are used as the principal examples. We indicate the pitfalls attendant upon the use of dissipation mechanisms based only upon their mathematical convenience and discuss alternatives which originate from more fundamental physical considerations. The last section introduces two mechanisms, namely; thermoelastic damping and shear diffusion damping, which, to our knowledge, have not been discussed previously in this context.

* Supported in part by the United States Air Force under Grant AFOSR 89 - 0031. Duplication is permitted for U.S. Government purposes.

1. Damping: Experience and Inference

It is now commonly accepted that models used for simulation of mechanical vibrations in elastic systems must incorporate realistic energy dissipation mechanisms. This appreciation has derived from several sources, including the very small, but very important, vibrational damping rates expected in projected space structures and the realization (see [12],e.g.) that consistency of approximating numerical procedures relative to control criteria universally recognized as significant cannot generally be expected in the absence of such mechanisms.

Energy dissipation in elastic systems results from a wide variety of sources, both internal and external with respect to the systems themselves. External energy losses occur due to interraction with fluids, such as atmospheric gases, liquids, etc., in which the elastic structure may be immersed, or through interface with other physical systems, such as support structures. On the other hand, internal losses result from processes occurring within the elastic system itself, such as conversion of mechanical energy to heat energy by means of internal friction, differential temperatures and resultant conduction due to thermoelastic effects and many other causes. The external dissipation mechanisms are very important and often the dominant ones in every day experience; indeed, their ubiquitous character is a serious difficulty in laboratory experiments whose object is the measurement of internal damping effects. Nevertheless it is the internal dissipation mechanisms which are of the greatest scientific interest. This is in part due to the fact that in projected space structures, which do not interract with other structures and are not immersed in fluid media, internal damping is likely to be dominant, and in part due to the fact that external losses, by virtue of their very multiplicity, are rarely amenable to systematic mathematical analysis (see [10], however). In this article we will be exclusively concerned with internal dissipation and we will use the familiar Euler - Bernoulli, Rayleigh and Timoshenko models for lateral vibrations of a beam as vehicles for our discussion.

In all treatments of questions in natural science one must begin with what is empirically known through experience of various sorts. Ideally, since much experimentation has been done in engineering laboratories on damping rates, one could simply cite those studies. Unfortunately, it is frequently difficult for an applied mathematician to obtain data in this way in a form which he will find usable (or even readable). A perusal of the summary of such studies provided by Bert in [2] will allow the reader to see what we mean here. It is for these reasons that the author, with the support of the Air Force Office of Scientific Reseach, the National Science Foundation, and other agencies, has undertaken a number of studies on his own, using the equipment available in the MIPAC (Modelling, Information Processing and Control) Facility at the University of Wisconsin and, more recently, at Virginia Polytechnic Institute and State University. As they apply to lateral vibrations of thin beams we can summarize the results of this laboratory experience, which is described more fully in [10], as follows:

i) In the case of thin elastic beams formed from composite materials (in which class we include, e.g., boron - epoxy composites, fiberglass, hardwood, etc.) it is found that the damping exponent $r = r(\omega)$ associated with a natural mode of vibration of frequency ω is very nearly linearly proportional to ω. That is, the time behavior of such modes is like that of the function

$$\exp\left[-r(\omega)t + i\omega t \right],$$

with $r(\omega) \approx \gamma\ |\omega|$ for some γ independent of ω. There is some evidence that at very low frequencies $r(\omega)$ may be more accurately described by a function quadratic in ω (so that, overall, the pair $(\omega,\ r(\omega))$ describes a branch of a hyperbola), but this evidence is not yet conclusive.

ii) In the case of thin metallic (aluminum, steel, brass, etc.) beams, it is found that the damping rates at low and intermediate frequencies behave in much the same way as for their composite counterparts. At high frequencies the picture is less clear because the data on internal dissipation is corrupted by external losses of various kinds. There is some evidence, but again not altogether convincing, that $r(\omega)$ may tend to a positive constant as $|\omega| \to \infty$. This might correspond to a functional relationship such as

$$r(\omega) = \frac{\omega^2}{\alpha^2 + \beta^2\omega^2} \cdot \tag{1.01}$$

between $r(\omega)$ and ω - many other formulae would do equally well; this one has no particular physical significance.

It is not clear if there is any real distinction between the different types of beams noted in i) and ii) above. Those of class i) exhibit relatively large damping rates; high frequency modes are so highly damped that they are, in fact, difficult to observe at all. So it may simply be that the behavior described under i) is really just a restriction to relatively low frequency modes of a more complete relationship of the type discussed under case ii), as indicated in Figure 1. Resolution of this question must await further experimental studies.

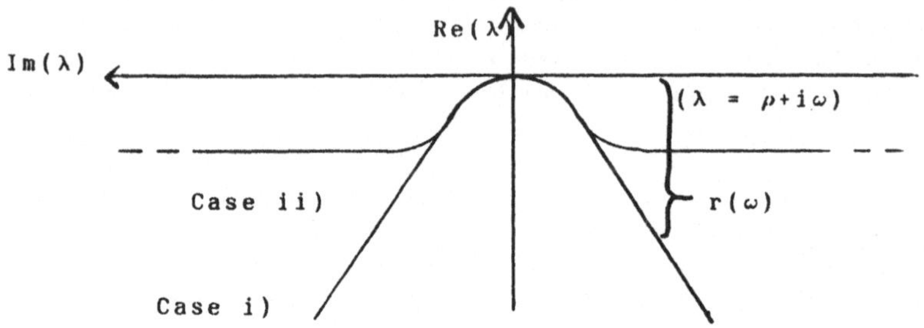

Figure 1: Damping rate versus frequency, Cases i) and ii).

Once a particular body of experience is at hand the question arises as to what may be inferred from the resulting data as far as what the structure of a mathematical model might be. We will see that this process is by no means automatic, especially when some of the data is uncertain, as it is in the cases described. What is needed before one even begins any such inference exercise is a correct appreciation of what a mathematical model is in the first place. Reality is too complex and multi-faceted for any mathematical formula to have the possibility of describing it in totality. The mathematical model purports only to simulate certain highly restricted aspects of the overall behavior of the physical system. Inevitably, some of the factors applying to the physical situation must be judged of minor importance, and then ignored, in developing any mathematical model of the system. If we keep this in mind we will realize that experimental results only suggest, rather than determine, the dissipation mechanisms which should be incorporated into a particular model, just as a relationship such as that shown in Figure 1 might be represented by (1.01) or, perhaps with equal validity, by

$$r(\omega) = \beta\,(\tan^{-1}\alpha\omega)^2, \qquad (1.02)$$

for appropriate positive α and β. In the case of imprecise measurements of $r(\omega)$ it might well be difficult to discriminate between the two.

It is clear, then, that there must be some guiding principle to enforce discipline on damping mechanism investigations. The one which we have chosen to espouse, here and in other articles on the same subject ([10], e.g.) is that the mechanisms deemed legitimate should make sense both mathematically and physically in the context appropriate to the overall model. That is, not only should the incorporation of the candidate dissipative mechanism result in a well-posed mathematical system, but it should also be explainable within the framework of mathematical physics common to all systems of the given class, including, for example, the corresponding conservative system. This may not be entirely unambiguous at this point but we will return to the question later with clarifying examples. We will also see that investigators may honestly differ on whether a proposed mechanism meets these criteria.

We can realize the dual objective of illustrating what we mean by model uncertainty and, at the same time, introduce some equations which will be important to us in the sequel, through inclusion of a short discussion of the common linear models for a vibrating beam. The oldest model, and still the most commonly used, is the so-called *Euler - Bernoulli* model. Letting w(x,t) denote lateral displacement of the elastic axis of the beam from its equilibrium position during an arbitrary motion, and letting

$$v(x,t) = \frac{\partial w}{\partial t}(x,t) ,$$

this model consists in the energy expression

$$\mathcal{E}(w,v) = \frac{1}{2} \int_0^L \left\{ \rho(x)(v(x,t))^2 + \left[EI(x)\frac{\partial^2 w}{\partial x^2}(x,t)\right]^2 \right\} dx \qquad (1.03)$$

and, requiring conservation of energy, the attendant partial differential equation

$$\rho(x) \frac{\partial^2 w}{\partial t^2} + \frac{\partial^2}{\partial x^2}\left[EI(x)\frac{\partial^2 w}{\partial x^2}\right] = 0 , \qquad (1.04)$$

whose solutions, obeying appropriate natural boundary conditions, may be seen to be conservative with respect to the energy (1.03). In (1.03) and (1.04) $\rho(x)$ is the linear mass distribution and EI(x) is the bending moment distribution.

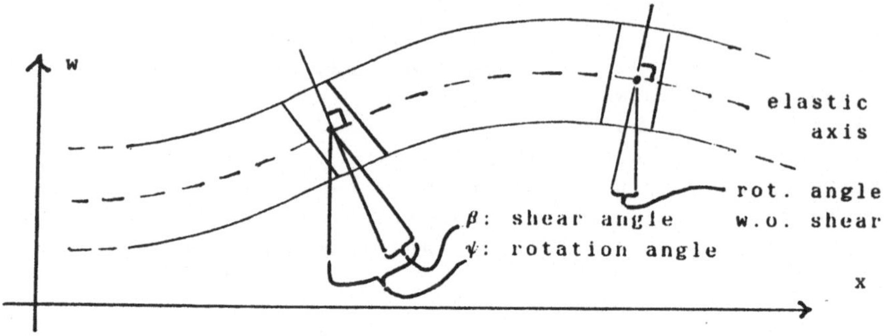

Figure 2: Beam elements; rotation and shear angles.

Referring now to Figure 2, we note that a real beam of finite thickness, may be seen as composed of a chain of rectangular "beam elements" of length δx, as shown. In the Euler - Bernoulli model only the linear momenta of beam elements in the w-direction is taken into account, the angular momenta are ignored. This reflects the operative assumption that the beam is very thin with respect to the wave length of any motions under consideration. If this is not a valid assumption the model (1.04) is no longer accurate, even for small amplitude motions. One may improve the performance of the model by augmenting the energy (1.03) to

$$\mathcal{R}(w,v) = \mathcal{E}(w,v) + \frac{1}{2}\int_0^L I_\rho(x)\,\frac{\partial v}{\partial x}(x,t)^2\,dx\,, \qquad (1.05)$$

where $I_\rho(x)$ is the mass moment of inertia about the elastic axis, axis, taken in the w direction, at the point x. Conservation requirements then lead to the modified *Rayleigh* equation

$$\rho(x)\frac{\partial^2 w}{\partial t^2} - \frac{\partial}{\partial x}\left[I_\rho(x)\frac{\partial^2 w}{\partial x\partial t}\right] + \frac{\partial^2}{\partial x^2}\left[EI(x)\frac{\partial^2 w}{\partial x^2}\right] = 0 \qquad (1.06)$$

and its accompanying natural boundary conditions.

While the angular momenta of beam elements has now been taken care of, we should note that we are still treating the individual beam elements as if they were perfectly rigid bodies - or, at any rate, deformable to orthogonal trapezoidal elements in bending motion. The linear model generally accepted as the most complete in the one-dimensional context is the *Timoshenko* model. Referring to Figure 2 again, we now admit the possibility that originally rectangular beam elements may be deformed into parallellograms, or skew trapezoids, through superposition of a shearing deformation onto those admitted earlier. In small amplitude motions the rotation of a beam element solely due to bending, measured by the angle $\tan^{-1}\frac{\partial w}{\partial x}$, can be adequately approximated by the partial derivative itself. Then the total rotation angle of a beam element centered at x is given by

$$\psi(x,t) = \frac{\partial w}{\partial x}(x,t) + \beta(x,t),$$

where, again as indicated in Figure 2, β is the *shear angle*. Now the appropriate quadratic energy expression is, with $\varphi = \frac{\partial \psi}{\partial t}$,

$$\mathcal{T}(w,v,\psi,\varphi) = \frac{1}{2}\int_0^L \Big\{ \rho(x)\ v(x,t)^2 + I_\rho(x)\ \varphi(x,t)^2$$

$$+ K(x)\left[\psi(x,t) - \frac{\partial w}{\partial x}(x,t)\right]^2 + EI(x)\ \frac{\partial \psi}{\partial x}(x,t)^2 \Big\}\ dx\ , \quad (1.07)$$

where K(x) is the *modulus of elasticity in shear* and the other coefficients are as defined earlier. Conservation of energy requirements now lead to the Timoshenko partial differential equations

$$\rho(x)\ \frac{\partial^2 w}{\partial t^2} + \frac{\partial}{\partial x}\left[K(x)\left[\psi - \frac{\partial w}{\partial x}\right]\right] = 0\ , \qquad (1.08)$$

$$I_\rho(x)\ \frac{\partial^2 \psi}{\partial t^2} - \frac{\partial}{\partial x}\left[EI(x)\frac{\partial \psi}{\partial x}\right] + K(x)\left[\psi - \frac{\partial w}{\partial x}\right] = 0\ , \qquad (1.09)$$

which are easily identified as a pair of coupled one dimensional wave equations. As discussed in [8], [9], (1.08),(1.09) is a *linear symmetric hyperbolic system* of partial differential equations in w and ψ. Such systems have very attractive mathematical properties, as developed in [4], e.g.. In the constant coefficient case one may check that the equation (1.08) can be differentiated with respect to x, solved for $\frac{\partial^2 \psi}{\partial x^2}$ amd substituted into the equation obtained from (1.09) by differentiating it twice with respect to x, thereby yielding the equivalent fourth order scalar equation

$$\rho\ \frac{\partial^2 w}{\partial x^2} - I_\rho\ \frac{\partial^2 w}{\partial t^2 \partial x^2} + EI\ \frac{\partial^4 w}{\partial x^4} + \frac{\rho}{K}\left[I_\rho\ \frac{\partial^4 w}{\partial t^4} - EI\ \frac{\partial^4 w}{\partial t^2 \partial x^2}\right] = 0\ ,$$

which can immediately be recognized as a singular perturbation of the Rayleigh model with ρ/K as the (assumed) small parameter.

2. Direct Dissipation Mechanisms.

In this section we propose to review a number of dissipation mechanisms which have been proposed for use to modify the energy-conservative equations (1.04),(1.06),(1.08),(1.09) so as to exhibit damping agreeing with laboratory experience. Some of these have been discussed in [10]; we will only summarize those cases here. The methods discussed here are *direct* in the sense that they involve direct insertion into the equations of motion of terms representing the forces which result in the dissipative behavior under discussion. The present section of our paper will, for brevity, deal only with the constant coefficient versions of the equations indicated.

The mathematical prototype for frequency proportional damping is the "square root" modelled, initially proposed in [3], which has the general form, assuming the conservative system is written as $\ddot{x} + Ax = 0$,

$$\ddot{x} + 2\gamma A^{1/2}\dot{x} + Ax = 0,$$

where γ is a (usually small) positive constant and $A^{1/2}$ is the non-negative square root of the non-negative self-adjoint operator A. The eigenvalues for any system of this sort are given by

$$\sigma_k = (-\gamma \pm i(1-\gamma^2)^{1/2})\, \omega_k , \quad k = 1,2,3,\ldots,$$

where ω_k is the natural frequency of the k-th mode of the conservative system. We see therefore that, with $r(\omega)$ as in §1,

$$r(\omega) = \frac{\gamma}{(1-\gamma^2)^{1/2}} |\omega|, \quad -\infty < \omega < \infty , \tag{2.01}$$

is the exact damping exponent versus frequency relationship for such a system. In the case of the constant coefficient Euler - Bernoulli beam, wherein

$$A = \frac{EI}{\rho}\frac{\partial^4}{\partial x^4},$$

we have shown in [11] that $A^{1/2}$ is a negative multiple of the second derivative operator for precisely those boundary conditions which admit purely trigonometric eigenfunctions; in other cases $A^{1/2}$ has a much more complicated expression. In the trigonometric cases just indicated, the square root model for the Euler - Bernoulli beam can be written in the form (redefining γ in an inessential way)

$$\rho \frac{\partial^2 w}{\partial t^2} - 2\gamma \frac{\partial^3 w}{\partial t \partial x^2} + EI \frac{\partial^4 w}{\partial x^4} = 0 , \qquad (2.02)$$

the boundary conditions remaining the same as in the conservative model. This equation is particularly appealing because the added damping term can be interpreted as a force proportional to the "bending rate" of the beam at the point in question.

In general, (2.02) does not yield dissipative behavior for the energy (1.03) if the boundary conditions imposed are taken from the set of natural boundary conditions for the conservative system. An example is provided in [10]. What is natural, then, is to look for modified boundary conditions for which the system is dissipative. In the indicated reference we note that if each occurrence of $EI \frac{\partial^3 w}{\partial x^3}$ in the boundary conditions is replaced by

$$-2\gamma \frac{\partial^2 w}{\partial t \partial x} + EI \frac{\partial^3 w}{\partial x^3} , \qquad (2.03)$$

we obtain

$$\frac{d\mathcal{E}}{dt} = -2\gamma \int_0^L \left(\frac{\partial^2 w}{\partial t \partial x} \right)^2 dx \leqq 0 . \qquad (2.04)$$

Thus we have a dissipative system, as desired. But we now find that we have gone too far because, in the case of a free beam, or a beam free at one end and hinged (simply supported) at the other, rotational inertial motions occur and we find that we have energy dissipation in these motions when (2.04) is valid. One physical interpretation of this may be found in the fact that our model is defined over an x-interval of fixed length L and the length of the curve representing the deformed elastic axis during an inertial motion

$$w(x,t) = w_0 + w_1 t + w_2 x + w_3 t x$$

changes length during the motion if $w_3 \neq 0$; the loss of energy in such a motion may be attributed to resistance to "stretching" of the beam. It is logical to follow up such an insight by adding a corresponding term to the potential energy, which might have the form $E\,(w(L,t) - w(0,t))^2$ for some $E > 0$. The idea is that the annihilator of the dissipation form (as a form in $\partial w/\partial t$) should be included in that of the potential energy (as a form in w itself). Alternatively, but in line with the same principle, we may weaken the dissipation form to

$$\frac{d\mathcal{E}}{dt} = -2\gamma \int_0^L \left[\frac{\partial^2 w}{\partial t \partial x} \right]^2 dx + \frac{2\gamma}{L} \left[\frac{\partial w}{\partial t}(L,t) - \frac{\partial w}{\partial t}(0,t) \right]^2, \quad (2.05)$$

leaving the potential energy alone, if we replace $EI\,\frac{\partial^3 w}{\partial x^3}$ wherever it occurs in boundary terms by

$$-2\gamma \frac{\partial^2 w}{\partial t \partial x} + EI \frac{\partial^3 w}{\partial x^3} - \frac{2\gamma}{L} \left[\frac{\partial w}{\partial t}(L,t) - \frac{\partial w}{\partial t}(0,t) \right]. \qquad (2.06)$$

The dissipation form on the right hand side of (2.05) is clearly contrived. It does result in dissipative behavior which is inactive during all inertial motions but it is difficult to justify it in physical terms. According to our earlier stated guidelines, therefore, it must remain a sort of *gedankexperiment* rather than a mechanism to be taken seriously. Nevertheless, it was this mathematical model which led the author to the main dissipation mechanism discussed in [10], which has been referred to as the *spatial hysteresis* model of internal damping.

S. Hansen has shown in his thesis, [7], that the model (2.02), with boundary conditions modified in either of the ways described above, has damping rates which are, *asymptotically*, linear functions of $|\omega|$, as in (1.01); they do not have this exact form for all ω as in the square root model. (This is also a good point for the author to acknowledge valuable conversations with Dr. Hansen in regard to the boundary conditions and modified potential energy forms discussed above.)

The spatial hysteresis approach to internal damping is based on the assumption that a beam element located at the point x in the interval $[0,L]$ experiences a torque due to differential rotation of that element relative to elements located at other points ξ which are, more or less, depending on the precise assumptions, located "near" x. Specifically, that torque is assumed to have the form

$$\tau(x,t) = 2\int_0^L h(x,\xi)\left[\frac{\partial^2 w}{\partial t\partial x}(x,t) - \frac{\partial^2 w}{\partial t\partial x}(\xi,t)\right]d\xi \qquad (2.07)$$

for an appropriate interraction kernel $h(x,\xi)$ which is non-negative and symmetric: $h(x,\xi) = h(\xi,x)$, a consequence of Newton's second law. What then results (see [10] for details) is the equation

$$\rho\frac{\partial^2 w}{\partial t^2} - 2\frac{\partial}{\partial x}\int_0^L h(x,\xi)\left[\frac{\partial^2 w}{\partial t\partial x}(x,t) - \frac{\partial^2 w}{\partial t\partial x}(\xi,t)\right]d\xi + EI\frac{\partial^4 w}{\partial x^4} = 0. \qquad (2.08)$$

The boundary conditions are modified from those applying to the conservative case by replacing $EI\frac{\partial^3 w}{\partial x^3}(X,t)$, where it occurs, by

$$EI\frac{\partial^3 w}{\partial x^3}(X,t) - 2\int_0^L h(X,\xi)\left[\frac{\partial^2 w}{\partial t\partial x}(X,t) - \frac{\partial^2 w}{\partial t\partial x}(\xi,t)\right]d\xi,$$

X being the boundary point in question (0 or L). We show in [10], and this is further developed in [I], that this damping mechanism can be considered a perturbation of the classical Kelvin - Voigt mechanism; the damping exponent versus frequency relationship is quadratic at low frequencies, as in the Kelvin-Voigt case but, unlike that case, is asymptotically similar to (1.01) as $|\omega| \to \infty$. Recent identification studies by H. T. Banks and his associates ([1]) seem to indicate that, with $h(x,\xi)$ of the form $\beta e^{-\sigma(x-\xi)^2}$, β and σ positive, excellent agreement is obtained between damping rates predicted by this model and those obtained from clamped composite material beams in the laboratory. The physical basis for this model, which is most persuasive in the case of composite materials in which relatively stiff fibers are imbedded in a more flexible matrix, is discussed in some detail in [10]. It is fair to say that this basis is not universally accepted and has been seriously questioned by some researchers most comfortable with the classical continuum mechanics approach.

All of the models discussed above, considered as modifications of the conservative Euler - Bernoulli model, produce an asymptotic damping exponent versus frequency relationship of the form (1.01). If one includes the rotational inertia of beam elements, the Rayleigh approach, the term $- I_\rho\frac{\partial^4 w}{\partial x^2\partial t^2}$ is added to the equations in question. The effect, in each case, is to produce damping rates $r(\omega)$ which are asymptotically a positive constant as $|\omega| \to \infty$. Incorporation into the Timoshenko model is only slightly more complicated and has the same effect; that system remains a linear symmetric hyperbolic system with the added damping terms and continues to correspond to a strongly continuous group rather than just a semigroup. Whether this situation should be considered more realistic than what results from the Euler - Bernoulli model depends on the application; additional laboratory experimental data is very much in order.

3. Indirect Dissipation Mechanisms.

The methods which we will introduce here are *indirect* in the sense that they involve coupling of the mechanical equations governing beam motion to related dissipative systems with their own dynamics, resulting in an overall system in which mechanical energy is dissipated.

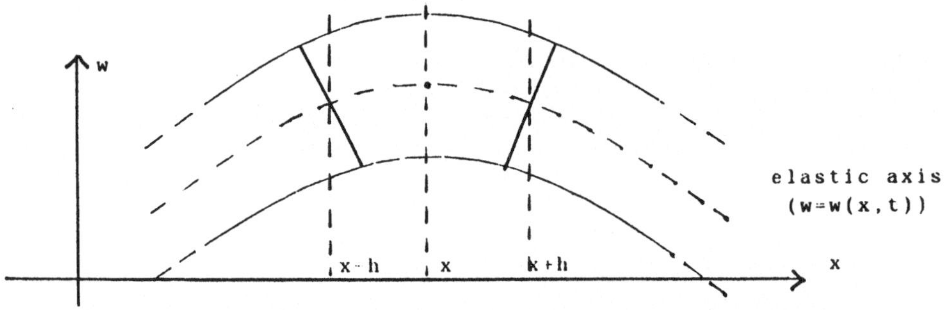

Figure 3: Schematic Representation for Thermoelastic Damping

Starting again with the Euler - Bernoulli model for beam motion (cf. (1.04)), we ask the reader to consider a cross-section of the beam as shown in Figure 3. An approximate trapezoid is bounded by the lines orthogonal to the elastic axis at x-h and x+h, where h is a small positive number, and by the upper and lower lateral surfaces of the beam. That curvilinear trapezoid is, in turn, divided into upper and lower halves by the elastic axis itself. Following private conversations with Prof. C. Zener and examination of his book [13], we are led to a thermoelastic dissipation mechanism involving transverse temperature differentials in the beam. For the purposes of this expository treatment it is enough to suppose that the absolute temperature in the upper trapezoidal region can be adequately approximated in mean by a function $T^+(x,t)$ and that in the lower region by $T^-(x,t)$, neglecting any more complex variations in the transverse direction. We probably should call the difference $\delta T(x,t)$ but, for notational simplicity, let us just write

$$T(x,t) = T^+(x,t) - T^-(x,t).$$

In the absence of other influences we should expect such a temperature differential to decay according to the simple law

$$\frac{\partial T}{\partial t}(x,t) = -k\, T(x,t) ,$$

assuming the beam to be sufficiently thin so that transverse conduction is much more rapid than that due to any longitudinal variations considered. However, the laws of thermoelasticity indicate a coupling of this process with the purely mechanical processes taking place in the beam, which we now proceed to elucidate.

Taking the thickness of the beam to be 2d, d > 0 but small, we may assume that for small h the volumes of the upper and lower trapezoids are represented by $V^+(x,t)dh$ and $V^-(x,t)dh$. Supposing a version of Boyle's law to be valid for solids as well as for the more familiar gas setting, we can see that in the absence of any sort of conduction we would have a simple relationship of the form

$$T(x,t) = \hat{K} \left[\frac{1}{V^+(x,t)} - \frac{1}{V^-(x,t)} \right] .$$

Letting $V(x,t) = V^+(x,t) - V^-(x,t)$ and supposing the variations in $V^+(x,t)$ and $V^-(x,t)$ to be very small relative to their mean values, which we take to be very nearly constant both in space and time, we can see that without conduction we should have an approximate relationship

$$\frac{\partial T}{\partial t}(x,t) = -\tilde{K}\, \frac{\partial V}{\partial t}(x,t) .$$

Reintroducing conductivity, the complete relationship becomes

$$\frac{\partial T}{\partial t}(x,t) = -\, k\, T(x,t) - \tilde{K}\, \frac{\partial V}{\partial t}(x,t) . \tag{3.01}$$

But, for small h, $V(x,t)$ is nearly proportional to $-\frac{\partial^2 w}{\partial x^2}(x,t)$, so we may replace (3.01) by

$$\frac{\partial T}{\partial t}(x,t) = -k\, T(x,t) + K \frac{\partial^3 w}{\partial t \partial x^2}(x,t) . \qquad (3.02)$$

Complementary to these temperature effects of mechanical motion, we now suppose that the total force couple about the elastic axis is modified from its usual Euler - Bernoulli value of EI $\frac{\partial^2 w}{\partial x^2}(x,t)$ to EI $\frac{\partial^2 w}{\partial x^2}(x,t) + M\, T(x,t)$, for some positive constant M. The work-energy relation for the beam now becomes

$$\frac{d\mathcal{E}}{dt} = \int_0^L \left\{ \rho \frac{\partial w}{\partial t}\frac{\partial^2 w}{\partial t^2} + \frac{\partial^2 w}{\partial t \partial x^2}\left[EI \frac{\partial^2 w}{\partial x^2} + M\, T(x,t) \right]\right\} dx ,$$

and, proceeding in the usual way, we derive the equation of motion for the beam (along with modified boundary conditions:

$$\rho \frac{\partial^2 w}{\partial t^2} + EI \frac{\partial^4 w}{\partial x^4} + L \frac{\partial^2 T}{\partial x^2} = 0 . \qquad (3.03)$$

The complete system then consists of (3.02) and (3.03), together with the modified boundary conditions, which we do not elaborate upon here.

We can replace the pair (3.02), (3.03) by a single equation in the following way. Differentiating (3.02) twice with respect to x and substituting into the equation obtained from (3.03) by applying to it the operator $\frac{\partial}{\partial t} + k\, I$, we arrive at

$$\rho \frac{\partial^3 w}{\partial t^3} + \rho k \frac{\partial^2 w}{\partial t^2} + (EI + LK) \frac{\partial^5 w}{\partial t \partial x^4} + EIk \frac{\partial^4 w}{\partial x^4} = 0 . \qquad (3.04)$$

To determine the rough form of the damping exponent as a function of the frequency we assume solutions

$$w(x,t) = e^{\lambda t} e^{i\nu x} \qquad (3.05)$$

(this is perfectly rigorous in the case of periodic boundary conditions and can be justified through use of the Fourier transform in the case of a beam of infinite length), thereby obtaining from (3.05) the equation

$$\rho\lambda^3 + \rho k\lambda^2 + (EI + LK)\lambda\nu^4 + EIk\nu^4 = 0 . \qquad (3.06)$$

This being a cubic with positive real coefficients, there must be a negative real root for all values of $\nu > 0$, reducing to $\lambda = -k$ when $\nu = 0$. Dividing the left hand side by $\lambda + k + \delta$, we obtain the remainder

$$(k+\delta)^2\rho\delta - LK(k+\delta)\nu^4 - EI\delta\nu^4$$

which vanishes if

$$\delta = \frac{LK(k+\delta)\nu^4}{\rho(k+\delta)^2 - EI\nu^4}.$$

In this case the quotient is

$$\rho\lambda^2 - \rho\delta\lambda + (EI + LK)\nu^4 - \rho\delta(k+\delta)$$

and we conclude that the complex roots have imaginary parts asymptotic to

$$\pm\left[\frac{EI + LK}{\rho}\right]^{1/2}\nu^2$$

as $\nu^2 \to \infty$ and imaginary parts equal to $\delta/2$, which tends to 0 as $\nu^2 \to 0$ and to $-LKk/2(EI + LK)$ as $\nu^2 \to \infty$. Near $\nu^2 = 0$ the real part is approximately a quadratic function of the imaginary part. Thus we obtain just the sort of behavior shown in Figure 1 of §1.

The real, negative exponents $\lambda = -(k+\delta)$, which approximate $-k(1 - LK/2(EI + LK))$ as $\nu^2 \to \infty$, correspond to *thermal relaxation* modes of the coupled system, wherein the beam is warped by a certain temperature differential distribution, the distortion of the mechanical structure and the transverse temperature differential decaying to zero together at the same exponential rate, i.e., like $e^{-(k+\delta)t}$. These nonvibratory modes are typical of indirect dissipation mechanisms and have also been observed in so-called "time hysteresis" damping schemes, involving PDE-delay equations, as studied in [5],[6], e.g.

The second indirect mechanism which we will consider is the *shear diffusion* model. It takes as its starting point the Timoshenko model, which is usually written in the form (1.08),(1.09). However, we choose here to use the variables w and $\beta = \psi - \partial w/\partial x$ rather than w and ψ because the damping mechanism is particularly dependent upon the evolution of the shear angle, β. In the variables w and β the energy is (cf. §1 for definition of constants)

$$\mathcal{T} = \frac{1}{2}\int_0^L \left\{\rho\left[\frac{\partial w}{\partial t}\right]^2 + K\beta^2 + I_\rho\left[\frac{\partial \beta}{\partial t} + \frac{\partial^2 w}{\partial t \partial x}\right]^2 + EI\left[\frac{\partial \beta}{\partial x} + \frac{\partial^2 w}{\partial x^2}\right]\right\} dx ,$$

from which we obtain a modified form of the Timoshenko equations (assuming constant coefficients here)

$$\rho\frac{\partial^2 w}{\partial t^2} - I_\rho\left[\frac{\partial^3 \beta}{\partial t^2 \partial x} + \frac{\partial^2 w}{\partial t^2 \partial x^2}\right] + EI\left[\frac{\partial^3 \beta}{\partial x^3} + \frac{\partial^4 w}{\partial x^4}\right] = 0 , \tag{3.07}$$

$$I_\rho\left[\frac{\partial^2 \beta}{\partial t^2} + \frac{\partial^3 w}{\partial t^2 \partial x}\right] + K\beta - EI\left[\frac{\partial^2 \beta}{\partial x^2} + \frac{\partial^3 w}{\partial x^3}\right] = 0 . \tag{3.08}$$

In (3.08) all terms except the first can be regarded as *shearing* forces to which the shear angle β responds through the action of the first term. In the shear diffusion model we suppose that a further viscous force affects the evolution of β:

$$I_\rho\left[\frac{\partial^2 \beta}{\partial t^2} + \frac{\partial^3 w}{\partial t^2 \partial x}\right] + 2\sigma\frac{\partial \beta}{\partial t} + K\beta - EI\left[\frac{\partial^2 \beta}{\partial x^2} + \frac{\partial^3 w}{\partial x^3}\right] = 0 . \tag{3.09}$$

Most commonly we assume I_ρ to be very small relative to the other constants present and we neglect the first term. What is then left is a diffusion process for β, hence the name. Applying $\partial/\partial x$ to (3.09) and adding the result to (3.07), that equation is replaced by

$$\rho\frac{\partial^2 w}{\partial t^2} + 2\sigma\frac{\partial^2 \beta}{\partial t \partial x} + K\frac{\partial \beta}{\partial x} = 0 , \tag{3.10}$$

and, assuming I_ρ very small, as we indicated, the differentiated equation (3.09) becomes

$$2\sigma\frac{\partial^2 \beta}{\partial t \partial x} + K\frac{\partial \beta}{\partial x} - EI\left[\frac{\partial^3 \beta}{\partial x^3} + \frac{\partial^4 w}{\partial x^4}\right] = 0 . \tag{3.11}$$

Comparing (3.10) and (3.11), we have

$$\rho\frac{\partial^2 w}{\partial t^2} + EI\left[\frac{\partial^3 \beta}{\partial x^3} + \frac{\partial^4 w}{\partial x^4}\right] = 0 . \tag{3.12}$$

Now we differentiate (3.10) twice with respect to x to obtain

$$\rho \frac{\partial^2 w}{\partial t^2 \partial x^2} + 2\sigma \frac{\partial^4 \beta}{\partial t \partial x^3} + K \frac{\partial^3 \beta}{\partial x^3} = 0 , \qquad (3.13)$$

solve (3.12) for $\frac{\partial^3 \beta}{\partial x^3}$ and substitute in (3.13) to finally obtain an equation in the lateral displacement w:

$$\frac{2\sigma}{K} \frac{\partial}{\partial t}\left[\rho \frac{\partial^2 w}{\partial t^2} - \frac{\rho EI}{2\sigma} \frac{\partial^3 w}{\partial t \partial x^2} + EI \frac{\partial^4 w}{\partial x^4}\right] + \rho \frac{\partial^2 w}{\partial t^2} + EI \frac{\partial^4 w}{\partial x^4} = 0. \quad (3.14)$$

Again, to get a rough idea of the damping versus frequency relationship, we consider solutions of the form (3.05), leading to the polynomial equation

$$\frac{2\sigma\rho}{K} \lambda^3 + \left[\rho + \frac{\rho EI}{K}\nu^2\right] \lambda^2 + \frac{2\sigma EI}{K}\nu^4 \lambda + EI\nu^4 = 0. \quad (3.15)$$

Dividing by the leading coefficient and renaming the resulting coefficients, we can take this equation to have the form

$$\lambda^3 + 2\gamma\nu^2\lambda^2 + \alpha^2\nu^4 + \varepsilon(\lambda^2 + \alpha^2\nu^4) = 0 , \qquad (3.16)$$

where all coefficients shown are positive. Letting $\lambda = \tilde{\lambda}\nu^2$, dividing the resulting equation by ν^4, equation (3.16) can, in turn, be replaced by

$$\tilde{\lambda}^2 + 2\gamma\tilde{\lambda} + \alpha^2 - \frac{2\gamma\varepsilon\tilde{\lambda}}{\nu^2\tilde{\lambda}+\varepsilon} = 0 \qquad (3.17)$$

For large values of λ we can expand the last term as a power series in $\varepsilon/\tilde{\lambda}\nu^2$:

$$\tilde{\lambda}^2 + 2\gamma\tilde{\lambda} + \alpha^2 - \frac{2\gamma\varepsilon}{\nu^2} + \frac{2\gamma\varepsilon^2}{\tilde{\lambda}\nu^4} + \cdots = 0$$

and conclude that, as $\nu^2 \to \infty$,

$$\lambda \approx \left[-\gamma \pm i\sqrt{\alpha^2 - \gamma^2}\right]\nu^2 ,$$

which shows the negative real part of λ to be asymptotically a multiple of the imaginary part, hence that the damping exponent is proportional to frequency at the high end of the spectrum. On the other hand, for small values of λ we can see that

$$\frac{2\gamma\varepsilon\tilde{\lambda}}{\nu^2\tilde{\lambda}+\varepsilon} = 2\gamma\tilde{\lambda} - \frac{2\gamma\nu^2\tilde{\lambda}^2}{\varepsilon} + \frac{2\gamma\nu^4\tilde{\lambda}^3}{\varepsilon^2} + \cdots ,$$

and we see that we can use

$$\tilde{\lambda}^2 + \alpha^2 + \frac{2\gamma\nu^2\tilde{\lambda}^2}{\varepsilon} - \frac{2\gamma\nu^4\tilde{\lambda}^3}{\varepsilon^2} + \cdots = 0 . \qquad (3.18)$$

To zero order we have $\tilde{\lambda}^2 = -\alpha^2$. Substituting this value for $\tilde{\lambda}^2$ in the last two terms, we obtain the equation

$$\tilde{\lambda}^2 + \frac{2\gamma\nu^4\alpha^4}{\varepsilon^2}\tilde{\lambda} + \alpha^2\left[1 - \frac{2\gamma\nu^2}{\varepsilon}\right] + \cdots = 0 .$$

From this we can see that

$$\lambda = \tilde{\lambda} \approx \left[-\gamma\alpha^2\nu^4 \pm i\sqrt{\alpha^4(1 - 2\gamma\nu^2/\varepsilon)^2 - \gamma^2\alpha^4\nu^8}\right]\nu^2 + \cdots$$

and hence that, at the low end of the spectrum, the damping exponent is proportional to the *cube* of the frequency - a somewhat unexpected result! However, the overall damping exponent versus frequency curve will still look much like what we have shown in Figure 1 of §1. It is also possible to see that for each value of ν^2 there is a real λ asymptotically tending to $-2\gamma\nu^2$ as $\nu^2 \to \infty$, corresponding to what we might call "shear relaxation modes".

The combined effect of the two kinds of damping considered in this section will yield a damping exponent with quadratic dependence on frequency at the low end and linear dependence at the high end of the spectrum. In between one might, depending on the relative strengths of the two types, obtain a curve resembling the one shown in Figure 4, below. Damping versus frequency relationships of this type have often been observed experimentally, and we make some remarks on this in [10] in connection with the spatial hysteresis model, but it is fair to add that those results involve external losses. It may well be that the interraction kernel $h(x,\xi)$ of that method can be specified so as to produce a similar relationship, which would provide an argument in favor of that procedure which some of its critics would find more palatable than those offered so far. The whole question of the nature of internal damping processes remains in its infancy, insofar as its mathematical development is concerned; any results at this stage must be considered tentative and subject to revision in the light of future developments.

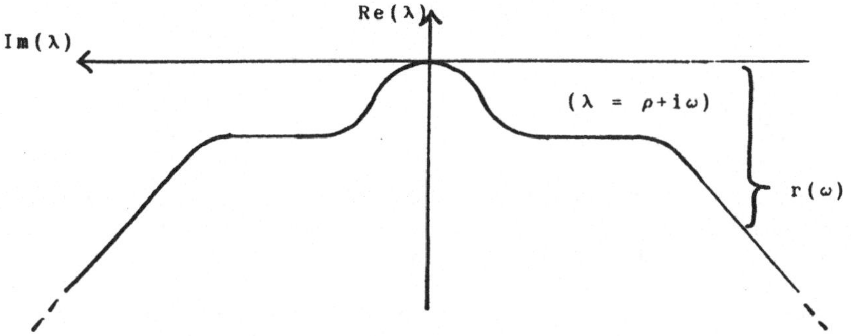

Figure 4: Combined Thermoelastic and Shear Diffusion

Finally, we note that both of the methods introduced in this section, which we have developed as partial differential equations of higher order than the original conservative model, can equally well be treated as PDE - Delay equations in which the order of the derivatives appearing is not raised, but at the expense of introducing a time dependent "memory" state. For the thermoelastic damping mechanism (cf.(3.04)) that equation takes the form, for an applied force f(x,t) satisfying appropriate decay requirements as $t \to -\infty$,

$$\rho \frac{\partial^2 w}{\partial t^2} + EI \frac{\partial^4 w}{\partial x^4} + LK \int_{-\infty}^{t} e^{-k(t-s)} \frac{\partial^5 w}{\partial t \partial x^4}(x,s)\, ds = f(x,t). \quad (3.19)$$

In the case of the shear diffusion mechanism the comparable equation is

$$\rho \frac{\partial^2 w}{\partial t^2} + EI \frac{\partial^4 w}{\partial x^4} - \frac{\rho EI}{2\sigma} \int_{-\infty}^{t} e^{\frac{K(t-s)}{2\sigma}} \frac{\partial^4 w}{\partial t^2 \partial x^2}(x,s)\, ds = f(x,t). (3.20)$$

In the case of both of the equations (3.19),(3.20) the homogeneous equation is not equivalent to (3.04),(3.14), respectively, because of the rapid growth of solutions of the homogeneous equation as $t \to -\infty$. These functional equations have much in common with a class of systems studied by Hannsgen, Renardy and Wheeler in [5],[6], relative to which additional developments are to be expected.

References

[1] Banks, H. T.: *Parameter estimation in partial differential equations; Computational methods*, to appear.

[2] Bert, C. W.: *Material damping: an introductory review of mathematical models, measures and experimental techniques.* J. Sound & Vib. 29 (1973), pp. 129 - 153.

[3] Chen, G., and D. L. Russell: *A mathematical model for linear elastic systems with structural damping.* Quart. Appl. Math., January, 1982.

[4] Courant, R. and D. Hilbert: *Methods of Mathematical Physics: Vol. II: Partial Differential Equations.* Interscience Pub. Co., New York, 1962

[5] Hannsgen, K., and R. Wheeler: *Time delays and boundary feedback stabilization in one-dimensional viscoelasticity.* Proc. 1986 Vorau Symposium on Identification and Control of Distributed Parameter Systems, Birkhauser, Pub.

[6] Hannsgen, K., Y. Renardy and R. Wheeler: *Effectiveness and robustness with respect to time delays of boundary and distributed stabilization in one-dimensional viscoelasticity.* SIAM J. On Cont. Opt., 26(1988), pp. 1200 - 1234.

[7] Hansen, S. W.: *General Input Elements.* Thesis, University of Wisconsin, Madison, December 1988.

[8] Russell, D. L.: *Quadratic performance criteria in boundary value control of linear symmetric hyperbolic systems.* SIAM J. Control, 11 (1973), pp. 475 - 509.

[9] Russell, D. L.: *Controllability and stabilizability theory for linear partial differential equations: recent progress and open questions.* SIAM Review, 4 (1978), pp. 639 - 739.

[10] Russell, D. L.: *On mathematical models for the elastic beam with frequency-proportional damping,* to appear in SIAM "Frontiers in Applied Mathematics" volume, H.T. Banks, Ed.

[11] Russell, D. L.: *On the positive square root of the fourth derivative operator,* to appear in *Quart. Appl. Math.*

[12] Russell, D. L.: *Some remarks on transfer function methods for infinite dimensional linear systems,* to appear.

[13] Zener, C.M.: *Elasticity and Anelasticity of Metals,* University of Chicago Press, 1948

PART 2

INVITED LECTURES

ON RIBLETS IN LAMINAR FLOWS

O. Pironneau: University of Paris 6 and INRIA, France.
INRIA 78153 Le Chesnay, FRANCE
G. Arumugam: University of Madurai, India

ABSTRACT:

Riblets are little groves dug on the surface of flying or swimming solid bodies in order to reduce their drag. Why riblets work is not known but it is believed that it is a turbulent process because they trap the main eddies that develop near the solid body surfaces.

This report deals with the efficiency of riblets in laminar viscous flows. We formulate the problem as a distributed parameter optimum design problem for Poiseuille flow, show that even without turbulence there are optimum riblets and compute them numerically by the Finite Element Method and the techniques of optimal control.

1. INTRODUCTION

Some recent experiences on sail boats (the Stars and Stripes) and airplanes have shown that their aerodynamism could be improved to some extent by digging tiny groves, called riblets, on their surface in the direction of the flow. Typically, these riblets have thickness and width less than a millimeter. Walsh [7] (see also Savill[9]) investigated experimentally the drag reduction that could be expected on a flat plate and found it to be around 5 % for saw-tooth shaped riblets of height half the logarithmic boundary layer thickness and period twice its height. He tried other periods and other shapes but found these characteristics to be the best, within measurement accuracy.

The explaination of this phenomenon is not clear but Walsh assumes that it is a turbulent phenomenon; horse-shoe structures and elongated eddies created by the plate are trapped between the teeth of the ribblets thus reducing the drag . But since the riblets are very small they lie well inside the laminar boundary layer and therefore, a priori, one could try to corolate the phenomenon to Poiseulle flow. The aim of this note is to elucidate this last point: ***could the drag of a flat plate in a plane Poiseuille laminar flow be improved by digging small periodic riblets on its surface in the direction of the flow?***

To answer this question we formulate it as an optimization problem with respect to the shape of the domain of a partial differential equation (optimal shape design [2], [5], [6],[8]). As we shall see there is a problem however with the modelling of the boundary layer so that one should not draw dramatic conclusion too hastily from our results. Nevertheless, by using the tools of calculus of variation in the context of optimal shape design we are able to prove that there are indeed better shapes than the flat plate for Poiseuille flow. The analysis is based on the derivation of first and second order optimality conditions and numerical solution of these on computer. The shape obtained is not as angular as the one suggested by Walsh but the height to length ratio is comparable; the improvement predicted is much more (around 15%) but this is probably because the displacement of the boundary layer thickness due to the riblet is not taken into account in this study. Thus our results indicate that one does not need to invoke the capturing of turbulent bursts by riblets to explain their drag reduction effect; to be sure this study should nevertheless be completed by a numerical simulation of the full Navier-Stokes equation in the proposed geometry.

Plain Poiseuille flow is modelled by a simple Laplace equation for the component of the velocity, u, which is parallel to the flat plate in the direction x_3 of the flow.

$$-\Delta u = k \quad in \quad \Omega \tag{1}$$

$$u|_\Sigma = 0, \quad \frac{\partial u}{\partial n}|_S = 0 \quad (\partial\Omega = \Sigma \cup S) \tag{2}$$

where Ω is the computational domain limited on the top by the flate plate with riblets dug on it (it is easier for us to set the cartesian reference frame so that the fluid is below the plate), on the bottom by the end of the viscous boundary layer where the Poiseuille approximation stops to be valid and the sides by the fact that there is a periodic array of riblets so one needs only to consider one cell. Here k is the pressure gradient in x_3 divided by the viscosity , $\frac{\partial u}{\partial n}$ is the normal derivative of u on $\partial\Omega$, the boundary of Ω; Σ is the riblets surface and S the remaining boundary of Ω, that is, the vertical and the bottom boundaries (figure 1). At the end of the boundary layer one should really put a matching condition with the external flow. The condition $\frac{\partial u}{\partial n}|_S = 0$ means that the flow is uniform, it is a classical matching condition but still it is rather arbitrary; this is certainly the weakest point of the paper because our results depend upon the position of the end of the boundary layer.

The boundary condition on the vertical parts of S implies periodicity when Σ has an axis of symetry (the equation for Σ is even in x_1).

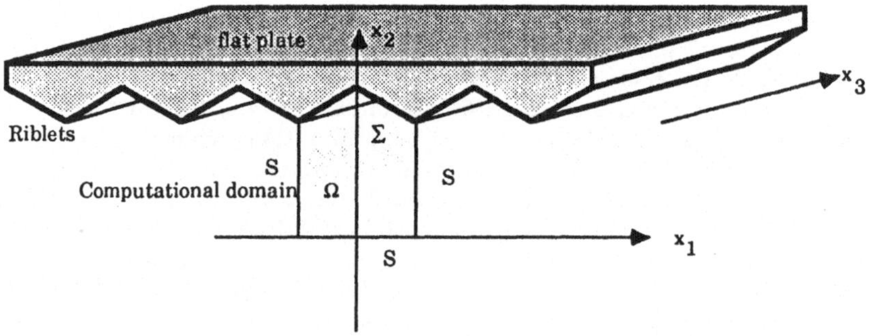

Figure 1
Computational domain; the riblets shown are on the lower side of the flat plate and the fluid flows below the flat plate normally to the cross section shown. The computational domain is limited by the riblet surface Σ, the end of

the boundary layer (horizontal part of S) and the periodicity (the vertical part of S)

2. FORMULATION OF THE OPTIMIZATION PROBLEM

The force dF that the fluid exerts on the surface element $d\gamma$ is

$$dF = \frac{\nu}{2}(0,0,\frac{\partial u}{\partial n})d\gamma \tag{3}$$

Therefore the total drag in the direction x_3 has modulus :

$$D = -\frac{\nu}{2}\int_{\Sigma}\frac{\partial u}{\partial n}d\gamma \tag{4}$$

So the riblets optimization problem is

$$\min_{\Sigma\in W}\{-\int_{\Sigma}\frac{\partial u}{\partial n}d\gamma; u \quad solution \quad of \quad (1),(2); \quad \int_{\Omega} u dx = d\}. \tag{5}$$

where W is the set of admissible curves Σ. One must give d or k otherwise the solution is $\Sigma = S$; it is reasonable to give the flux d and let the pressure drop k be adjusted so as to give d.

Let

$$v = \frac{u}{k}, \tag{6}$$

It can be shown easily (see [1]) that (1), (2), (5) become

$$-\Delta v = 1 \quad in \quad \Omega; v|_{\Sigma} = 0, \quad \frac{\partial v}{\partial n}|_{S} = 0 \tag{7}$$

$$\max_{\Sigma\in W}\{\frac{1}{|\Omega|}\int_{\Omega}|\nabla v|^2 dx : v \quad solution \quad of \quad (7)\}; \tag{8}$$

A relation of the type :

$$|\Omega| = f(\Sigma) \tag{9}$$

is also needed because if Σ is the flat plate (no riblets) then by sending the flat plate at infinity the drag on it tends to zero and we get an optimum with an infinite maximum in (8). The most simple relation of the type (9) is , for example, to say that the average height of the boundary layer is constant :

$$\int_{\Sigma} x_2 d\gamma = constant \tag{10}$$

or, in other words:

$$|\Omega| = a, \quad constant. \tag{11}$$

3. MATHEMATICAL ANALYSIS OF THE PROBLEM

3.1. Existence of the solution

We can apply the results of D.Chenais[3] but we can also get rid of regularity assumptions on Σ by applying the method used in [6] .

Let

$$V_d(\Omega) = \{v \in H^1(\Omega) : \quad v|_\Sigma = 0\} \tag{12}$$

Then (7)-(8) is also :

$$\max_{\Sigma \in W, v \in V_d(\Omega)} \frac{1}{2} \int_\Omega (|\nabla v|^2 - v) dx \tag{13}$$

and ***this problem has at least one solution*** when W is a set of all boundaries determining Ω in the allowable bounded open domain and such that $|\Omega| \geq a$.

$$W = \{\Sigma : |\Omega| \geq a, \quad \Omega \subset O\} \tag{14}$$

For the proof , see [6, Theorem 3, p. 37].

3.2.Optimality Conditions

Let Σ be a solution of (8) and let Ω be the corresponding domain. Let Σ' a neighbouring boundary of Σ be defined by

$$\Sigma' = \{x + \lambda\alpha(x)n(x) : x \in \Sigma\} \tag{15}$$

and Ω' be the corresponding domain. Here $n(x)$ is normal to Σ at x, $\alpha(x)$ is a regular function and λ is a scalar, very small. Let $v^{\Omega'}, v^\Omega$ denote the solutions of (7) in Ω' and Ω and let $\delta v = v^{\Omega'} - v^\Omega$, Finally let S' be the boundary of $\Omega' - \Sigma'$. Now we have

$$-\Delta\delta v = 0 \quad in \quad \Omega' \cap \Omega, \quad \frac{\partial \delta v}{\partial n}|_{S \cap S'} = 0, \tag{16}$$

But, $v^{\Omega'}(x) = v^\Omega(x) + \delta v(x)$ and

$$0 = v^{\Omega'}(x + \lambda\alpha n) = v^{\Omega'}(x) + \lambda\alpha\frac{\partial v^{\Omega'}}{\partial n} + o(\lambda), \quad \forall x \in \Sigma, \tag{17}$$

so $\delta v(x) = -\lambda\alpha\frac{\partial v^{\Omega'}}{\partial n} + o(\lambda) \quad \forall x \in \Sigma$ Therefore, $v'_\alpha = lim_{\lambda\to 0}\frac{v^{\Omega'} - v^\Omega}{\lambda}$ verifies

$$-\Delta v'_\alpha = 0 \quad in \quad \Omega, \quad v'_\alpha|_\Sigma = -\alpha\frac{\partial v}{\partial n}, \quad \frac{\partial v_{\alpha'}}{\partial n}|_S = 0. \tag{18}$$

The criteria (8) is

$$J(\Sigma) = \frac{1}{|\Omega|}\int_\Omega |\nabla v|^2 dx \tag{19}$$

If Σ is a solution, then it has to verify :

$$0 \geq \delta J = J(\Sigma') - J(\Sigma) = \frac{1}{|\Omega'|}\int_{\Omega'} |\nabla(v+\delta v')|^2 dx - \frac{1}{|\Omega|}\int_{\Omega} |\nabla v|^2 dx$$

$$= \frac{1}{|\Omega'|}\int_{\Omega'} |\nabla v|^2 dx - \frac{1}{|\Omega|}\int_{\Omega} |\nabla v|^2 dx + \frac{2}{|\Omega|}\int_{\Omega} \nabla v \nabla \delta v dx + o(\lambda)$$

$$= \frac{2}{|\Omega|}\int_{\Omega} \nabla v \nabla \delta v dx - \frac{\lambda}{|\Omega|^2}\int_{\Sigma} \alpha d\gamma \int_{\Omega} |\nabla v|^2 dx + \frac{\lambda}{|\Omega|}\int_{\Sigma} \alpha |\nabla v|^2 d\gamma + o(\lambda) \tag{20}$$

since

$$\int_{\delta\Omega} f(x)dx = \int_{\Omega'\backslash\Omega'\cap\Omega} f(x)dx - \int_{\Omega\backslash\Omega'\cap\Omega} f(x)dx = \int_{\Sigma} \lambda \alpha f(x) d\gamma + o(\lambda)$$

That is for all admissible α , we have

$$\frac{2}{|\Omega|}\int_{\Omega} \nabla v \nabla v'_{\alpha} dx - \frac{1}{|\Omega|^2}\int_{\Sigma} \alpha d\gamma \int_{\Omega} |\nabla v|^2 dx + \frac{1}{|\Omega|}\int_{\Sigma} \alpha |\nabla v|^2 d\gamma \leq 0 \tag{21}$$

The first integral is zero because

$$\int_{\Sigma} \nabla v \nabla v'_{\alpha} dx = \int_{\Omega} -\Delta v'_{\alpha} v dx + \int_{\partial\Omega} \frac{\partial v'_{\alpha}}{\partial n} v d\gamma = \int_{\Sigma} \frac{\partial v'_{\alpha}}{\partial n} v d\gamma = 0 \tag{22}$$

with (7), the equation (21) becomes :

$$\int_{\Sigma} \alpha |\frac{\partial v}{\partial n}|^2 d\gamma \leq \frac{\int_{\Sigma} \alpha d\gamma}{|\Omega|}\int_{\Omega} |\nabla v|^2 dx \quad \forall \alpha \quad admissible \tag{23}$$

In particular if W is given by (14) with O sufficiently large, the set of admissible α is $\{\alpha : \int_{\Sigma} \alpha d\gamma = 0\}$ and (23) implies that $|\partial v/\partial n|$ is constant on Σ.

The previous computation is somewhat formal but it can be justified with regularity assumptions on Σ and v, either directly (see [6]) or via the mapping technique of Murat-Simon [5] (see also Cea-Haug [2], Zolezio [8], ...) Let us sum up the results in the following proposition :

Proposition 1 *If Σ is a regular solution of (8) with (11) such that v is in $H^2(\Omega)$ then*

$$|\frac{\partial v}{\partial n}| = \quad constant \quad on \quad \Sigma \tag{24}$$

3.3.Search for an analytic solution

In both of the above cases, the problem is to find, for some given constant c, a domain Ω such that

$$-\Delta v = 1, \quad v|_\Sigma = 0, \quad \frac{\partial v}{\partial n}|_\Sigma = c, \quad \frac{\partial v}{\partial n}|_S = 0 \tag{25}$$

Notice before going any further that any stationnary point of (13), local maximums included, will satisfy (25), so a check on the sign of the second derivative of (13) will be necessary to check that we have a local minimum or a local maximum.

One trivial solution of the problem is : *the flat plate* In fact, if $\Omega =]-l, l[\times]0, L[$ then the solution of (25) is $v = -(x_2^2 - L^2)/2$ corresponding to $c = -3d/2lL^2$, which gives the drag .

$$D_p = \frac{3d\nu}{2L^2} = \frac{3d\nu}{2l^2}(\frac{l}{L})^2 \tag{26}$$

There is another trivial solution for the problem (25) :the *semi-circle* . If $\Omega = \{x : |x| < l, \ x_2 < 0\}$ $S \neq 0$, then (25) rewritten in polar coordinates admits a solution

$$v(r, \theta) = -(r^2 - l^2)/2,$$

with $c = -8d/\pi l^3$ and the drag

$$D_c = \frac{4d\nu}{l^2}. \tag{27}$$

Since $|\Omega|$ has to be the same in both the cases, we have $l/L = 4/\pi$. So $D_p/D_c = 6/\pi^2 \cong 0.6$, which says that *the flat plate is 40% better than the semi-cylinder.* So the semi-circle is only a stationary point of the problem (8) and not a minimum.

To know whether the flat plate is a minimum let us calculate the second variation of the criteria J with respect to $\alpha(x)$. That is, calculate $d^2J/d\lambda^2|_{\lambda=0}$. If this quantity is positive, then J is locally convex with respect to α. Let us do that calculation for the (8) with (11).

If $v^{\Omega'} = v^\Omega + \lambda v'_\alpha + \lambda^2/2 \ v''_{\alpha\alpha}$ then ,

$$J''_{\lambda\lambda} = \frac{d^2}{d\lambda^2}\int_{\Omega(\lambda)} |\nabla v(\lambda)|^2 dx = 2\int_\Omega |\nabla v'_\alpha|^2 dx + 2\int_\Omega \nabla v . \nabla v''_{\alpha\alpha} dx \tag{28}$$

$$+\frac{d^2}{d\lambda^2}\int_{\Omega(\lambda)} |\nabla v|^2 dx + 4\int_\Sigma \alpha \nabla v'_\alpha . \nabla v d\gamma,$$

But

$$\frac{d^2}{d\lambda^2}\int_{\Omega(\lambda)} |\nabla v|^2 dx = \int_\Sigma \alpha^2 \frac{\partial |\nabla v|^2}{\partial n} d\gamma$$

Indeed let us calculate the second variation of $\int_\Omega f(x, y) dx dy$ when $\Omega = \{(x, y) : -l \leq x \leq l; 0 \leq y \leq L(x)\}$ is replaced by $\Omega' = \{(x, (1 + \beta(x))y) : (x, y) \in \Omega\}$.

$$\int_{\Omega'} f(x, y) dx dy = \int_\Omega f(x, y(1 + \beta(x)))(1 + \beta(x)) dx dy$$

$$= \int_{\Omega} f(x,y)dxdy + \int_{\Omega} \beta(x)(f + yf'_{,y})dxdy + \int_{\Omega} \beta^2(x)(f'_y y + f'_{yy}\frac{y^2}{2})dxdy$$

But

$$\int_{\Omega} \beta y f'_{,y} dxdy = \int_{-l}^{l} \beta dx \int_{O}^{L} y f'_{,y} dy = \int_{-l}^{l} Lf(x, L(x))\beta(x)dx - \int_{\Omega} f\beta dxdy$$

and

$$\frac{1}{2}\int_{-l}^{l} \beta^2 f'_{yy} y^2 dxdy = \int_{-l}^{l} \beta^2 [f'_y(x, L(x))\frac{L(x)^2}{2} - \int_{O}^{L} f'_{,y} y dy]dx.$$

so

$$\int_{\Omega'} f dx = \int_{\Omega} f dx + \int_{\Sigma} fL\beta d\gamma + \int_{\Sigma} \frac{(\beta L)^2}{2} f'_{,y} d\gamma$$

which implies when $\lambda\alpha = L\beta$ and Ω is a rectangle that $\frac{d^2}{d\lambda^2}\int_{\Omega} f dx = \int_{\Sigma} \alpha^2 \frac{\partial f}{\partial n} d\gamma$.

Now since $v|_{\Sigma} = 0$, $\Delta v''_{\alpha\alpha} = 0$ and $\partial v''_{\alpha\alpha}/\partial n\ |_S = 0$, we also have: $\int_{\Omega} \nabla v \nabla v''_{\alpha\alpha} dx = 0$.

So

$$J''_{\lambda\lambda} = 2\int_{\Omega} |\nabla v'_{\alpha}|^2 dx + 4\int_{\Sigma} \alpha(\frac{\partial v'_{\alpha}}{\partial n})(\frac{\partial v}{\partial n})d\gamma + \int_{\Sigma} \alpha^2 \frac{\partial |\nabla v|^2}{\partial n} d\gamma \tag{29}$$

From (18) we have $v'_{\alpha} = -\alpha \partial v/\partial n$ and so

$$\int_{\Sigma} \alpha \frac{\partial v}{\partial n}\frac{\partial v'_{\alpha}}{\partial n} d\gamma = -\int_{\Sigma} v'_{\alpha} \frac{\partial v'_{\alpha}}{\partial n} d\gamma = -\int_{\partial\Omega} v'_{\alpha} \frac{\partial v'_{\alpha}}{\partial n} d\gamma = -\int_{\Omega} |\nabla v'_{\alpha}|^2 dx; \tag{30}$$

Finally we have

$$J''_{\lambda\lambda} = -2\int_{\Omega} |\nabla v'_{\alpha}|^2 dx + \int_{\Sigma} \alpha^2 \frac{\partial |\nabla v|^2}{\partial n} d\gamma = 2\int_{\Sigma} \alpha \frac{\partial v}{\partial n}\frac{\partial v'_{\alpha}}{\partial n} d\gamma + \int_{\Sigma} \alpha^2 \frac{\partial |\nabla v|^2}{\partial n} d\gamma. \tag{31}$$

In the case of *the flat plate* $v = (L^2 - x_2^2)/2$ and it is not easy to decide the sign of J''. So, since Σ has to be periodic we can develop $\alpha(x)$ in Fourier series. As we shall see in §4.1 the angle of contact between Σ, S has to be at 90 degrees, thus we should have $d\alpha(\pm l)/dx_1 = 0$ and so the Fourier series expansion for α is

$$\alpha(x_1) = \sum_{k=1}^{\infty} \alpha_k cos(k\pi\frac{x_1}{l}).$$

From (18) we find that

$$v'_{\alpha} = L\sum_k \alpha_k cos(k\pi\frac{x_1}{l})ch(k\pi\frac{x_2}{l})/ch(k\pi\frac{L}{l})$$

So

$$J''_{\lambda\lambda} = \int_{-l}^{l} -2\alpha L^2 \frac{\pi}{l}\sum_k k\alpha_k cos(k\pi\frac{x_1}{l})\tanh(k\pi\frac{L}{l}) + 2\int_{-l}^{l} \alpha^2 L dx$$

$$= 2Ll\sum_k \alpha_k^2[-\pi\frac{L}{l}k\tanh(k\pi\frac{L}{l}) + 1]$$

The above expression change its sign according to the sign of $-\pi\frac{L}{l}k\tanh(k\pi\frac{L}{l}) + 1$. Summing up the above result we have the following

Theorem 1

if $\pi L/l > z$, the root of $1 - x\tanh x$ then the flat plate is a local maximum of J for a given area because $J''_{\lambda\lambda} < 0$. Otherwise the flat plate is not a local maximum of J and a better shape of same area can be found.

Remark 1:

A similar calculation can be done for the semicercle; it is found that

$$J'' = 128\pi l\sum_{k\geq 1}\alpha^2(k-1).$$

So the semicercle is also not a local maximum of J.

4. SEARCH FOR A NUMERICAL SOLUTION

Before solving (8) numerically to discretize the domain correctly it is better to know the general shape of Σ, a priori, in particular, the corners if they exist. It is shown in [1] that Σ cannot have any corner except may be cusps and that the connection between S and Σ is at right angle.

Let $\{T_h\}_h$be a family of triangulation of Ω, indexed by the size of the largest edge of the element i.e. a set of triangles T_k with the following properties:

- $T_k \cap T_l = \emptyset$ or 1 edge or 1 vertex (k $\neq$ l)
- $\Omega_h = \cup T_k$; $\partial\Omega_h$ and $\partial\Omega$ have the same corners.
-no triangle should have an angle which tends to zero or π when h tends to zero.

Let V_h be the space of interpolation functions of degree 1 and which are zero valued on Σ_h :

$$V_h = \{w_h \text{ continuous, affine on } T_k\ \forall k, \text{ zero on } \Sigma_h\} \tag{33}$$

Let $v_h \in V_h$ be a solution of

$$\int_{\Omega_h} \nabla v_h \nabla w_h dx = \int_{\Omega_h} w_h dx \quad \forall w_h \in V_h, \tag{34}$$

Now (8) in discrete version becomes

$$\max_{\Sigma_h \in W}\{\frac{1}{|\Omega_h|}\int_{\Omega_h}|\nabla v_h|^2 dx : \quad v_h \quad solution \quad of \quad (34)\} \tag{35}$$

Proposition 2

Problem (35) has at least one solution. There exists a subsequence of a sequence of solutions of (35) which converges to a solution of (8) when h tends to zero.

Proof: Cf [6, Theorem 8 p120]

In fact the problem (35) is an optimization problem on the position of the vertices $\{q^i\}_{1...N}$ of the triangulation.

$$\max_{\{q_i\}\in Q}\{J(q^1..q^N)=\frac{1}{|\Omega_h|}\int_{\Omega_h}|\nabla v_h|^2dx:\quad v_h \quad solution \quad of \quad (34)\} \tag{36}$$

Gradients of J with respect to q_j have been computed and a computer program was written to solve the problem by a conjugate gradient method (see [1] for more details).

The results are shown on figure 2; 3 cases have been computed, each with same area 0.03.

In the first case the riblet width is 0.5; after 60 iterations a shape is obtained (figure 6) with drag equal to:

$$\frac{d}{2}\nu|\frac{\Omega|}{J_{60}}=d\nu\frac{0.03.10^4}{2*0.58}=250d\nu$$

givin a drag per unit width of $500d\nu$. This should be compared with the drag per unit width of a flat plate which is equal to $834d\nu$.

In the second case the riblet width is 3/4; it gives a drag equal to $520d\nu$ and a drag per unit width of $600d\nu$.

Finally in the third case computed the width is 1.2, the drag is $790d\nu$ and the drag per unit width $660d\nu$. *Therefore the first shape (figure 2) is the best.* It would have been better to produce a fourth computation with a riblet steeper than the first shape but for such shapes the computations are not so stable and it is more difficult to converge.

Figure 2

Top

Results after 60 iterations starting from a rectangle (L=0.06, l=1/2). The cost function $\int_\Omega |\nabla v|^2 dx$ *has initial value* $= 0.36 * 10^-4$ *and final value* $= 0.58 * 10^-4$. *The figure on the left shows the shape at iteration 60; the values of the inverse of the cost function at each iteration are shown in the center, the value of* $\partial v/\partial n$ *on* Σ *is shown on the right.*

middle

Same as top but with another width: L=0.04, l=3/4; giving $J_{60} = 0.29 * 10^-4$.

Bottom

Same as Top but L=0.025, l=1.2; giving $J_{60} = 0.6 * 10^-5$

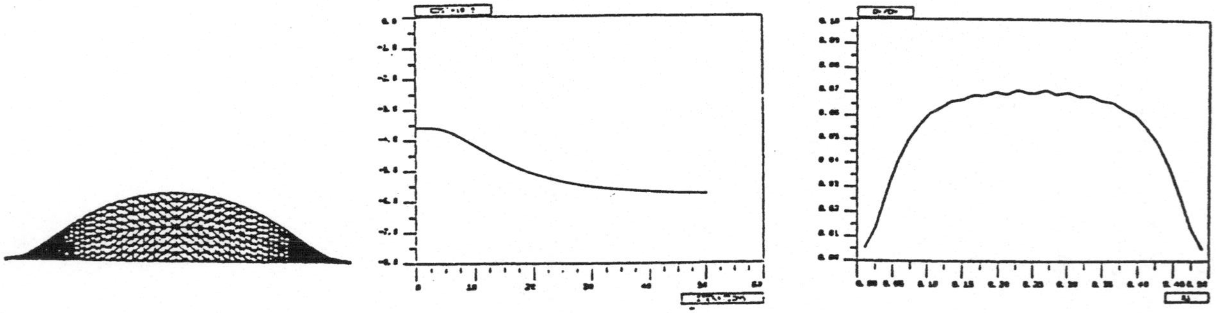

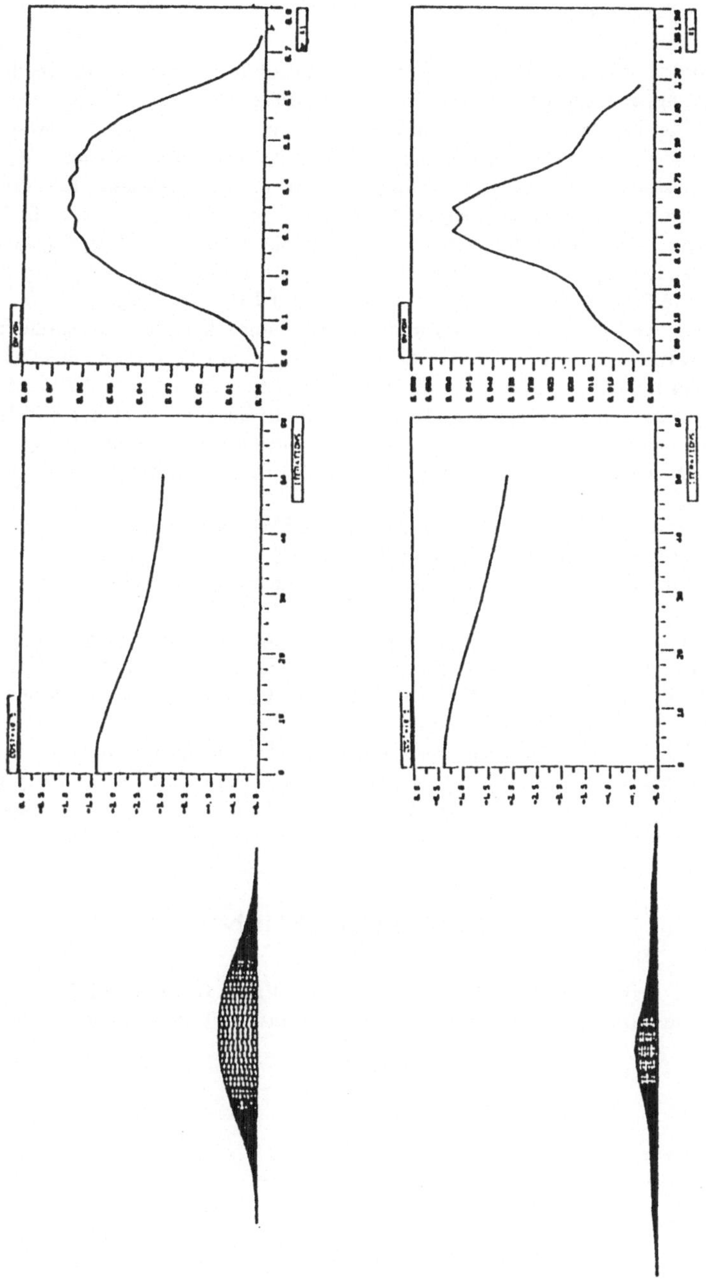

5. CONCLUSION

Assuming that the ribblets are in the viscous sublayer of the boundary layer so that the Poiseuille approximation is valid , we have found that there are better shapes than the flat plate. This fact can be proved analytically by studying the second order optimality conditions of the optimization problem corresponding to the minimization of the drag with respect to the shape of the riblet for a fixed boundary layer thickness. The shape found is shown on figure 2; it is a smooth shape, in height approximately half the thickness of the boundary layer. Conversely for a given boundary layer thickness L and a given period l for the riblets there is also an optimal shape; 3 such shapes have been computed and are shown on figure 2; this shape is the flat plate itself if L/l is too large.

This gives the following picture for riblets all along a flat plate: at some cross section perpandicular to the flow direction the absolute riblet shape is chosen. Since the boundary layer thichness is known at this section, this fixes the spanwise periodicity of the riblets; then upstream of that cross section the riblets should become smoother and thinner (one of the other optimal shapes for a given L and l) because the boundary layer is thinner; down stream of the chosen cross section the ribblet should steepen till the critical ration L/l is reached.

If we compare these conclusions witht those of Walsh we note that:

-1. the same aspect ratio, height/width is found

-2. the same ratio riblet height/ boundary layer height is found

but

-3. in our case it is the viscous boundary layer and not the log layer

-4. our riblets gives an improvement of 15% instead of the expected 8%

-5. our conclusions assume that the boundary layer thickness is independant of the riblets shape.

The authors believe that, because of item 5, the conclusions reached in this paper should not be taken for granted but nevertheless this works indicates clearly that it may be possible to improve the drag of the flat plate *in the laminar regime* by digging smooth tiny ribs on its surface.

ACKNOWLEDGEMENTS

We thank Prof.J.Cousteix, Prof.P.Perrier and Prof. R. Scott for the important discussions on the modelling of the problem and the interpretation of the results.

REFERENCES

[1] G. Arumugam, O. Pironneau: On the problem of riblets as a drag reduction device. (to appear)

[2] J. Cea, E.V. Haug: ***Optimization of distributed parameter structures***, Alphen aan der Rijn. Amsterdam 1981.

[3] D. Chenais: On the existence of solution in a domain identification problem. J. Math Anal Appl. **52** (2) (1975)

[4] J. Cousteix: cours de mécanique des fluides, école supérieure d'aéronauti- que.

[5] F. Murat, J. Simon: Studies in optimal shape design problem; in ***Lecture notes in computer sciences*** J. Cea de. Vol 41. Springer 1976.

[6] O. Pironneau: Optimal shape design for elliptic systems. Springer lecture notes in computational physics. New York 1984.

[7] M. Walsh: Drag characteristics of V-groove and transverse curvature riblets. Proc. Symposium on viscous drag reduction. Dallas, Texas, nov 1979.

[8] J.P. Zolezio: The material derivative method for optimal shape design problems. In J. Cea, E.V. Haug: ***Optimization of distributed parameter structures***, Alphen aan der Rijn. Amsterdam 1981.

[9] A. Savill, T. Truong, I. Ryhming: Report on the first european meeting on drag reduction. ***Journal de Mecanique thorique et appliqu.*** **7**, 4, pp353-378. 1988.

ITERATED HOMOGENIZATION AND THE EFFECTIVE PROPERTIES OF POLYCRYSTALS

Marco Avellaneda
Courant Institute of Mathematical Sciences
New York, U.S.A.

1. A ***single-phase polycrystal*** is a composite material formed by an aggregate of grains of different shapes and orientations of a single homogeneous crystal. Examples of such materials include metals, compacted powders and certain solidified gases [1].

The electrical conductivity tensor of the homogeneous crystal is denoted by a 3×3 symmetric matrix

(1)
$$\sigma_0 = \begin{pmatrix} \sigma_1 & 0 & 0 \\ 0 & \sigma_2 & 0 \\ 0 & 0 & \sigma_3 \end{pmatrix},$$

where

$$0 < \sigma_1 \leq \sigma_2 \leq \sigma_3 < \infty$$

are the three principal conductivities. We consider σ_0 as given (usually σ_0 is determined from the atomic structure of the pure crystal or from direct measurement). The ***local*** (i.e. fluctuating) ***conductivity tensor*** can be written as

(2)
$$\sigma(x) = \mathbf{R}^T(x)\sigma_0\mathbf{R}(x),$$

where $\mathbf{R}(x)$ is a field of rotations reflecting the variations of crystal orientation from point to point. We assume that the material is ***homogeneous on a macroscopic scale***. Then, the local relation between current and electric fields (Ohm's law)

(3)
$$\vec{j}(x) = \sigma(x)\vec{e}(x)$$

implies an effective relation between the average fields

(4)
$$< j(x) > = \sigma^* < e(x) >,$$

where σ^* is, by definition, the ***effective*** or ***homogenized*** conductivity tensor of the polycrystal. If Q denotes a ***representative volume element*** of material, σ^* is given by:

(5)
$$\sigma^* \cdot \vec{e}_0 = \fint_Q \sigma(x)(\nabla\phi(x) + \vec{e}_0)dx, \ \ \vec{e}_0 \in \mathbb{R}^3,$$

where $\nabla\phi$ is the ***fluctuating part of the electric field***, satisfying the partial differential equation

(6)
$$\begin{aligned} &\nabla \cdot \sigma(x)(\nabla\phi(x) + \vec{e}_0) = 0 \text{ in } Q \\ &\fint_Q \phi(x)dx = 0, \\ &\phi \text{ is } Q-\text{periodic}. \end{aligned}$$

Problem of interest: to determine the set of all possible σ^* as $R(x)$ varies over all possible fields of rotations, corresponding to different phase geometries.

In this paper, I shall give a partial answer to this problem and discuss also some aspects of the analogous problem for the *elastic constants*. These results were obtained in collaboration with K. Lurie, A. Cherkaev (Leningrad), and Graeme Milton (New York) [2,3].

2. One of the few phase-geometries for which σ^* is exactly computable is the *laminar geometry*, in which the crystallites are thin, parallel, periodically distributed planar slices. Consider two matrices

$$(7)\qquad \sigma_1 = \begin{pmatrix} \sigma_1^1 & 0 & 0 \\ 0 & \sigma_1^2 & 0 \\ 0 & 0 & \sigma_1^3 \end{pmatrix}, \quad \sigma_2 = \begin{pmatrix} \sigma_2^1 & 0 & 0 \\ 0 & \sigma_2^2 & 0 \\ 0 & 0 & \sigma_2^3 \end{pmatrix},$$

and set

$$(8)\qquad \sigma(x) = \sigma^1\chi(x_1) + \sigma^2(1-\chi(x_1)),$$

where χ is the characteristic function

$$(9)\qquad \chi(t) = \begin{cases} 1 & 0 \le t < p \\ 0 & p \le t < 1 \\ \chi(t+1) & t \in \mathbb{R} \end{cases},$$

and $p \in (0,1)$. Then, the corresponding effective tensor is given by

$$(10)\qquad \sigma^* = \begin{bmatrix} \sigma_1^* & 0 & 0 \\ 0 & \sigma_2^* & 0 \\ 0 & 0 & \sigma_3^* \end{bmatrix}$$

with

$$(11)\qquad \begin{aligned} \sigma_1^* &= \left(\frac{p}{\sigma_1^1} + \frac{(1-p)}{\sigma_2^1}\right)^{-1} \\ \sigma_2^* &= p\sigma_1^2 + (1-p)\sigma_2^2 \\ \sigma_3^* &= p\sigma_1^3 + (1-p)\sigma_2^3. \end{aligned}$$

3. Information on σ^* will be sought in the form of *inequalities* or *bounds* correlating the eigenvalues of σ^*. These bounds should depend on $\sigma_1, \sigma_2, \sigma_3$ (the given data). In obtaining sharp inequalities we (loosely) follow the following strategy.

(I) an *inequality* is derived from variational considerations or from a comparison principle;

(II) we analyze the behavior of the electric fields which saturate the bound of step (I) and derive *optimality conditions*,

(III) An *optimal microgeometry* (compatible with the inequality of (I)) is constructed based on the *optimality conditions* of (II).

4. Iterated Homogenization.

A key tool for constructing a wide class of computable microstructures is *iterated homogenization*, which corresponds physically to the concept of an *effective medium approximation* [4,5,6]. (See also [7, Ch. 1, §8].)

Suppose that the crystallites within the representative volume element Q fall under two categories: some are of size $O(1)$ and others of size $O(\epsilon)$ ($\epsilon << 1$) with respect to Q. Then, the local conductivity is of the form

$$\sigma_\epsilon = \sum(x, \frac{x}{\epsilon}). \tag{12}$$

We may assume, to fix ideas, that $\sum(x,y)$ is Q−periodic in both arguments. Define the inhomogeneous tensor $\sigma^{(1)}(x)$, and the uniform tensor $\sigma^{(2)}$ by

$$\sigma^{(1)}(x) = \text{ homogenized tensor of } \sum(x,\cdot), \tag{13}$$

(where x is kept *fixed*)

$$\sigma^{(2)}(x) = \text{ homogenized tensor of } \sigma^{(1)}(\cdot). \tag{14}$$

Then, it can be shown [7] that

$$\lim_{\epsilon \downarrow 0} \sigma_\epsilon^* = \sigma^{(2)}. \tag{15}$$

Physically, this means that the aggregate of $O(\epsilon)$ grains surrounding the $O(1)$ grains is replaced by an *equivalent effective medium*. This has been a long-standing procedure used in order to obtain approximate formulas for σ^* [4,5 and references therein]. It is now understood that these formulas are *exact* for appropriate phase-geometries with *widely separated scales*. Moreover, and perhaps somewhat surprisingly, some of these materials provide microstructures which achieve *optimal bounds* for σ^*.

5. Voigt's bound [2,8,9,10] is obtained by taking $\phi \equiv 0$ in (5). Accordingly, for each $\vec{e}_0 \in \mathbb{R}^3$, we have

$$\begin{aligned} \vec{e}_0^T \cdot \sigma^* \cdot \vec{e}_0 &= \inf_\phi \fint_Q \vec{e}(x) \cdot \sigma(x) \cdot \vec{e}(x) dx \qquad (\vec{e}(x) = \nabla\phi + \vec{e}_0) \\ &\leq \fint_Q \vec{e}_0 \cdot \sigma(x) \cdot \vec{e}_0, \end{aligned} \tag{16}$$

so that, summing up (16) over an orthonormal basis of $\mathbb{R}^3$,

$$\text{trace } \sigma^* \leq \textit{trace } \sigma_0. \tag{17}$$

Since

$$\textit{trace } \sigma^* = \inf_{\substack{\langle e(x) \rangle = I \\ \nabla \times e(x) = 0}} \fint_Q \textit{trace}[\mathrm{e}^T(x) \cdot \sigma(x) \cdot \mathrm{e}(x)] dx, \tag{18}$$

Voigt's bound is attained if and only if the tensor-valued electric field $\vec{e}(x)$ satisfies $\mathbf{e}(x) = I$ almost everywhere. Consequently, div $\sigma(x) = \operatorname{div}(\sigma(x)I) = 0$, so that the field $\mathbf{R}(x)$ satisfies the *optimality condition:*

(19) $$\operatorname{div}[\mathbf{R}^T(x)\sigma_0\mathbf{R}(x)] = 0.$$

Let us analyze this condition for *piecewise constant* fields $\mathbf{R}(x)$. Suppose that $\operatorname{div}[\mathbf{A}_1\chi_E(x) + \mathbf{A}_2(1 - \chi_E(x))] = 0$, where $\mathbf{A}_i$ are two constant matrices and χ_E is the characteristic function of a set E. This is possible if and only if

(20) $$(\mathbf{A}_1 - \mathbf{A}_2)\vec{n}(x) = 0, \ x \in \partial E,$$

where $\vec{n}(x)$ is the unit normal to ∂E. We conclude that (i) $\vec{n}(x)$ must be (piecewise) constant, and (ii) that it must coincide with a null eigendirection of $\mathbf{A}_1 - \mathbf{A}_2$. Thus, laminar microstructures are good candidates for optimal phase-geometries (saturating (17)). In fact, set

(21) $$\sigma(x) = \begin{bmatrix} \sigma_1 & 0 & 0 \\ 0 & \sigma_2 & 0 \\ 0 & 0 & \sigma_3 \end{bmatrix} \chi(x_1) + \begin{bmatrix} \sigma_1 & 0 & 0 \\ 0 & \sigma_3 & 0 \\ 0 & 0 & \sigma_2 \end{bmatrix} (1 - \chi(x_1)),$$

with χ as in (9). Then, condition (20) is satisfied. The corresponding σ^* is also diagonal in this frame of reference, and, from (11),

(22) $$\sigma_1^* = \sigma_1, \ \sigma_2^* = p\sigma_2 + (1-p)\sigma_3, \ \sigma_3^* = p\sigma_3 + (1-p)\sigma_2.$$

It is convenient to visualize the effective tensors σ^* in $(\sigma_1^*, \sigma_2^*, \sigma_3^*)$–space. Clearly, we have

(23) $$\sigma_1 \le \sigma_i^* \le \sigma_3, \ \ i = 1, 2, 3,$$

which restricts $(\sigma_1^*, \sigma_2^*, \sigma_3^*)$ to a *cube*. The bound (17) corresponds to

(21) $$\sigma_1^* + \sigma_2^* + \sigma_3^* \le \sigma_1 + \sigma_2 + \sigma_3$$

Formula (22) shows that the hexagon in which the plane $\sigma_1^* + \sigma_2^* + \sigma_3^* = \sigma_1 + \sigma_2 + \sigma_3$ intersects the boundary of the cube (23) can be swept by effective tensors corresponding to laminates, as we vary the volume fraction p. But we also know more: namely that if $(\alpha_1, \beta_1, \gamma)$ and $(\alpha_2, \beta_2, \gamma)$ are two points on this hexagon, then, by forming the laminate

(25) $$\sigma(x) = \begin{bmatrix} \alpha_1 & 0 & 0 \\ 0 & \beta_1 & 0 \\ 0 & 0 & \gamma \end{bmatrix} \chi(x_3) + \begin{bmatrix} \alpha_2 & 0 & 0 \\ 0 & \beta_2 & 0 \\ 0 & 0 & \gamma \end{bmatrix} (1 - \chi(x_3)),$$

we obtain an effective tensor σ^*, such that $(\sigma_1^*, \sigma_2^*, \sigma_3^*)$ describes, as we vary $p \in (0,1)$, the segment joining $(\alpha_1, \beta_1, \gamma)$ and $(\alpha_2, \beta_2, \gamma)$. We conclude that the ***interior*** of the hexagon also corresponds to realizable materials (laminates of laminates). In particular, the bound

(26) $$\sigma^* \le \frac{\sigma_1 + \sigma_2 + \sigma_3}{3},$$

for isotropic polycrystals is sharp. This result was obtained in 1976 by K. Schulgasser [10].

Note: For *uniaxial* single-crystals, the above mentioned hexagon reduces to a triangle.

6. A new optimal lower bound.

The Reuss bound [9,10]

$$\text{trace}(\sigma^*)^{-1} \leq \text{trace}(\sigma_0)^{-1} \tag{27}$$

leads to the optimality condition

$$\nabla \times [\mathbf{R}^T(x)(\sigma_0)^{-1}\mathbf{R}(x)] = 0, \tag{28}$$

which implies that the difference between the resistivities of neighboring crystallites must form a matrix of *rank 1*. It can be seen that this condition *cannot* be satisfied except in trivial cases. This gives evidence pointing to the suboptimality of (27). Schulgasser [10,11] showed in fact that this was the case and conjectured a possible better bound for *isotropic* polycrystals with *uniaxial single-phase* ($\sigma_1 = \sigma_2$ or $\sigma_2 = \sigma_3$). We shall derive a new *comparison principle* for this problem based on a *null-Lagrangian* which leads to an optimal bound. In the uniaxial case, this bound coincides with Schulgasser's conjecture.

Let $I_2(M)$ denote, for 3×3 matrices M, the function

$$I_2(M) = \text{trace Cof}(M) = \lambda_1\lambda_2 + \lambda_1\lambda_3 + \lambda_2\lambda_3, \tag{29}$$

where λ_i, $i = 1,2,3$ are the (possibly complex) eigenvalues of M. It can be shown using Plancherel's formula that

$$⨍_Q I_2(\nabla\vec{\phi}(x) + M)dx = I_2(M), \tag{30}$$

for all $M \in \mathbb{R}^{3\times 3}$ and all $\vec{\phi} \in H^1(Q, \mathbb{R}^3)$ which are Q-periodic. Null-Lagrangians have been studied systematically by Ball and coworkers [12] and Murat [13].

The additional fact that will be used about $I_2(M)$ is that, if σ is a positive constant such that

$$\begin{bmatrix} \sigma_1 & -\sigma/2 & -\sigma/2 \\ -\sigma/2 & \sigma_2 & -\sigma/2 \\ -\sigma/2 & -\sigma/2 & \sigma_3 \end{bmatrix} \geq 0, \tag{31}$$

then

$$\text{trace}[M^T\mathbf{R}^T\sigma_0\mathbf{R}M] \geq \sigma I_2(M). \quad \text{(See [2])}. \tag{32}$$

Let σ_* denote the largest positive constant satisfying (31). Combining (30) and (32), we conclude that, for all M,

$$\begin{aligned} \text{trace}[M^T\sigma^* M] &= \inf_{<\nabla\vec{\phi}> = I} \int_Q \mathit{trace}[(\nabla\vec{\phi}(x)^T\sigma(x)(\nabla\vec{\phi}(x)]dx \\ &\geq \inf_{<\nabla\vec{\phi}> = I} \sigma_* \int_Q I_2(\nabla\vec{\phi}(x))dx \\ &= \sigma_* I_2(M). \end{aligned} \tag{33}$$

This inequality is equivalent to

$$\sum_{i=1}^{3} \frac{\sigma_i^* - \sigma_s}{2\sigma_i^* + \sigma_s} \geq 0. \tag{34}$$

It can be shown from (31) that the constant σ_s is the unique positive root of the equation

$$\sum_{i=1}^{3} \frac{\sigma_i - \sigma_s}{2\sigma_i + \sigma_s} = 0. \tag{35}$$

Inequality (34), with σ_s given by (35), constitutes a new bound on the effective tensor σ^* of a single-phase polycrystal. In the case of an *isotropic* polycrystal, this bound becomes simply

$$\sigma^* \geq \sigma_s. \tag{36}$$

For *uniaxial* single-crystals, σ_s can be written more explicitly in terms of σ_1, σ_2, σ_3:
(I) $\sigma_1 = \sigma_2 < \sigma_3$:

$$\sigma_s = \frac{-\sigma_3 + \sqrt{\sigma_3^2 + 8\sigma_1\sigma_3}}{2}; \tag{37}$$

(II) $\sigma_1 < \sigma_2 = \sigma_3$:

$$\sigma_s = \frac{-\sigma_1 + \sqrt{\sigma_1^2 + 8\sigma_1\sigma_3}}{2}. \tag{38}$$

(See [2] for details.)

7. Attainability of the new bound.

The bound (34) is saturated if the tensor electric field $\nabla\phi(x)$ satisfies

$$\text{trace}[(\nabla\vec{\phi}(x))^T\sigma(x)(\nabla\vec{\phi}(x))] = \sigma_s I_2(\nabla\vec{\phi}(x)). \tag{39}$$

It can be shown that this relation implies that $\nabla\vec{\phi}(x) = (\nabla\vec{\phi}(x))^T$ and that, if $\alpha_1(x)$, $\alpha_2(x)$, $\alpha_3(x)$ denote the real eigenvalues of $\nabla\vec{\phi}(x)$, one has also

$$\sigma(x) = \mathbf{R}^T(x)\sigma_0\mathbf{R}(x); \quad \nabla\vec{\phi}(x) = \mathbf{R}^T(x)\begin{bmatrix} \alpha_1(x) & 0 & 0 \\ 0 & \alpha_2(x) & 0 \\ 0 & 0 & \alpha_3(x) \end{bmatrix}\mathbf{R}(x), \tag{40}$$

and

$$\begin{bmatrix} \sigma_1 & -(1/2)\sigma_s & (-1/2)\sigma_s \\ (-1/2)\sigma_s & \sigma_2 & (-1/2)\sigma_s \\ (-1/2)\sigma_s & (-1/2)\sigma_s & \sigma_3 \end{bmatrix}\begin{bmatrix} \alpha_1(x) \\ \alpha_2(x) \\ \alpha_3(x) \end{bmatrix} = 0, \tag{41}$$

for almost all $x \in Q$. Thus, the eigenvalues of $\nabla\vec{\phi}$ satisfy

$$\frac{\alpha_i(x)}{\alpha_j(x)} = \frac{2\sigma_j + \sigma_s}{2\sigma_i + \sigma_s}, \quad 1 \leq i < j \leq 3. \tag{42}$$

This optimality condition for the tensor electric field is realized in the uniaxial case by a *composite sphere assemblage*. To fix ideas we consider the case $\sigma_1 < \sigma_2 = \sigma_3$. First, define on the unit sphere of $\mathbb{R}^3$ the conductivity tensor

$$\hat{\sigma}(x) = \sigma_1 \frac{x}{|x|} \otimes \frac{x}{|x|} + \sigma_3 (I - \frac{x}{|x|} \otimes \frac{x}{|x|}), \quad |x| \leq 1. \tag{42}$$

Next, consider a covering of the cell Q by disjoint spheres S_i with centers x_i and radii r_i ($i = 1, 2, 3, \ldots$); and deifne

$$\sigma(x) = \hat{\sigma}(\frac{x - x_i}{r_i}), \quad \text{if } x \in Q \cap S_i. \tag{43}$$

Schulgasser [11] proved that the effective conductivity associated with (43) is given by (38). It can also be checked that the resulting elastic field satisfies the optimality conditions (40)-(42).

In [2], we showed that an *infinite-rank laminate* also realizes the bound $\sigma^* = \sigma_*$. This geometry is composed of successive, randomly oriented, infinitesimal inclusions of laminar crystallites of σ_0 in which the normal to the laminate is aligned with the axis of symmetry of the single crystal.

In a more recent development, Milton and E. Nesi [14] proved that the bound $\sigma^* = \sigma_*$ is also achieved when $\sigma_1 < \sigma_2 < \sigma_3$. Their microstructure is built by infinitely many successive laminations with suitable directions and volume fractions chosen at each step. The article [2] and recent unpublished work of Nesi and Milton indicate that the bound (34) is saturated for various effectively *anisotropic* composites as well. For example if σ_0 is uniaxial, simple laminates of $\sigma_* I$ and σ_0 with layers orthogonal to the axis of symmetry of the single crystal produce *optimal uniaxial materials* for which equality holds in (34).

8. One can also study the *elastic properties* of polycrystals [3]. (This is in fact a classical problem dating back to Voigt [8].) We consider as data the *elasticity* or *stiffness tensor* of the single crystal:

$$\mathbf{C}^{(0)} = (C^{(0)}_{ijkl}). \tag{44}$$

The local elasticity tensor $\mathbf{C}(x)$ of the polycrystal is then given by:

$$C_{ijkl}(x) = C^{(0)}_{\alpha\beta\gamma\delta} \mathbf{R}_{i\alpha}(x) \mathbf{R}_{j\beta}(x) \mathbf{R}_{k\gamma}(x) \mathbf{R}_{l\delta}(x). \tag{45}$$

We have investigated the problem of optimal bounds on the effective properties of *isotropic configurations*, for which $\mathbf{C}^*$ takes the form

$$C^*_{ijkl} = (\kappa^* - \frac{2}{3}\mu^*)\delta_{ij}\delta_{kl} + \mu^*(f_{ik}f_{jl} + f_{il} + f_{jk}). \tag{46}$$

The *effective bulk modulus* κ^* satisfies

$$\kappa^* = \frac{1}{9} I : \mathbf{C}^* : I \tag{47}$$

and measures the energy per unit volume stored by the polycrystal under a uniform *pressure field*. The Voigt-Reuss bounds for κ^* [3,9] are

$$\kappa^R \leq \kappa^* \leq \kappa^V, \tag{48}$$

with

$$\kappa^V = \frac{1}{9} I : C^{(0)} : I = \frac{1}{9} C^{(0)}_{iijj} \tag{49}$$

and

$$\kappa^R = I : (C^0)^{-1} : I]^{-1} = (S^{(0)}_{iijj})^{-1}, \tag{50}$$

where $\sigma^{(0)} = (\mathbf{C}^{(0)})^{-1}$ is the compliance of the single crystal.

An analysis of the upper Voigt bound shows that it will be saturated if and only if the *strain field* is uniform, i.e.

$$\mathrm{e}(x) = I, \quad a.e. \quad x \in Q. \tag{51}$$

The corresponding stress field is

$$\begin{aligned} \tau(x) &= \mathbf{C}(x)\epsilon(x) \\ &= \mathbf{R}(x)[\mathbf{C}^{(0)} : \mathbf{R}^T(x)\mathrm{e}(x)\mathbf{R}(x)]\mathbf{R}^T(x) \\ &= \mathbf{R}(x)(\mathbf{C}^{(0)} : I)\mathbf{R}^T(x) \end{aligned} \tag{52}$$

From div $\tau(x) = 0$, $\mathbf{R}(x)$ must satisfy for $\mathbf{R}(x)$:

$$\mathrm{div}[\mathbf{R}(x)(\mathbf{C}^{(0)} : I)\mathbf{R}^T(x)] = 0, \tag{53}$$

and, since the polycrystal is isotropic,

$$\fint_Q \mathbf{R}(x)(\mathbf{C}^{(0)} : I)\mathbf{R}^T(x)\, dx = 3\kappa^V I. \tag{54}$$

Examining equations (53), (54) we see that the problem of optimality of the Voigt bound $\kappa^* \leq \kappa^V$ is equivalent to the problem of optimality of the bound (26) on the conductivity of isotropic polycrystal; the conductivity tensor σ^0 being replaced by the *compressibility tensor* $\mathbf{C}^{(0)} : I$. We conclude therefore that $\kappa^* \leq \kappa^V$ is an *optimal* bound, in the same manner as we proved in section 5 the optimality of $\sigma^* \leq (1/3)(\sigma_1 + \sigma_2 + \sigma_3)$. The microgeometry realizing the bound (asymptotically) is a rank-two laminate [3].

9. An analysis of the Reuss bound shows that $\kappa^* = \kappa^R$ if and only if the stress tensor $\tau(x)$ is uniform, i.e.

$$\tau(x) = I, \quad x \ a.e. \tag{55}$$

As in the previous section, we can use this to derive a corresponding optimality condition on the field of orientations $\mathbf{R}(x)$. This condition is:

$$\mathbf{R}(x)(\mathbf{S}^{(0)} : I)\mathbf{R}^T(x) = \frac{1}{2}[\nabla\vec{u} + (\nabla\vec{u})^T], \tag{56}$$

for some displacement field $\vec{u}(x)$. Isotropy implies the supplementary condition

$$\fint_Q \mathbf{R}(x)(\mathbf{S}^{(0)} : I)\mathbf{R}^T(x)dx = \frac{1}{3\kappa^R} I. \tag{57}$$

In [3] it is shown that there exists a rank-two laminate geometry which is macroscopically isotropic and which achieves the Reuss bound. The argument is similar to the one for Voigt's bound. We refer the reader to this article for further details.

Thus, there is a lack of symmetry between the lower bound on the conductivity and the lower bound on the bulk modulus of isotropic polycrystals. This lack of symmetry is reflected in the fact that the optimality condition (28) for the Reuss bound for conductivity (which is not realizable) is equivalent to

$$\mathbf{R}^T(x)\sigma_0^{-1}\mathbf{R}(x) = \nabla\vec{u}, \text{ for some } \vec{u}, \tag{58}$$

which is more stringent than (56).

10. The analysis of *field optimality* conditions to construct microstructures and the new comparison principles based on null-Lagrangians are, in my belief, powerful tools for studying composite materials. Many other results along these lines can be proved by these methods and various open problems (e.g., optimal bounds on the *shear modulus* μ^*, study of "multiphase" polycrystals) remain to be analyzed. I hope to report on progress in this area in the near future.

REFERENCES

[1] H. B. HUNTINGTON, *The Elastic Constants of Crystals*, Solid State Reprints, Academic Press, NY, 1958.

[2] M. AVELLANEDA, A. V. CHERKAEV, K. A. LURIE, G. W. MILTON, *On the effective conductivity of polycrystals and a three-dimensional phase-interchange inequality*, J. App. Phys., May 15, 1988.

[3] M. AVELLANEDA, G. W. MILTON, Optimal bounds on the effective bulk modulus of polycrystals, to appear in SIAM Jour. Appl. Math.

[4] G. W. MILTON, *Microgeometries corresponding exactly with effective medium theories*, in: Physics and Chemistry of Porous Media, Johnson and Sen, Eds., AIP Conf. # 107, American Institute of Physics, NY, 1984.

[5] G. W. MILTON, *The coherent potential approximation is a realizable effective medium scheme*, Comm. Math. Physics, 99, 463-500, 1985.

[6] M. AVELLANEDA, *Iterated homogenization, differential effective medium theory, and applications*, Comm. in Pure and Appl. Math., Vol. XL, 527-554, 1987.

[7] A. BENSOUSSAN, J. L.LIONS, G. PAPANICOLAOU, *Asymptotic Analysis of Periodic Structures*, North-Holland, 1978.

[8] W. VOIGT, *Lehrbuch der Krystallphysik*, Teubner, Berlin, 1910.

[9] J. R. WILLIS, *Variational and related methods for the overall properties of composites*, in: Advances in Applied Mechanics 21, Yih, ed., Academic Press, New York, 1981.

[10] K. SCHULGASSER, *Bounds on the conductivity of statistically isotropic polycrystals*, J. Phys. C, 10, 1977, p. 407-417.

[11] K. SCHULGASSER, *Sphere assemblage model for polycrystals and symmetric materials*, J. Appl. Phys. 54, 1982, p. 1380.

[12] J. BALL, J. C. CURRIE and P. J. OLVER, *Null Lagrangians, weak continuity and variational problems of arbitrary order*, Jour. Func. Anal. 41, 1981.

[13] F. MURAT, *Compacité par compensation: condition nécessaire et suffisante de continuité sous hypothèse de rang constant*, Ann. Sc. Norm. Sup. Pisa 8, 1981, p. 69.

[14] G. W. MILTON and E. NESI, private communication on *cyclic microstructures*.

SOME REMARKS ON THE EXACT CONTROLLABILITY IN A NEIGHBOURHOOD OF THE BOUNDARY OF A PERFORATED DOMAIN

Doina CIORANESCU
Laboratoire d'Analyse Numérique
Université Pierre et Marie Curie
Tour 55-65, 4 place Jussieu
75252 PARIS CEDEX 05, France.

Patrizia DONATO
Dipartimento di Matematica e Applicazioni
Università di Napoli
Via Mezzocannone 8
80134 NAPOLI, Italy.

O.INTRODUCTION

Let us consider in a bounded domain of $\mathbb{R}^N$, $N \geq 2$, a system described by the wave equation. In J.L. Lions [5] it is proved that, given a Dirichlet boundary condition, one can act on the system by a control applied in a neighborhood of the boundary. This control, obtained by HUM (Hilbert Uniqueness Method) is an exact one, i. e. the system is driven to rest in finite time.

We treat here the case of a domain Ω perforated by periodic holes , the period being ε. The holes are completely contained in Ω. We consider mixed boundary conditions : Dirichlet on the exterior boundary $\partial\Omega$ and Neumann on the boundary ∂S_ε of the holes. In the first part of this work (§1- §3), for any fixed $\varepsilon > 0$, we construct by HUM an exact control v_ε with support in a neighborhood of the whole boundary $\partial\Omega \cup \partial S_\varepsilon$. Let ω be a neighborhood of the exterior boundary and ω_ε a neighborhood of the boundary of the holes (see figure 1). The thickness δ of ω is supposed to be independent of ε. Obviously the thickness $\delta(\varepsilon)$ of ω_ε depends on the size $r(\varepsilon)$ of the holes, hence on ε. The control v_ε we construct has its support in $\omega \cup \omega_\varepsilon$.

In the second part of the work we study the behaviour of the control v_ε as $\varepsilon \to 0$ (which implies, of course that $\delta(\varepsilon)$ and $r(\varepsilon) \to 0$, whereas the number of the holes tends to infinity). Two natural questions arise: can $r(\varepsilon)$ and $\delta(\varepsilon)$ be choosen in such a way that v_ε converges to some v with support in ω ? And if so, is this limit an exact control for a system posed in whole Ω ?

We give here a partial answer to these questions. More precisely , we show that for $N > 4$ the answer is a positive one : there exists r_0 such that if $r(\varepsilon) \geq r_0$ then v_ε converges to a control v applied in ω. For $N \leq 3$, in general, whatever is the choice of $r(\varepsilon)$ and $\delta(\varepsilon)$, our control v_ε diverges. The question for N=4 is still open. These results are proved by using a priori estimates which are of order $\delta^{-3}(\varepsilon)\, O(1)$ and in our approach these estimates are optimal.

1. STATEMENT OF THE PROBLEM.

Let Ω be a bounded open set of $\mathbb{R}^N$, $N \geqslant 2$, with boundary $\partial\Omega$ sufficiently smooth. Denote

$$\Gamma_0 = \{ x \in \partial\Omega \,/\, x.\nu > 0 \} ; \qquad \Gamma_0^* = \partial\Omega \setminus \overline{\Gamma}_0 ,$$

where ν is the exterior normal to Ω.

Let $\mathcal{V}$ be a neighborhood of Γ_0 of thickness δ and set

$$\omega = \mathcal{V} \cap \Omega .$$

Introduce the representative cell of $\mathbb{R}^N$: $Y = [0, \ell_1[\times .. \times [0, \ell_N[$, and denote by S an open subset of Y, with smooth boundary ∂S, such that $\overline{S} \subset Y$.

Denote by $\tau(r(\varepsilon)\,\overline{S})$ the set of all translated images of $r(\varepsilon)\overline{S}$ of the form $\varepsilon k_\ell + r(\varepsilon)\overline{S}$, $k \in \mathbb{Z}^N$, $k_\ell = (k_1 \ell_1, ..., k_N \ell_N)$, where $r(.)$ is a positive function such that $r(\varepsilon) \to 0$ if $\varepsilon \downarrow 0$.

We denote by S_ε the set of the holes contained in $\Omega \,/\, \overline{\omega}$ and define :

$$\Omega_\varepsilon = \Omega \setminus S_\varepsilon .$$

This means that the domain $\Omega \setminus \overline{\omega}$ is ε-periodically perforated by holes of size $r(\varepsilon)$.

Let $\delta(.)$ be a positive function, such that $\delta(\varepsilon) \to 0$ if $\varepsilon \downarrow 0$. Let

$$\omega_\varepsilon = \{ x \in \Omega_\varepsilon \,/\, \text{dist}\,(x, \partial S_\varepsilon) \leqslant \delta(\varepsilon) \}.$$

If, for example, the origin belongs to Ω we can have the following situation :

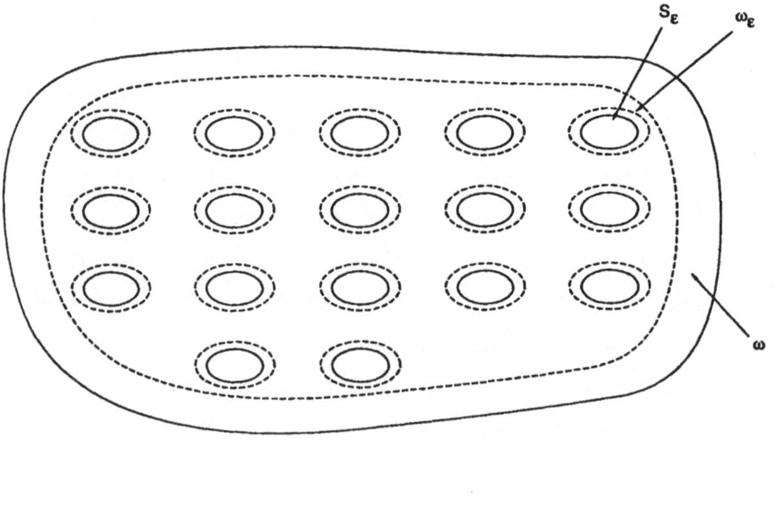

Figure 1

Introduce the space

$$V_\varepsilon = \{ \varphi \,/\, \varphi \in H^1(\Omega_\varepsilon),\ \varphi|_{\partial\Omega} = 0 \}$$

provided with the norm

$$\|\varphi\|_{V_\varepsilon} = \|\nabla\varphi\|_{(L^2(\Omega_\varepsilon))^N}$$

which is equivalent to the $H^1(\Omega_\varepsilon)$-norm with a Poincaré constant independent of ε.

In the following we use the notations :

$$\frac{\partial}{\partial\nu} = \partial_\nu\ ,\quad \frac{\partial}{\partial x_k} = \partial_{x_k}\ ,\quad \frac{\partial}{\partial t} = \partial_t .$$

Consider now the problem :

$$\begin{cases} z_\varepsilon'' - \Delta z_\varepsilon = v_\varepsilon & \text{in } \Omega_\varepsilon \times (0,T) \\ z_\varepsilon = 0 & \text{on } \partial\Omega \times (0,T) \\ \partial_\nu z_\varepsilon = 0 & \text{on } \partial S_\varepsilon \times (0,T) \\ z_\varepsilon(0) = z_\varepsilon^0 & \text{in } \Omega_\varepsilon \\ z_\varepsilon'(0) = z_\varepsilon^1 & \text{in } \Omega_\varepsilon\ . \end{cases} \tag{1}$$

Assume that the initial data are such that

$$\begin{cases} z_\varepsilon^0 \in L^2(\Omega_\varepsilon) \\ z_\varepsilon^1 \in V_\varepsilon' . \end{cases} \tag{2}$$

By the transposition method, if $v_\varepsilon \in [H^1(0,T;V_\varepsilon)]'$ then problem (1) has a unique solution $z_\varepsilon \in L^2(\Omega_\varepsilon \times (0,T))$, i.e. z_ε satisfies :

$$\int_0^T\!\!\int_{\Omega_\varepsilon} z_\varepsilon\, f\, dx\, dt + (z_\varepsilon^0, \theta_\varepsilon'(0)) - \langle z_\varepsilon', \theta_\varepsilon(0)\rangle = \langle v_\varepsilon, \theta_\varepsilon\rangle \tag{3}$$

for all $f \in L^1(0,T;L^2(\Omega))$, where θ_ε is solution of the transposed problem :

$$\begin{cases} \theta_\varepsilon'' - \Delta\theta_\varepsilon = f & \text{in } \Omega_\varepsilon \times (0,T) \\ \theta_\varepsilon = 0 & \text{on } \partial\Omega \times (0,T) \\ \partial_\nu \theta_\varepsilon = 0 & \text{on } \partial S_\varepsilon \times (0,T) \\ \theta_\varepsilon(T) = \theta_\varepsilon'(T) = 0 & \text{in } \Omega_\varepsilon\ . \end{cases}$$

We refer the reader for the details to J.L.Lions–E.Magenes [7].

Moreover we have the following estimate (see J.L.Lions [5]) :

$$\| z_\varepsilon \|_{L^\infty(0,T;L^2(\Omega_\varepsilon))} + \| z'_\varepsilon \|_{L^\infty(0,T;H^{-1}(\Omega_\varepsilon))} \leqslant c \{ \| z^0_\varepsilon \|_{L^2(\Omega_\varepsilon)} + \| z^1_\varepsilon \|_{H^{-1}(\Omega_\varepsilon)} +$$

$$+ \| v_\varepsilon \|_{[H^1(0,T;V_\varepsilon)]'} \} \tag{4}$$

with a constant c independent of ε.

First we fix ε and by HUM we construct an exact control for (1), i.e. a function v_ε such that there exists T (independent of ε) for which :

$$z_\varepsilon(T) = z'_\varepsilon(T) = 0 .$$

We shall require to this control to have its support in $\omega \cup \omega_\varepsilon$.

Next we study the behaviour of v_ε and z_ε as $\varepsilon \to 0$.

2. A PRIORI ESTIMATES.

We prove here some a priori estimates which enable us to apply HUM to our problem. In this section ε is fixed.

Let us consider the equation :

$$\begin{cases} \varphi''_\varepsilon - \Delta \varphi_\varepsilon = 0 & \text{in } \Omega_\varepsilon \times (0,T) \\ \varphi_\varepsilon = 0 & \text{on } \partial\Omega \times (0,T) \\ \partial_\nu \varphi_\varepsilon = 0 & \text{on } \partial S_\varepsilon \times (0,T) \\ \varphi_\varepsilon(0) = \varphi^0_\varepsilon & \text{in } \Omega_\varepsilon \\ \varphi'_\varepsilon(0) = \varphi^1_\varepsilon & \text{in } \Omega_\varepsilon \end{cases} \tag{5}$$

with $\varphi^0_\varepsilon \in V_\varepsilon$ and $\varphi^1_\varepsilon \in L^2(\Omega_\varepsilon)$. By classical results (5) has a unique solution $\varphi_\varepsilon \in L^\infty(0,T;V_\varepsilon)$ with $\varphi'_\varepsilon \in L^\infty(0,T;L^2(\Omega_\varepsilon))$.

In the following E^ε_0 denotes the energy related to problem (5):

$$E_0^\varepsilon = \frac{1}{2}\int_{\Omega_\varepsilon}(|\varphi_\varepsilon^1|^2 + |\nabla\varphi_\varepsilon^0|^2)\,dx.$$

Further for any ψ such that $\nabla\psi \in [L^1(\partial\Omega_\varepsilon)]^N$, we denote by ∇_σ the 1st order tangential operator on $\partial\Omega_\varepsilon$ defined by :

$$\partial_{x_i}\psi = \nu_i\,\partial_\nu\psi + (\nabla_\sigma)_i\,\psi.$$

Lemma 2.1. *Let* q *be a vector field of* $[C^1(\mathbb{R}^N \times (0,T))]^N$. *Then if* φ_ε *is the solution of* (5) *the following identity holds :*

$$\frac{1}{2}\int_0^T\int_{\partial\Omega} q_k\nu_k\,|\partial_\nu\varphi_\varepsilon|^2\,d\sigma\,dt = (\varphi_\varepsilon'(t), q_k\,\partial_{x_k}\varphi_\varepsilon(t))\Big|_0^T + \frac{1}{2}\int_0^T\int_{\Omega_\varepsilon}(|\varphi_\varepsilon'|^2 - |\nabla\varphi_\varepsilon|^2)\,\partial_{x_k}q_k\,dx\,dt -$$

$$-\frac{1}{2}\int_0^T\int_{\partial S_\varepsilon} q_k\nu_k\,|\varphi_\varepsilon'|^2\,d\sigma\,dt + \frac{1}{2}\int_0^T\int_{\partial S_\varepsilon} q_k\nu_k\,|\nabla\varphi_\varepsilon|^2\,d\sigma\,dt + \int_0^T\int_{\Omega_\varepsilon}(\partial_{x_j}\varphi_\varepsilon)(\partial_{x_j}q_k)(\partial_{x_k}\varphi_\varepsilon)\,dx\,dt -$$

$$-\int_0^T\int_{\Omega_\varepsilon}\varphi_\varepsilon'\,q_k'\,\partial_{x_k}\varphi_\varepsilon\,dx\,dt. \qquad (6)$$

Proof. We follow here the multipliers method, see J.L.Lions [5]. Multiply equation (5) by $q_k\,\partial_{x_k}\varphi_\varepsilon$. Integrating by parts we have :

$$0 = (\varphi_\varepsilon'(t), q_k\,\partial_{x_k}\varphi_\varepsilon(t))\Big|_0^T - \int_0^T\int_{\Omega_\varepsilon}\varphi_\varepsilon'\,q_k'\,\partial_{x_k}\varphi_\varepsilon\,dx\,dt - \int_0^T\int_{\Omega_\varepsilon}\varphi_\varepsilon'\,q_k\,\partial_{x_k}\varphi_\varepsilon'\,dx\,dt +$$

$$+\int_0^T\int_{\Omega_\varepsilon}\partial_{x_j}\varphi_\varepsilon\,\partial_{x_j}(q_k\,\partial_{x_k}\varphi_\varepsilon)\,dx\,dt - \int_0^T\int_{\partial\Omega} q_k\,\partial_{x_k}\varphi_\varepsilon\,\partial_\nu\varphi_\varepsilon\,d\sigma\,dt.$$

But

$$\int_0^T\int_{\Omega_\varepsilon}\varphi_\varepsilon'\,q_k\,\partial_{x_k}\varphi_\varepsilon'\,dx\,dt = \frac{1}{2}\int_0^T\int_{\Omega_\varepsilon} q_k\,\partial_{x_k}(|\varphi_\varepsilon'|^2)\,dx\,dt =$$

$$= -\frac{1}{2}\int_0^T\int_{\Omega_\varepsilon}|\varphi_\varepsilon'|^2\,\partial_{x_k}q_k\,dx\,dt + \frac{1}{2}\int_0^T\int_{\partial S_\varepsilon} q_k\nu_k\,|\varphi_\varepsilon'|^2\,dx\,dt$$

and, from the boundary conditions:

$$\int_0^T\int_{\Omega_\varepsilon} \partial_{x_j}\varphi_\varepsilon\, \partial_{x_j}(q_k\, \partial_{x_k}\varphi_\varepsilon)\, dx\, dt =$$

$$= \int_0^T\int_{\Omega_\varepsilon} (\partial_{x_j}\varphi_\varepsilon)(\partial_{x_j} q_k)(\partial_{x_k}\varphi_\varepsilon)\, dx\, dt + \frac{1}{2}\int_0^T\int_{\Omega_\varepsilon} q_k\, \partial_{x_k}(|\nabla\varphi_\varepsilon|^2)\, dx\, dt =$$

$$= \int_0^T\int_{\Omega_\varepsilon} (\partial_{x_j}\varphi_\varepsilon)(\partial_{x_j} q_k)(\partial_{x_k}\varphi_\varepsilon)\, dx\, dt - \frac{1}{2}\int_0^T\int_{\Omega_\varepsilon} |\nabla\varphi_\varepsilon|^2\, \partial_{x_k} q_k\, dx\, dt +$$

$$+ \frac{1}{2}\int_0^T\int_{\partial\Omega} q_k\nu_k\, |\partial_\nu\varphi_\varepsilon|^2\, d\sigma\, dt + \frac{1}{2}\int_0^T\int_{\partial S_\varepsilon} q_k\nu_k\, |\nabla\varphi_\varepsilon|^2\, d\sigma\, dt.$$

On the other hand, since $\partial_{x_k}\varphi_\varepsilon = \nu_k\, \partial_\nu\varphi_\varepsilon$ on $\partial\Omega$, it follows

$$\int_0^T\int_{\partial\Omega} q_k\, \partial_{x_k}\varphi_\varepsilon\, \partial_\nu\varphi_\varepsilon\, d\sigma\, dt = \int_0^T\int_{\partial\Omega} q_k\nu_k\, |\partial_\nu\varphi_\varepsilon|^2\, d\sigma\, dt.$$

Therefore :

$$0 = (\varphi_\varepsilon'(t),\, q_k\, \partial_{x_k}\varphi_\varepsilon(t))\Big|_0^T - \int_0^T\int_{\Omega_\varepsilon} \varphi_\varepsilon'\, q_k'\, \partial_{x_k}\varphi_\varepsilon\, dx\, dt + \frac{1}{2}\int_0^T\int_{\Omega_\varepsilon} |\varphi_\varepsilon'|^2\, \partial_{x_k} q_k\, dx\, dt -$$

$$- \frac{1}{2}\int_0^T\int_{\partial S_\varepsilon} q_k\nu_k\, |\varphi_\varepsilon'|^2\, d\sigma\, dt + \int_0^T\int_{\Omega_\varepsilon} (\partial_{x_j}\varphi_\varepsilon)(\partial_{x_j} q_k)(\partial_{x_k}\varphi_\varepsilon)\, dx\, dt -$$

$$- \frac{1}{2}\int_0^T\int_{\Omega_\varepsilon} \partial_{x_k} q_k\, |\nabla\varphi_\varepsilon|^2\, dx\, dt - \frac{1}{2}\int_0^T\int_{\partial\Omega} q_k\nu_k\, |\partial_\nu\varphi_\varepsilon|^2\, d\sigma\, dt + \frac{1}{2}\int_0^T\int_{\partial S_\varepsilon} q_k\nu_k\, |\nabla\varphi_\varepsilon|^2\, d\sigma\, dt$$

which proves the result.

■

Lemma 2.2. *There exists* T_0 *independent of* ε *such that for any* $T > T_0$ *and for* $(\varphi_\varepsilon^0, \varphi_\varepsilon^1)$ *in* $V_\varepsilon \times L^2(\Omega_\varepsilon)$ *we have :*

$$(2/R)\,(T - T_0)\, E_0^\varepsilon \leqslant \int_0^T\int_{\Gamma_0} |\partial_\nu\varphi_\varepsilon|^2\, d\sigma\, dt + (1/R)\left| \int_0^T\int_{\partial S_\varepsilon} x_k\nu_k(|\nabla_\sigma\varphi_\varepsilon|^2 - |\varphi_\varepsilon'|^2)\, d\sigma\, dt \right|$$

where R *is of same order as the diameter of* Ω.

Proof. Choose $q_k = x_k$ in (6). Set $X = (\varphi_\varepsilon'(t),\, x_k\, \partial_{x_k}\varphi_\varepsilon(t))\Big|_0^T$. Then :

$$0 = X - \frac{1}{2}\int_0^T\int_{\partial\Omega} x_k\nu_k |\partial_\nu\varphi_\varepsilon|^2 \, d\sigma\, dt + \frac{1}{2}\int_0^T\int_{\partial S_\varepsilon} x_k\nu_k(|\nabla_\sigma\varphi_\varepsilon|^2 - |\varphi_\varepsilon'|^2)\, d\sigma\, dt + \tag{7}$$

$$+ \frac{1}{2}(N-1)\int_0^T\int_{\Omega_\varepsilon}(|\varphi_\varepsilon'|^2 - |\nabla\varphi_\varepsilon|^2)\, dx\, dt + \frac{1}{2}\int_0^T\int_{\Omega_\varepsilon}(|\varphi_\varepsilon'|^2 + |\nabla\varphi_\varepsilon|^2)\, dx\, dt.$$

Moreover we have :

$$\frac{1}{2}\int_0^T\int_{\Omega_\varepsilon}(|\varphi_\varepsilon'|^2 + |\nabla\varphi_\varepsilon|^2)\, dx\, dt = T\, E_0^\varepsilon. \tag{8}$$

Denote $Y = (\varphi_\varepsilon', \varphi_\varepsilon)\Big|_0^T$. Multiplying (5) by φ_ε it comes easily that :

$$\int_0^T\int_{\Omega_\varepsilon} |\nabla\varphi_\varepsilon|^2 dx\, dt = \int_0^T\int_{\Omega_\varepsilon} |\varphi_\varepsilon'|^2\, dx\, dt - Y.$$

Hence by (7) and (8) we deduce that :

$$0 = X + \frac{1}{2}(N-1)\, Y + T\, E_0^\varepsilon - \frac{1}{2}\int_0^T\int_{\partial\Omega} x_k\nu_k |\partial_\nu\varphi_\varepsilon|^2\, d\sigma\, dt +$$

$$+ \frac{1}{2}\int_0^T\int_{\partial S_\varepsilon} x_k\nu_k(|\nabla_\sigma\varphi_\varepsilon|^2 - |\varphi_\varepsilon'|^2)\, d\sigma\, dt$$

which implies

$$T\, E_0^\varepsilon \leqslant |X| + \frac{1}{2}(N-1)|Y| + \frac{1}{2}\int_0^T\int_{\partial\Omega} x_k\nu_k |\partial_\nu\varphi_\varepsilon|^2\, d\sigma\, dt +$$

$$+ \frac{1}{2}\left|\int_0^T\int_{\partial S_\varepsilon} x_k\nu_k(|\nabla_\sigma\varphi_\varepsilon|^2 - |\varphi_\varepsilon'|^2)\, d\sigma\, dt\right|$$

where $R = \max_{x\in\Omega_\varepsilon} |x|$.

The definition of Γ_0 and Γ_0^* yields :

$$\int_0^T\int_{\partial\Omega} x_k\nu_k |\partial_\nu\varphi_\varepsilon|^2\, d\sigma\, dt \leqslant \frac{1}{2} R \int_0^T\int_{\Gamma_0} |\partial_\nu\varphi_\varepsilon|^2 d\sigma\, dt.$$

On the other hand $|X| \leqslant 2R\, E_0^\varepsilon$. Further if $(1/\mu_0)$ is the Poincaré constant of Ω , $|Y| \leqslant E_0^\varepsilon / \mu_0$.

Then :

$$[T - 2R - \tfrac{1}{2}(N-1)\,\mu_0^{-1}]\,E_0^\varepsilon \leq \tfrac{1}{2}R\int_0^T\int_{\Gamma_0}|\partial_\nu\varphi_\varepsilon|^2\,d\sigma\,dt +$$

$$+\tfrac{1}{2}\Big|\int_0^T\int_{\partial S_\varepsilon} x_k\nu_k(|\nabla_\sigma\varphi_\varepsilon|^2 - |\varphi_\varepsilon'|^2)\,d\sigma\,dt\Big|$$

which gives the claimed result if we choose $T_0 = 2R + \tfrac{1}{2}(N-1)\ \mu_0^{-1}$. ∎

Lemma 2.3. *We have :*

$$\tfrac{1}{2}\int_0^T\int_{\Gamma_0}|\partial_\nu\varphi_\varepsilon|^2\,d\sigma\,dt \leq E_0^\varepsilon + c\int_0^T\int_\omega(|\nabla\varphi_\varepsilon|^2 + |\varphi_\varepsilon'|^2)\,dx\,dt \tag{9}$$

where c *is a positive constant independent of* ε.

Proof . To prove (9) we need only to take in (6) : $q(x,t) = h(x)$, with $h(x) \in [C^1(\mathbb{R}^N)]^N$ satisfying

$$h\,.\,\nu = 1 \quad \text{on } \partial\Omega, \quad \text{supp}\,h \subset \omega, \quad \|h\|_{L^\infty(\omega)} \leq 1.$$

The existence of such a field is showed in J.L.Lions [5]. The constant c depends on R and δ. ∎

Lemma 2.4. *We have*

$$\tfrac{1}{2}\Big|\int_0^T\int_{\partial S_\varepsilon} x_k\nu_k(|\nabla_\sigma\varphi_\varepsilon|^2 - |\varphi_\varepsilon'|^2)\,d\sigma\,dt\Big| \leq R\,E_0^\varepsilon + c\,\delta^{-1}(\varepsilon)\int_0^T\int_{\omega_\varepsilon}(|\nabla\varphi_\varepsilon|^2 + |\varphi_\varepsilon'|^2)\,dx\,dt \tag{10}$$

where c *is a positive constant independent of* ε.

Proof. To prove the result it is enough to choose in (6) the field $q_k^\varepsilon(x,t) = x_k\,w_\varepsilon(x)$, where $w_\varepsilon \in C^1(\mathbb{R}^N)$ satisfies

$$\text{supp}\,w_\varepsilon \subset \omega_\varepsilon, \qquad w_\varepsilon = 1 \text{ on } \partial S_\varepsilon, \qquad 0 \leq w_\varepsilon \leq 1. \tag{11}$$

To construct w_ε let us consider a neighbourhood $\mathcal{U}_\varepsilon$ of S of thickness $\delta(\varepsilon)/r(\varepsilon)$. We introduce a function $h_\varepsilon(x)$ satisfying (see J.L.Lions [4], Chap. 1, Lemma 7.2) :

$$\begin{cases} \text{supp}\, h_\varepsilon \subset \mathcal{U}_\varepsilon, & h_\varepsilon = 1 \text{ on } \partial S, \\ 0 \leqslant h_\varepsilon \leqslant 1, & \|\nabla h_\varepsilon\|_{L^\infty(\mathbb{R}^N)} \leqslant c\, r(\varepsilon)\, \delta^{-1}(\varepsilon). \end{cases} \tag{12}$$

Set $w_\varepsilon(x) = h((x-\varepsilon k)/r(\varepsilon))$, $k \in \mathbb{Z}^N$. Hence w_ε verifies (11). Moreover :

$$\|\nabla w_\varepsilon\|_{L^\infty(\mathbb{R}^N)} \leqslant c\, \delta^{-1}(\varepsilon). \tag{13}$$

Using (12) and (13) into (6) we obtain the result.

In the special case of spheric holes one can also choose as w_ε the test function explicitely constructed in D.Cioranescu – F.Murat [2]. ■

Lemma 2.5. *There exists* T_1 *independent of* ε *such that for* $T < T_1$ *we have :*

$$\int_0^T\int_{\Gamma_0} |\partial_\nu \varphi_\varepsilon|^2\, d\sigma\, dt + \left|\int_0^T\int_{\partial S_\varepsilon} x_k \nu_k (|\nabla_\sigma \varphi_\varepsilon|^2 - |\varphi_\varepsilon'|^2)\, d\sigma\, dt\right| \leqslant$$

$$\leqslant c\int_0^T\int_\omega (|\nabla \varphi_\varepsilon|^2 + |\varphi_\varepsilon'|^2)\, dx\, dt + c\,\delta^{-1}(\varepsilon)\int_0^T\int_{\omega_\varepsilon} (|\nabla \varphi_\varepsilon|^2 + |\varphi_\varepsilon'|^2)\, dx\, dt \tag{14}$$

where c *is a positive constant independent of* ε.

Proof. Denote

$$\mathcal{J}_1 = \int_0^T\int_{\Gamma_0} |\partial_\nu \varphi_\varepsilon|^2\, d\sigma\, dt$$

$$\mathcal{J}_2 = \left|\int_0^T\int_{\partial S_\varepsilon} x_k \nu_k (|\nabla_\sigma \varphi_\varepsilon|^2 - |\varphi_\varepsilon'|^2)\, d\sigma\, dt\right|.$$

From lemmas 2.4 and 2.5 we know that :

$$\mathcal{J}_1 + \mathcal{J}_2 \leqslant 2(1+R)\, E_0^\varepsilon + c\int_0^T\int_\omega (|\nabla \varphi_\varepsilon|^2 + |\varphi_\varepsilon'|^2)\, dx\, dt +$$

$$+ c\,\delta^{-1}(\varepsilon)\int_0^T\int_{\omega_\varepsilon} (|\nabla \varphi_\varepsilon|^2 + |\varphi_\varepsilon'|^2)\, dx\, dt.$$

By lemma 2.2 it follows that

$$2(1+R)\,E_0^\varepsilon \leqslant R(1+R)\,(T-T_0)^{-1}\,\mathcal{J}_1 + (1+R)\,(T-T_0)^{-1}\,\mathcal{J}_2 .$$

Hence

$$[1-R(1+R)\,(T-T_0)^{-1}]\mathcal{J}_1 + [1-(1+R)\,(T-T_0)^{-1}]\,\mathcal{J}_2 \leqslant$$

$$\leqslant c\int_0^T\!\!\int_\omega (|\nabla\varphi_\varepsilon|^2 + |\varphi_\varepsilon'|^2)\,dx\,dt + c\,\delta^{-1}(\varepsilon)\int_0^T\!\!\int_{\omega_\varepsilon}(|\nabla\varphi_\varepsilon|^2 + |\varphi_\varepsilon'|^2)\,dx\,dt$$

which concludes the proof if $T_1 = \max\{R(1+R)+T_0\,,\ 1+R+T_0\}$. ■

Remark. For the sake of simplicity all integrals are written between 0 and T. Actually they could, as well, have been written between τ and $T-\tau$ with $\tau > 0$ and sufficiently small. More precisely, by using (8), inequality (14) in lemma 2.5 becames :

$$\int_\tau^{T-\tau}\!\!\int_{\Gamma_0}|\partial_\nu\varphi_\varepsilon|^2\,d\sigma\,dt + \Big|\int_\tau^{T-\tau}\!\!\int_{\partial S_\varepsilon} x_k\nu_k(|\nabla_\sigma\varphi_\varepsilon|^2 - |\varphi_\varepsilon'|^2)\,d\sigma\,dt\Big| \leqslant$$

$$\leqslant c\int_\tau^{T-\tau}\!\!\int_\omega(|\nabla\varphi_\varepsilon|^2 + |\varphi_\varepsilon'|^2)\,dx\,dt + c\,\delta^{-1}(\varepsilon)\int_\tau^{T-\tau}\!\!\int_{\omega_\varepsilon}(|\nabla\varphi_\varepsilon|^2 + |\varphi_\varepsilon'|^2)\,dx\,dt. \qquad (15)$$

Similarly if τ is choosen to have $T-2\tau > T_0$, from lemma 2.2 we get :

$$(2/R)\,(T-T_0)\,E_0^\varepsilon \leqslant \int_\tau^{T-\tau}\!\!\int_{\Gamma_0}|\partial_\nu\varphi_\varepsilon|^2\,d\sigma\,dt +$$

$$+ (1/R)\,\Big|\int_\tau^{T-\tau}\!\!\int_{\partial S_\varepsilon} x_k\nu_k(|\nabla_\sigma\varphi_\varepsilon|^2 - |\varphi_\varepsilon'|^2)\,d\sigma\,dt\Big|. \qquad (16)$$

■

Denote by $\omega_{\varepsilon/2}$ the set obtained by substituting $\delta(\varepsilon)$ by $\delta(\varepsilon)/2$ in the definition of ω_ε.

Lemma 2.6. *The following estimate holds :*

$$\int_\tau^{T-\tau}\!\!\int_{\omega_{\varepsilon/2}}|\nabla\varphi_\varepsilon|^2\,dx\,dt \leqslant C\,\delta^{-2}(\varepsilon)\int_0^T\!\!\int_{\omega_\varepsilon}(|\varphi_\varepsilon|^2 + |\varphi_\varepsilon'|^2)\,dt\,dx \qquad (17)$$

where C is a constant independent of ε.

Proof. By the same argument as in the proof of Lemma 2.5, we construct a function $\bar{w}_\varepsilon$ in $C^1(\mathbb{R}^N)$ satisfying:

$$\operatorname{supp} \bar{w}_\varepsilon \subset \omega_\varepsilon, \qquad \bar{w}_\varepsilon = 1 \text{ on } \omega_{\varepsilon/2}, \qquad 0 \leqslant \bar{w}_\varepsilon \leqslant 1. \tag{18}$$

Multiply equation (5) by $t(T-t)\,\bar{w}_\varepsilon^2 \varphi_\varepsilon$. We have:

$$\int_\tau^{T-\tau}\int_{\omega_{\varepsilon/2}} |\nabla \varphi_\varepsilon|^2 \,dx\,dt \leqslant \int_0^T\int_{\omega_\varepsilon} t(T-t)\,\bar{w}_\varepsilon^2 |\nabla\varphi_\varepsilon|^2\,dx\,dt \leqslant$$

$$\leqslant \left| \int_0^T\int_{\omega_\varepsilon} (T-2t)\,\bar{w}_\varepsilon^2\, \varphi_\varepsilon\, \varphi_\varepsilon' \,dx\,dt \right| + \bar{C} \int_0^T\int_{\omega_\varepsilon} \bar{w}_\varepsilon^2 \,|\varphi_\varepsilon'|^2\,dx\,dt \,+$$

$$+\, \bar{C}\int_0^T\int_{\omega_\varepsilon} |\, t(T-t)\,\bar{w}_\varepsilon\, \varphi_\varepsilon\, \nabla\bar{w}_\varepsilon\, \nabla\varphi_\varepsilon \,|\,dx\,dt .$$

Then (17) follows by using Young inequality in the last term, the definition (18) of $\bar{w}_\varepsilon$ and the estimate: $\|\nabla \bar{w}_\varepsilon\|_{L^\infty(\mathbb{R}^N)} \leqslant c\,\delta^{-1}(\varepsilon)$. ■

Proposition 2.7. *There exist* T_1, C *and* C_1 *independent of* ε *such that :*

$$\begin{aligned} C\,E_0^\varepsilon &\leqslant \int_0^T\int_\omega (|\varphi_\varepsilon|^2 + |\varphi_\varepsilon'|^2)\,dx\,dt + \delta^{-3}(\varepsilon)\int_0^T\int_{\omega_\varepsilon}(|\varphi_\varepsilon|^2 + |\varphi_\varepsilon'|^2)\,dt\,dx \leqslant \\ &\leqslant C_1 \delta^{-3}(\varepsilon)\,E_0^\varepsilon \end{aligned} \tag{19}$$

for all $T > T_1$.

Proof. The estimate (19) is a simple consequence of (15), (16), (17) and of the following inequality :

$$\int_\tau^{T-\tau}\int_\omega |\nabla\varphi_\varepsilon|^2\,dx\,dt \leqslant C(\omega)\int_0^T\int_\omega(|\varphi_\varepsilon|^2 + |\varphi_\varepsilon'|^2)\,dx\,dt$$

proved in J.L.Lions [5]. Here $C(\omega)$ is a constant dependent on δ^{-1} and independent of ε.

■

3. CONSTRUCTION OF THE EXACT CONTROL - HUM .

Consider again problem (1) under assumption (2). By making use of HUM we construct now a control v_ε. Let us point out that the control we obtain is not regular. For the case of domains without holes, such controls are introduced by J.L.Lions [5] and in general they are not distributions.

Let φ_ε be solution of (5) and ψ_ε solution of the backward problem :

$$\begin{cases} \psi_\varepsilon'' - \Delta\psi_\varepsilon = (-\varphi_\varepsilon'' + \varphi_\varepsilon)\,\chi_\omega + \delta^{-3}(\varepsilon)\,(-\varphi_\varepsilon'' + \varphi_\varepsilon)\,\chi_{\omega_\varepsilon} & \text{in } \Omega_\varepsilon \times (0,T) \\ \psi_\varepsilon = 0 & \text{on } \partial\Omega \times (0,T) \\ \partial_\nu \psi_\varepsilon = 0 & \text{on } \partial S_\varepsilon \times (0,T) \\ \psi_\varepsilon(T) = \psi_\varepsilon'(T) = 0 & \text{in } \Omega_\varepsilon \end{cases} \tag{20}$$

where $-\varphi_\varepsilon''\,\chi_\omega$, resp. $-\varphi_\varepsilon''\,\chi_{\omega_\varepsilon}$, are defined by

$$< -\varphi_\varepsilon''\,\chi_\omega\,, \Psi > = \int_0^T\!\!\int_\omega \varphi_\varepsilon'\,\Psi'\,dx\,dt,$$

respectively

$$< -\varphi_\varepsilon''\,\chi_{\omega_\varepsilon}\,, \Psi > = \int_0^T\!\!\int_{\omega_\varepsilon} \varphi_\varepsilon'\,\Psi'\,dx\,dt$$

for all $\Psi \in H^1(0,T;L^2(\Omega_\varepsilon))$. Therefore $-\varphi_\varepsilon''\,\chi_\omega$, resp. $-\varphi_\varepsilon''\,\chi_{\omega_\varepsilon}$, $\in [H^1(0,T;L^2(\Omega_\varepsilon))]'$.

The solution ψ_ε is defined by the transposition method, as the solution z_ε of problem (1). The only difference is that now the test function θ_ε in (3) satisfies the forward equation (transposed of (20)) with homogeneous initial data.

Let us define the following space:

$$F_\varepsilon = V_\varepsilon \times L^2(\Omega_\varepsilon)$$

and consider the operator $\Lambda_\varepsilon \in \mathcal{L}(F_\varepsilon\,, F_\varepsilon')$ defined by

$$\Lambda_\varepsilon\{\varphi_\varepsilon^0, \varphi_\varepsilon^1\} = \{\psi_\varepsilon'(0), -\psi_\varepsilon(0)\}.$$

By (5) and (20) we have :

$$< \Lambda_\varepsilon \{ \varphi_\varepsilon^0, \varphi_\varepsilon^1 \}, \{ \varphi_\varepsilon^0, \varphi_\varepsilon^1 \} > =$$

$$= \int_0^T \int_\omega (| \varphi_\varepsilon |^2 + | \varphi_\varepsilon' |^2) dx\, dt + \delta^{-3}(\varepsilon) \int_0^T \int_{\omega_\varepsilon} (| \varphi_\varepsilon |^2 + | \varphi_\varepsilon' |^2)\, dx\, dt. \tag{21}$$

Fix ε. According to (19) it turns out that the right hand term in (21) is a norm on F_ε, equivalent to the inner one, i.e. to E_0^ε. Hence Λ_ε is an isomorphism between F_ε and F_ε'. Therefore the equation

$$\Lambda_\varepsilon \{ \varphi_\varepsilon^0, \varphi_\varepsilon^1 \} = \{ z_\varepsilon^1, -z_\varepsilon^0 \} \tag{22}$$

has an unique solution. Set

$$v_\varepsilon = (-\varphi_\varepsilon'' + \varphi_\varepsilon)\, \chi_\omega + \delta^{-3}(\varepsilon)\, (-\varphi_\varepsilon'' + \varphi_\varepsilon)\, \chi_{\omega_\varepsilon}$$

which is an element of $[H^1(0,T; L^2(\Omega_\varepsilon))]'$. The uniqueness theorem implies that $z_\varepsilon = \psi_\varepsilon$ so that v_ε is the exact control for system (1).

4. LIMIT FOR $\varepsilon \to 0$.

In the following we denote by $\tilde{v}$ the zero extension to Ω for any function v defined on Ω_ε. We suppose :

$$\left| \begin{array}{lll} \text{i).} & \tilde{z}_\varepsilon^0 \rightharpoonup z^0 & \text{in } L^2(\Omega) \text{ weak} \\ \text{ii).} & \left| \begin{array}{l} \| z_\varepsilon^1 \|_{V_\varepsilon'} \leq \text{const.(independently of } \varepsilon) , \\ z_\varepsilon^1 \rightharpoonup z^1 \end{array} \right. & \\ & & \text{in } H^{-1}(\Omega) \text{ weak} \end{array} \right. \tag{23}$$

$$r(\varepsilon) = C\,\varepsilon^{\alpha}, \quad \delta(\varepsilon) = c\,\varepsilon^{\alpha} \quad \text{with } C, c > 0. \tag{24}$$

Theorem 4.1. *Assume* (23) *and* (24). *If* $N > 4$ *and* $\alpha \geqslant (N+2)/(N-4)$ *then :*

$$\begin{cases} \tilde{\varphi}_\varepsilon\,(\chi_\omega + \delta^{-3}(\varepsilon)\,\chi_{\omega_\varepsilon}) \rightharpoonup \varphi\,\chi_\omega & \text{in } L^1(\Omega \times (0,T)) \text{ weak} \\ \tilde{\varphi}_\varepsilon'\,(\chi_\omega + \delta^{-3}(\varepsilon)\,\chi_{\omega_\varepsilon}) \rightharpoonup \varphi'\,\chi_\omega & \text{in } L^1(\Omega \times (0,T)) \text{ weak.} \end{cases} \tag{25}$$

Moreover the function

$$v = (-\varphi'' + \varphi)\,\chi_\omega \tag{26}$$

is the exact control given by HUM for the problem

$$\begin{cases} z'' - \Delta z = v & \text{in } \Omega \times (0,T) \\ z = 0 & \text{on } \partial\Omega \times (0,T) \\ z(0) = z^0 & \text{in } \Omega \\ z'(0) = z^1 & \text{in } \Omega. \end{cases} \tag{27}$$

Furthermore

$$\tilde{z}_\varepsilon \rightharpoonup z \qquad \text{in } L^2(0,T; L^2(\Omega)) \text{ weak}. \tag{28}$$

Proof in the case $\alpha > (N+2)/(N-4)$.

The assumption (23) on the size of $r(\varepsilon)$ implies that the holes are very small compared to the period. Clearly in this case

$$\chi_{S_\varepsilon} \to 0 \qquad \text{in } L^p(\Omega) \text{ strong, } p < +\infty.$$

From the definition of the operator Λ_ε and from the equation (22) it follows that

$$\begin{cases} \|\varphi_\varepsilon^0\|_{V_\varepsilon} \leq c \\ \\ \|\varphi_\varepsilon^1\|_{L^2(\Omega_\varepsilon)} \leq c \end{cases} \tag{29}$$

where c is independent of ε.

Hence, up to a subsequence :

$$\begin{cases} \tilde{\varphi}_\varepsilon^0 \rightharpoonup \varphi^0 & \text{in } L^2(\Omega) \text{ weak} \\ \\ \tilde{\varphi}_\varepsilon^1 \rightharpoonup \varphi^1 & \text{in } L^2(\Omega) \text{ weak.} \end{cases} \tag{30}$$

Consequently, there exists an extension operator $P_\varepsilon \in \mathcal{L}(L^\infty(0,T;H^k(\Omega_\varepsilon));L^\infty(0,T;H^k(\Omega)))$, $k = 0,1$, such that

$$\begin{cases} P_\varepsilon\varphi_\varepsilon \rightharpoonup \varphi & \text{in } L^\infty(0,T;H_0^1(\Omega)) \text{ weak *} \\ \\ (P_\varepsilon\varphi_\varepsilon)' \rightharpoonup \varphi' & \text{in } L^\infty(0,T;L^2(\Omega)) \text{ weak *.} \end{cases} \tag{31}$$

The existence of such extension operators follows from C.Conca-P.Donato [3] and D.Cioranescu-P.Donato [1].

Since we are in the case of a domain with "small" holes ($r(\varepsilon) << \varepsilon$) the limit function φ satisfies :

$$\begin{cases} \varphi'' - \Delta\varphi = 0 & \text{in } \Omega \times (0,T) \\ \varphi = 0 & \text{on } \partial\Omega \times (0,T) \\ \varphi(0) = \varphi^0 & \text{in } \Omega \\ \varphi'(0) = \varphi^1 & \text{in } \Omega. \end{cases} \tag{32}$$

We show now that $\varphi^0 \in H_0^1(\Omega)$. To prove it, let Q_ε be an extension operator in $\mathcal{L}(V_\varepsilon;H_0^1(\Omega))$ (see [3]). We have from (27) :

$$Q_\varepsilon\varphi_\varepsilon^0 \rightharpoonup \varphi^0 \qquad \text{in } H_0^1(\Omega) \text{ weak.}$$

But

$$\tilde{\varphi}_\varepsilon^0 = Q_\varepsilon \varphi_\varepsilon^0 \ \chi_{\Omega_\varepsilon}$$

where

$$\chi_{\Omega_\varepsilon} \rightharpoonup 1 \qquad \text{in } L^\infty(\Omega) \text{ weak*}$$

so that $\varphi^0 = \varphi_*^0$.

In the next step of the proof we establish convergences (25). From (31) we deduce that

$$P_\varepsilon \varphi_\varepsilon \rightharpoonup \varphi \qquad \text{in } L^\infty(0,T; L^{2^*}(\Omega)) \text{ weak *}$$

where $2^* = 2N/(N-2)$. Hypothesis (24) implies that, up to a subsequence :

$$\delta^{-3}(\varepsilon)\ \chi_{\omega_\varepsilon} \to 0 \qquad \text{in } L^{(2^*)'}(\Omega) \text{ strong}$$

where $(2^*)' = 2N/(N+2)$. Indeed, a simple computation yields :

$$\int_\Omega (\ \delta^{-3}(\varepsilon)\ \chi_{\omega_\varepsilon})^{(2^*)'} dx \ \simeq \ \varepsilon^{[\alpha N(N-4)/(N+2)\ ]-N}\ O(1). \tag{33}$$

Finally, to prove (26)-(28) we need to pass to the limit in problem (1) (and (20)). Let us recall that z_ε is defined by the transposition method so that we have to pass to the limit in the identity (3). The result comes easily from the estimate (4), hypothesis (23) and convergences (25). ■

To prove the result in the case $\alpha > (N+2)/(N-4)$ we shall make use of a result due to L.Tartar [9], that we state here together with its proof for the reader's convenience.

Lemma 4.2.(Tartar). *Assume* $N \geqslant 3$. *Let* $\{u_\varepsilon\}$ *be a bounded sequence in* $H^1(\Omega)$ *such that*

$$u_\varepsilon \rightharpoonup u \qquad \textit{in } H_0^1(\Omega) \textit{ weak.}$$

Let $\{g_\varepsilon\}$, $g_\varepsilon \geq 0$, *be a bounded sequence in* $L^{(2^*)'}(\Omega)$ *such that*

$$\begin{cases} g_\varepsilon \rightharpoonup g & \text{in } L^{(2^*)'}(\Omega) \text{ weak} \\ |g_\varepsilon|^{(2^*)'} \rightharpoonup h & \text{weakly* in the sense of measures} \end{cases} \tag{34}$$

with $h \in L^1(\Omega)$. *Then :*

$$g_\varepsilon u_\varepsilon \rightharpoonup g\,u \qquad \text{in } \mathcal{D}'(\Omega).$$

Proof. Without loss of generality we can suppose that

$$u_\varepsilon \geq 0 \quad \text{and} \quad u_\varepsilon \rightharpoonup 0 \qquad \text{in } H_0^1(\Omega) \text{ weak.} \tag{35}$$

The hypotheses imply, as $g_\varepsilon u_\varepsilon$ is bounded in $L^1(\Omega)$, that, up to a subsequence

$$g_\varepsilon u_\varepsilon \rightharpoonup \mu \qquad \text{in } \mathcal{D}'(\Omega)$$

where $\mu \geq 0$ is a measure. Moreover since $|u_\varepsilon|^{(2^*)}$ is bounded in $L^1(\Omega)$ one has also (up to a subsequence) :

$$|u_\varepsilon|^{(2^*)} \rightarrow \nu \qquad \text{in } \mathcal{D}'(\Omega)$$

where ν is a bounded measure.

By the concentration-compactness principle of P.L.Lions (see lemma in the introduction of [8]) it follows that there exist $\{a_i\}_{i\in\mathbb{N}} \subset \Omega$, such that

$$\nu = \Sigma_{i\in\mathbb{N}}\, c_i\, \delta_{a_i} \tag{36}$$

where c_i are nonnegative constants verifying $\Sigma_{i\in\mathbb{N}}\, (c_i)^{2/2^*} < +\infty$ and δ_{a_i} is the Dirac mass at a_i. As ν is a bounded measure it results that $\Sigma_{i\in\mathbb{N}}\, c_i < +\infty$.

Let $\theta \in \mathcal{D}(\Omega)$. By Holder inequality :

$$\int_\Omega \theta\, g_\varepsilon u_\varepsilon\, dx \leqslant \left(\int_\Omega \theta\, |g_\varepsilon|^{(2^*)'} dx\right)^{1/(2^*)'} \left(\int_\Omega \theta\, |u_\varepsilon|^{2^*} dx\right)^{1/2^*}$$

hence, letting $\varepsilon \to 0$:

$$\int_\Omega \theta\ d\mu \leqslant \left(\int_\Omega \theta\, h\, dx\right)^{1/(2^*)'} \left(\int_\Omega \theta\, d\nu\right)^{1/2^*} \tag{37}$$

for $\theta \in \mathcal{D}(\Omega)$.

Let E a Borel set. Since its characteristic function χ_E can be approached by smooth functions (for example by taking the convolution of χ_E with mollifiers), (37) gives :

$$\mu(E) \leqslant \left(\int_E h\, dx\right)^{1/(2^*)'} (\nu(E))^{1/2^*} \tag{38}$$

We can take as E the Borel set $A = \cup_{i\in N} a_i$. Then from (36) we have $\nu(\Omega \setminus A) = 0$, so that from (38) we get

$$\mu(\Omega \setminus A) = 0, \tag{39}$$

therefore μ is an atomic measure.

On the other side, since $h \in L^1(\Omega)$ it follows that $\int_A h\, dx = 0$, which implies $\mu(A) = 0$. Now using (39) one deduces $\mu(\Omega) = 0$. The positivity of μ implies that $\mu = 0$. ■

Proof in the case $\alpha = (N+2)/(N-4)$.

Arguing as in the proof of Theorem 4.1 we obtain the convergences (31) with φ verifying (32). By the choice of α we deduce from (33) that $\delta^{-3}(\varepsilon)\ \chi_{\omega_\varepsilon}$ is bounded in $L^{(2^*)'}(\Omega)$. Then since

$$\delta^{-3}(\varepsilon)\ \chi_{\omega_\varepsilon} \to 0 \qquad \text{in } L^1(\Omega) \text{ strong}$$

we have

$$\delta^{-3}(\varepsilon)\ \chi_{\omega_\varepsilon} \to 0 \qquad \text{in } L^{(2^*)'}(\Omega) \text{ weak.}$$

Therefore, to obtain the result we apply Lemma 4.2 with $u_\varepsilon = \varphi_\varepsilon$ and $g_\varepsilon = \delta^{-3}(\varepsilon)\ \chi_{\omega_\varepsilon}$.

■

Remarks.

1. We notice that the limit control (26) which has its support in ω is the same as that obtained in J.L.Lions [5] when one wants to control in a neighbourhood of the boundary of a (non perforated) domain.

2. The theorem requires that $N > 4$. For $N \leqslant 4$ the control v_ε can diverge. Actually it diverges for $N \leqslant 3$ since in this case, for any choice of $r(\varepsilon)$ and $\delta(\varepsilon)$, $\delta^{-3}(\varepsilon)\ \chi_{\omega_\varepsilon}$ diverges in $L^1(\Omega)$.

For $N = 4$ we can have convergence of $\delta^{-3}(\varepsilon)\ \chi_{\omega_\varepsilon}$ in $L^1(\Omega)$ for suitable α but not in $L^{(2^*)'}(\Omega)$, as its norm in this space is not bounded whatever is the choice of $r(\varepsilon)$ and $\delta(\varepsilon)$. So, we cannot answer in this case.

The negative answer for $N \leqslant 3$, comes from the a priori estimates (19), which are of order $\delta^{-3}(\varepsilon)\ O(1)$. According to the method we used, this order is optimal (see also J.L.Lions [5], Chap.VII, §2.3). ■

REFERENCES

[1] D. Cioranescu, P. Donato. Exact internal controllability in perforated domains, to appear in Journal Math. Pures et Appl. 1988.

[2] D. Cioranescu, F. Murat. Un terme étrange venu d'ailleurs, dans *Nonlinear Partial Differential Equations and their Applications. Collège de France Seminar*, vol. II, 98-138, vol III, 157-178, Research Notes in Mathematics, Pitman, London, 1981.

[3] C. Conca, P. Donato. Non homogeneous Neumann's problems in domains with small holes, to appear in Modélisation Math. Analyse Numérique vol. 4 .1988.

[4] J. L. Lions. *Quelques méthodes de résolution de problèmes aux limites non linéaires*, Dunod, Paris 1969.

[5] J. L. Lions. *Controlabilité exacte et stabilisation des systèmes distribués.*, vol. I, à paraître dans RMA - Masson Paris 1988.

[6] J. L. Lions. Contrôlabilité exacte et homogénéisation (I), Asymptotic Analysis 1(1988), 3- 11.

[7] J. L. Lions , E. Magenes. *Problèmes aux limites non homogènes et applications* ., vol. 1 et 2. Dunod, Paris 1968.

[8] P. L. Lions. The concentration- compactness principle in the calculus of variations. The limit case. Part I, Rev. Mat. Iber. I, 1 (1985),145-200.

[9] L. Tartar. Personal communication .

SHAPE ANALYSIS AND OPTIMIZATION: LAGRANGIAN METHODS AND APPLICATIONS[1]

M.C. Delfour

Centre de recherches mathématiques et
Département de mathématiques et de statistique
Université de Montréal
C.P. 6128, Succursale A
Montréal (Québec), Canada, H3C 3J7.

J.P. Zolésio

Laboratoire de Physique Mathématique
Université des Sciences et Techniques
du Languedoc,
Pl. Eugène-Bataillon
34060 - Montpellier, Cédex, France.

ABSTRACT.

The computation of Shape Gradients is a central element in Shape Analysis and Optimization Problems. Various techniques are available to perturb the domain and the shape of the object under consideration. The deformation of the domain by a Velocity Field (the so-called Speed Method) provides an efficient way to get computable expressions for the Shape Gradient. Simple techniques using theorems on the differentiability of a Min Max with respect to a parameter and a Lagrangian formulation have recently been used. They provide a mathematical justification to many quick computations suggested by several authors for a large family of problems.

In this paper we present new results for the computation of Shape Gradients in cases where the application of theorems on the differentiability of a Min Max fails or is difficult to apply. Applications to nonlinear problems and variational inequalitites will be considered.

1. INTRODUCTION.

The Computation of Shape Gradients is a central element in Shape Optimization and Analysis Problems. Various Techniques are available to perturb the domain under consideration. In particular the use of a Velocity Field (also called Speed Method) provides an efficient way to develop a Shape calculus.

A direct approach which has been extensively used in the literature is the direct computation of the derivative of the state with respect to the domain. However this turns out to be only an intermediate computation which completely disappears from the final expression for the gradient upon introduction of an appropriate adjoint variable.

1 This research has been supported by NSERC Grants 0GP 008730 and INF 0007939 and a FCAR Grant for the Ministère de l'Education du Québec.

It is now well-known that the final expression can be obtained by introducing an appropriate Lagrangian where the adjoint variable coincides with the "Lagrange multiplier" associated with the state equation. This approach widely used in the Engineeering literature and quick computations can be performed (cf. J. CEA [1]) for most classical problems.

In this paper we go over several mathematical techniques to justify the Lagrangian approach in Shape Sensitivity Analysis. For convex cost functions and linear boundary value problems, everything can be justified by introducing the "Shape Lagrangian" and making use of theorems on the differentiability of a saddle point with respect to a parameter. These basic results can be extended to quasi convex cost functionals. (cf. DELFOUR and ZOLESIO [1, 2, 3, 4, 7]).

The Lagrangian approach is however more difficult to justify for nonlinear boundary value problems where the convexity is lost. Yet it is well known that the results are formaly the same as in the linear case. Variational inequalities provide another example which is not easily amenable to a Lagrangian formulation. In both cases a penalization method has been used to justify the Lagrangian approach. It is based on recent paper by DELFOUR and ZOLESIO [5, 6]. Of special interest is the case of variational equations where we manage to construct an adjoint variable which is the solution of a variational inequality. Since the underlying concepts are rather technical, we give an application to the control of variational inequalitites and a one dimensional example is completely worked out. The results are presented for convex cost functional, but they also readily extend to the quasi convex case.

2. SHAPE CALCULUS AND SHAPE LAGRANGIAN.

In this section we shall use the following simple illustrative example. Let Ω be an open domain in $\mathbb{R}^n$ with smooth boundary Γ and let $y = y(\Omega)$ be the solution of the boundary value problem

(1) $$-\Delta y + y = f \text{ in } \Omega, \frac{\partial y}{\partial n} = 0 \text{ on } \Gamma.$$

We want to compute the "derivative" of the cost function

(2) $$J(\Omega) = \int_\Omega |y(\Omega) - y_d|^2 dx$$

with respect to the domain Ω. The functions f and y_d are fixed and given a priori in $H^1(\mathbb{R}^n)$. $\mathbb{R}$ will denote the field of real numbers.

2.1. Velocity (or Speed) Method.

Domains Ω are generally not elements of a vector space and this requires the development of a "Shape calculus" to make sense of a "derivative" or a "gradient". Several techniques are available for this purpose. We shall use the "Velocity (or speed) Method" which contains the others as special cases. To define the "deformations" of the domain Ω we specify an artifical evolution time $t \geq 0$ and a time-dependent velocity field $V: [0, t^*] \times \mathcal{O} \to \mathbb{R}^n$ for a large enough open neighborhood $\mathcal{O}$ of Ω and an appropriate $t^* > 0$. The displacement of the point's X of $\mathbb{R}^n$ is governed by the differential equation

$$\frac{dx}{dt}(t) = V(t, x(t)) \ , \ t \geq 0 \ , \ x(0) = X \tag{3}$$

which generates a transformation $T_t(V)$ of $\mathbb{R}^n$ defined as

$$T_t(V)X = x(t) \ , \ t \geq 0 \ , \ X \in \mathbb{R}^n. \tag{4}$$

When V is specified we shall simply write T_t. Under the action of the field V, the domain Ω is transformed into a new domain $\Omega_t = T_t(\Omega)$. We now consider the solution y_t on Ω_t of the boundary value problem

$$-\Delta y_t + y_t = f \text{ in } \Omega_t \ , \ \frac{\partial y_t}{\partial n_t} = 0 \text{ on } \Gamma_t \tag{5}$$

(Γ_t, the boundary of Ω_t) and the associated cost function

$$J(\Omega_t) = \frac{1}{2} \int_{\Omega_t} |y_t - y_d|^2 dx. \tag{6}$$

The <u>Eulerian semi-derivative</u> of J at Ω for the velocity field V is defined as

$$dJ(\Omega; V) = \lim_{t \searrow 0} \frac{J(\Omega_t) - J(\Omega)}{t}. \tag{7}$$

In general the cost function J is an explicit function of the state variable y_t. Here

$$J(\Omega_t) = F(\Omega_t, Y_t) \tag{8}$$

where

$$F(\Omega, \varphi) = \frac{1}{2} \int_{\Omega} |\varphi - y_d|^2 dx \ , \ \forall \Omega \ , \ \forall \varphi. \tag{9}$$

A direct approach which is often followed in the literature is to use the derivative of the state with respect to t. Since the domain is moving there are at least two natural choices for that derivative. The first one is a "partial derivative" with respect to the parameter t at $t = 0$. It assumes that the state y_t on Ω_t has a smooth extension

$$Y : [0, t^*] \times \mathcal{O} \to \mathbb{R} \tag{10}$$

$$Y(t, \cdot)|_{\Omega_t} = y_t(\cdot) \tag{11}$$

where $t^* > 0$ and $\mathcal{O}$ is a large enough neighborhood of Ω. Then

$$Y'(x) = \lim_{\substack{t > 0 \\ t \to 0}} \frac{Y(t, x) - Y(0, x)}{t}. \tag{12}$$

The second "derivative" uses the "transported state"

$$y^t = y_t \circ T_t \tag{13}$$

on the fixed domain Ω. In this case all the differential quotients are defined on Ω and

$$\dot{y} = \lim_{t \searrow 0} \frac{y^t - y^0}{t}. \tag{14}$$

In both case it is necessary to study the properties of the state and find an equation for Y' or $\dot{y}$. In the end the final expression for $dJ(\Omega; V)$ will be constrained both by the state equation and the equation for the derivative of the state. In particular $dJ(\Omega; V)$ will not be explicitly expressed as a linear function of V

$$dJ(\Omega; V) = \langle \text{grad } J(\Omega), V \rangle. \tag{15}$$

At this stage, it is customary to a posteriori introduce an adjoint variable p which will explicitly give an expression for grad $J(\Omega)$ which only depends on y and p but not on the velocity field V. The "end result" is an expression which could have been directly obtained from a Lagrangian formulation without introducing or studying the derivative of the state Y' or $\dot{y}$.

2.2. Lagrangian method.

The solution y_t of the state equation (3) is the minimizing element of the quadratic energy functional

(13)
$$E(\Omega_t, \varphi) = \frac{1}{2}\int_{\Omega_t} \{ |\Delta\varphi|^2 + \varphi^2 - 2f\varphi\} \, dx$$

over the function space $H^1(\Omega_t)$. It is completely characterized by the variational equation

(14)
$$y_t \in H^1(\Omega_t), \; dE(\Omega_t, y_t; \varphi) = 0 \;, \quad \forall \varphi \in H^1(\Omega_t).$$

Moreover

(15)
$$J(\Omega_t) = F(\Omega_t, y_t),$$

where

(16)
$$F(\Omega_t, \varphi) = \frac{1}{2}\int_{\Omega} |\varphi - y_d|^2 dx.$$

This readily suggests to introduce the Lagrangian

(17)
$$G(\Omega_t, \varphi, \psi) = F(\Omega_t, \varphi) + dE(\Omega_t, \varphi; \psi)$$

since

(18)
$$J(\Omega_t) = \operatorname*{Inf}_{\varphi \in H^1(\Omega_t)} \; \operatorname*{Sup}_{\psi \in H^1(\Omega_t)} \; G(\Omega_t, \varphi, \psi).$$

We now require a theorem to differentiate a saddle point with respect to a parameter t.

2.3. Derivative of a saddle point with respect to a parameter.

For a saddle point where the spaces $H^1(\Omega_t)$ and $H^1(\Omega_t)$ depend on the parameter t, it is difficult to find a general theorem giving the expression for the derivative of $J(\Omega_t)$ with respect to t. When the spaces are independent of t we have a number of interesting results (cf. CORREA and SEEGER [1], DELFOUR and ZOLEZIO [2] and their bibliographies).

When the saddle point is unique we can use the following simple result.

__Theorem.__ Let A and B be arbitrary sets and L a map

(19)
$$L : [0, t^*] \times A \times B \to \mathbb{R}$$

for some $t^* > 0$. We make the following hypotheses:

H1. $\forall t \in [0, t^*]$, there exists a unique pair $(y^t, p^t) \in (A, B)$ such that for all $y \in A$ and $p \in B$

(20)
$$L(t, y, p^t) \le L(t, y^t, p^t) \le L(t, y^t, p).$$

H2. $\forall y \in A, p \in B, \partial_t L(t, \varphi, \psi), \forall t \in [0, t^*]$

H3. There exist topologies τ_A and τ_B on A and B such that

(i) $t \to (y^t, p^t)$ continuous at $t = 0$

(ii) $\forall p$, $(t, y) \to \partial_t L(t, y, p)$ is lower semi continuous

(iii) $\forall y$, $(t, p) \to \partial_t L(t, y, p)$ is upper semi continuous.

Then the function

(21) $$g(t) = L(t, y^t, p^t)$$

is differentiable from the right and

(22) $$dg(t) = \lim_{s>0\to 0} [g(t+s) - g(t)]/s = \partial_t L(t, y^t, p^t). \quad \blacklozenge$$

To get around the difficulty of the dependence of the space $H^1(\Omega_t)$ on t, we introduce a technique which is probably unique to Shape Analysis. We parametrize the elements of $H^1(\Omega_t)$ with the help of the diffeomorphism T_t

(23) $$\varphi \to \varphi \circ T_t^{-1} : H^1(\Omega) \to H^1(\Omega_t)$$

and introduce the Shape Lagrangian

(24) $$L(t, \varphi, \psi) = G(\Omega_t, \varphi \circ T_t^{-1}, \psi \circ T_t^{-1})$$

and necessarily

(25) $$J(\Omega_t) = \operatorname*{Inf}_{\varphi \in H^1(\Omega)} \ \operatorname*{Sup}_{\psi \in H^1(\Omega)} L(t, \varphi, \psi).$$

It turns out that the saddle point (y_t, p_t) of $G(\Omega_t, \cdot, \cdot)$ and (y_t, p_t) of $L(t, \cdot, \cdot)$ are related as follows:

(26) $$y^t = y_t \circ T_t \quad p^t = p_t \circ T_t$$

The final expression is given by

(27) $$dJ(\Omega; V) = \int_\Gamma [|y - y_d|^2 + \nabla y \cdot \nabla p + yp - fp] V(0) \cdot n d\Gamma$$

where

(28) $$-\Delta y + y = f \text{ in } \Omega , \quad \frac{\partial y}{\partial n} = 0 \text{ on } \Gamma,$$

(29) $$-\Delta p + p + 2(y - y_d) = 0 \text{ in } \Omega , \quad \frac{\partial p}{\partial n} = 0 \text{ on } \Gamma.$$

The purpose of the above example was to show the underlying philosophy behind the method. More details and extensions to non-differentiable convex cost function functionals can be found in DELFOUR and ZOLESIO [2]. Usually a non-differentiability will result in the non-uniqueness of solution to the adjoint system and the more general theorem of CORREA and SEEGER [1] is necessary. A version which is directly applicable to our problem along with a simple example can also be found in DELFOUR and ZOLESIO [2]. It is the same example as in part 2.2 but with the cost function

(30) $$J(\Omega_t) = F(\Omega_t, y_t)$$

where

(31) $$F(\Omega_t, \varphi) = \int_\Omega |\varphi - y_d| dx.$$

To deal with this example we introduce an extra variable

(32) $$\mu \in M_t = \{\mu \in L^2(\Omega_t) : |\mu(x)| \leq 1, \text{ a.e. in } \Omega\}$$

and express F as

(33) $$F(\Omega_t, \varphi) = \operatorname*{Sup}_{\mu \in M_t} \hat{F}(\Omega_t, \varphi, \mu)$$

where

$$\hat{F}(\Omega_t, \varphi, \mu) = \int_\Omega \mu\ (\varphi - y_d)\, dx .$$

Finally

(34) $$J(\Omega_t) = \operatorname*{Inf}_{\varphi \in H^1(\Omega_t)} \ \operatorname*{Sup}_{\substack{\psi \in H^1(\Omega_t) \\ \mu \in M_t}} \hat{G}(\Omega_t, \varphi, \mu, \psi)$$

where

(35) $$\hat{G}(\Omega_t, \varphi, \mu, \psi) = \hat{F}(\Omega_t, \varphi, \mu) + dE(\Omega_t, \varphi; \psi).$$

So the method can handle a whole family of problem which are amenable to the differentiability of a saddle point.

2.4. Extension to differentiable quasi convex cost functionals.

However the method is not limited to convex cost functionals. We can compute the derivative of concave cost functionals $F(t, \varphi)$ by applying the theorem to $-F(t, \varphi)$. So linear combinations of convex and concave functionals can be handled. In fact the whole class of quasi-convex functional can be handled.

A functional $F(\varphi)$ is quasi-convex if there exists a convex functional $K(\varphi)$ such that the functional

(36) $$(F + K)(\varphi) = F(\varphi) + K(\varphi)$$

is convex.

In particular any functional F defined on a Hilbert space H which is twice differentiable and for which

(37) $$\exists c > 0, \forall \psi \in H, d^2F(\varphi; \psi; \psi) + 2c \|\psi\|^2 \geq 0$$

is quasi-convex. This is equivalent to say that

(38) $$\exists c > 0, F(\varphi) + c \|\varphi\|^2$$

is convex. When F is only once differentiable the above condition is equivalent to

(39) $$\exists c > 0, \forall \varphi, \forall \psi, F(\psi) - F(\varphi) \geq dF(\varphi; \psi - \varphi) - c \|\psi - \varphi\|^2.$$

2.5. Conclusions.

In the light of the above discussion, the introduction of the Shape Lagrangian and the basic use of theorems on the differentiability of a saddle point with respect to a parameter makes it possible to consider

- convex non-differentiable cost functionals
- differentiable non-convex cost functionals

for most classical well posed linear boundary value problems. In particular this means that we can compute not only first but also second, third,.....derivatives without any technical difficulties.

3. A PENALIZATION METHOD FOR PROBLEMS GOVERNED BY NONLINEAR STATE EQUATIONS.

In section 2.4 we have indicated how to remove the convexity hypothesis on the cost functional $F(t, \varphi)$. In this section we consider state equations arising from the minimization of an energy functional $E(t, \varphi)$ over a convex closed subset K of a reflexive Banach space B

(1) $$E(t, y^t) = \inf_{\varphi \in B} E(t, \varphi),\ y^t \in B.$$

When $K = B$ and $E(t, \varphi)$ is non-quadratic in φ we obtain a nonlinear problem

(2) $$y^t \in B,\ dE(t, y^t; \varphi) = 0, \forall \varphi \in B;$$

when $K \neq B$ we obtain variational inequalities of the form

(3) $$y^t \in K,\ dE(t, y^t; \varphi - y^t) \geq 0,\ \forall \varphi \in K.$$

When $K = B$ the main difficulty in the application of the theorem of section 2 is the loss of convexity of the functional

(4) $$\varphi \rightarrow dE(t, \varphi; \psi).$$

For instance the convex functional

(5) $$E(\varphi) = \frac{1}{5}\int_\Omega |\varphi|^5 dx - \int_\Omega \varphi dx$$

leads to

(6) $$dE(\varphi; \psi) = \int_\Omega |\varphi|^3 \varphi\psi dx - \int_\Omega \psi dx.$$

This functional is convex when $\varphi \geq 0$ on Ω and concave when $\varphi \leq 0$ on Ω.

For variational inequalities the construction of the Lagrangian functional becomes more delicate. By adding one variable the problem.

(7) $$J(t) = F(t, y^t)$$

where y^t is the solution of (3) can be reformulated as an Inf Sup problem

(8) $$J(t) = \inf_{\varphi \in K} \sup_{\substack{\mu \geq 0 \\ \psi \in K}} \{F(t, \varphi) - \mu dE(t, \varphi; \psi - \varphi)\}$$

but the previous results are not readily applicable here.

3.1. A penalization method.

To get around this difficulty we have proposed to use the following penalization method which can be found in DELFOUR and ZOLESIO [5, 6]. Since the critical details and hypotheses can be found there we simply give here a formal description of the main steps.

Let $E : \mathbb{R}^+ \times K \to \mathbb{R}$ be an energy functional defined over a closed convex subset K of a Banach space B. For each t in an interval $[0, T]$, $T > 0$ we assume that

(9) $$\varphi \to E(t, \varphi)$$

is convex and continuous on K and that there exists a unique solution $y^t \in K$ to the minimization problem

(10) $$y^t \in K, \; E(t, y^t) = \inf_{\varphi \in K} E(t, \varphi) \overset{\Delta}{=} e(t)$$

and that y^t is the unique solution of the variational inequality

(11) $$y^t \in K, \; dE(t, y^t; \varphi - y^t) \geq 0 \; , \quad \forall \varphi \in K,$$

where

(12) $$dE(t, \varphi; \psi) = \lim_{\substack{\theta > 0 \\ \theta \to 0}} [E(t, \varphi + \theta\psi) - E(t, \varphi)]/\theta.$$

Associate with the solution y^t of (10) the cost function

(13) $$J(t) = F(t, y^t)$$

for some given cost functional

(14) $$F = \mathbb{R}^+ \times K \to \mathbb{R}.$$

The main objective was to show that, under appropriate hypotheses, the cost function $J(t)$ can be expressed in the form

(15) $$J(t) = J(0) + \int_0^t f(s)ds$$

for some function f in $L^\infty(0, T)$ which will be characterized in terms of the state y^t and the solution p^t of an appropriate unilateral problem for each t. When f belongs to $C^0[0, T]$, J belongs to $C^1[0, T]$ and

(16) $$dJ(t) = \lim_{\substack{s > 0 \\ s \to 0}} [J(t + s) - J(t)]/s = f(t).$$

The basic idea is to introduce a family of minimization problems indexed by $\varepsilon > 0$

(17) $$J_\varepsilon(t) = \inf_{\varphi \in K} F_\varepsilon(t, \varphi)$$

where

(18) $$F_\varepsilon(t, \varphi) = F(t, \varphi) + \frac{1}{\varepsilon} [E(t, \varphi) - E(t, y^t)].$$

Then, under appropriate hypotheses,

(19) $$y^t_\varepsilon \to y^t \text{ in } B \text{ and } J_\varepsilon(t) \to J(t) \text{ as } \varepsilon \to 0.$$

So we introduce the function

(20) $$p^t_\varepsilon = (y^t_\varepsilon - y^t)/\varepsilon$$

and show that

(21) $$p^t_\varepsilon \to p^t \text{ in } V \text{ (weak) as } \varepsilon \to 0,$$

where V is Hilbert space such that $B \subset V$.

We now use the problem indexed by $\varepsilon > 0$. The minimizing element y_ε^t of (17) is characterized by

(22) $$y_\varepsilon^t \in K,\ dF(t, y_\varepsilon^t ; \varphi - y_\varepsilon^t) + \frac{1}{\varepsilon} dE(t, y_\varepsilon^t ; \varphi - y_\varepsilon^t) \geq 0,\ \forall \varphi \in K.$$

So we can use the theorem on the derivative of a Min with respect to a parameter t

(23) $$dJ_\varepsilon(t) = \partial_t F_\varepsilon(t, y_\varepsilon^t) = \partial_t F(t, y_\varepsilon^t) + \frac{1}{\varepsilon} \partial_t [E(t, y_\varepsilon^t) - E(t, y^t)]$$

and the identity

(24) $$J_\varepsilon(t) = J_\varepsilon(0) + \int_0^t dJ_\varepsilon(s) ds.$$

As ε goes to zero we obtain

(25) $$J(t) = J(0) + \int_0^t f(s) ds$$

where $f \in L^\infty(0, T)$ is given by

(26) $$f(t) = \partial_t F(t, y^t) + \partial_t dE(t, y^t ; p^t)$$

where

(27) $$y^t \in K,\ dE(t, y^t ; \varphi - y^t) \geq 0,\ \forall \varphi \in K$$

and, under hypothesis (H) given below,

(28) $$p^t \in S(t),\ \forall \psi \in S(t),\ dF(t, y^t ; \psi - p^t) + d^2E(t, y^t ; \psi - p^t ; p^t) \geq 0,$$

where

(29) $$S(t) = T_K(y^t) \cap \nabla E(t, y^t)^\perp$$

(30) $$T_K(y^t) = V - \text{closure } \{\lambda(\varphi - y^t) : \varphi \in K,\ \lambda \geq 0\}$$

(31) $$\nabla E(t, y^t)^\perp = V - \text{closure } \{\psi \in B : dE(t, y^t ; \psi) = 0\}.$$

and

(H) $$S(t) = \overline{co}\, A(t)$$

(32) $$A(t) = \left\{ \psi \in V \;\middle|\; \begin{array}{l} \exists \{\varphi_\varepsilon : \varepsilon > 0\} \subset K,\ \psi_\varepsilon = (\varphi_\varepsilon - y^t)/\varepsilon \text{ such that} \\ \psi_\varepsilon \to \psi \text{ in } V \text{ (weak) as } \varepsilon \to 0 \text{ and} \\ \lim\limits_{\varepsilon \searrow 0} \frac{1}{\varepsilon} dE(t, y^t ; 0, \psi_\varepsilon) = 0. \end{array} \right\}$$

Hypothesis (H) is quite abstract. Fortunately it is always verified when the so-called F. MIGNOT [1]'s hypothesis is verified since it can be easily shown that

(33) $$\mathbb{R}^+(K - y^t) \cap \nabla E(t, y^t)^\perp \subset A(t) \subset S(t).$$

3.2. Limiting behaviour as t goes to zero.

The results from the previous section only indicate that $J(t)$ is differentiable almost everywhere. One would like to say something about the existence of $dJ(0)$. This requires the existence of the limit

(34) $$f(s) \to f^* \text{ as } s \to 0$$

and to obtain a complete characterization of $dJ(0) = f^*$ in term of y and the adjoint variable we must be able to say something about the limit points of p^t as t goes to zero. It turns out that for t in a neighborhood of $t = 0$, the p^t's are bounded in the Hilbert space V. So we can extract subsequences such that

(35) $$p^{t_n} \to p^* \text{ in } V(\text{weak}).$$

The difficult part is to characterize p^*. In general p^* is not the solution of

(36) $$p \in S(0),\ \forall \psi \in S(0),\ dF(0,y;\psi - p) + d^2E(0,y;\psi - p;p) \geq 0$$

since the sets $S(t)$ do not "converge" to $S(0)$. In the case $K = B$, most of the above difficulties dissappear and we recover the expected results under reasonable assumptions.

3.3. Application to the control of variational inequalitites.

Since the previous results are quite abstract, it is useful to specialize them for a simple example which brings out the important features. To further simplify we shall even focus our attention of a classical control problem which is much simpler than the Shape problem since the control enters linearily in the variational inequality.

We need the following ingredients

$H = L^2(\Omega)$ with the usual norm $|\cdot|$ and inner product $(\cdot,\cdot)$

$V = H_0^1(\Omega)$ with the usual norm $\|\cdot\|$ and inner product $((\cdot,\cdot))$

$a : V \times V \to \mathbb{R}$ bilinear continous and coercive

$K = \{\varphi \in V:\ \varphi \geq 0 \text{ a.e. in } \Omega\}$

$f \in H$, $v \in H$ (control variable).

Consider the solution y of the variational inequality

(37) $$y = y(v) \in K,\ a(y, \varphi - y) \geq (f + v, \varphi - y),\ \forall \varphi \in K$$

and the associated cost function

(38) $$J(v) = \frac{1}{2}\{|y(v) - z_d|^2 + |v|^2\}.$$

The corresponding energy and cost functionals are

(39) $$E(t,\varphi) = \frac{1}{2}a(\varphi,\varphi) - (f + u + tv, \varphi)$$

(40) $$F(t,\varphi) = \frac{1}{2}\{|\varphi - z_d|^2 + |u + tv|^2\}.$$

Here F. MIGNOT [1]'s condition holds and from the previous section

(41) $$J(u + tv) = J(u) + \int_0^t (u + sv - p^s, v)ds$$

where

(42) $$y^s \in K,\ a(y^s, \varphi - y^s) \geq (f + u + sv, \varphi - y^s),\ \forall \varphi \in K$$

(43) $$p^s \in S(s),\ a(p^s, \psi - p^s) + (y^s - z_d, \psi - p^s) \geq 0,\ \forall \psi \in S(s)$$

It can also be shown that

(44) $$\|p^t\| \leq \frac{2}{\alpha}(t|v| + |y^0 - z_d|).$$

So there are weak limit points of $\{p^t\}$ in V. Let

(45) $$p = p(u, v) \in V$$

be such that

(46) $$\exists \{p^{t_n}\},\ p^{t_n} \in S(t_n) \text{ such that } p^{t_n} \to p \text{ in } V \text{ (weak).}$$

As a result

(47) $$a(p, p) + (y^0 - z_d, p) \le 0$$

At the minimum

(48) $$\exists u^* \in H,\ J(u^*) = \inf_{v \in H} J(v)$$

So for a fixed v, denote by $p(u^*, v)$ a weak limit point of $p^t = p(u^* + tv)$. Then

(49) $$J(u^* + tv) - J(u^*) = \int_u^t (u^* + sv - p(u^* + sv), v)ds$$

which implies that

(50) $$(u^* - p(u^*, v), v) \ge 0 \forall v \in H.$$

Remark 3.1. At this stage it is important to notice that the limit points $p(u^*,v)$ are dependent on u^* and the direction v. This is the generalization of the adjoint variable in differentiable problems. So far we have not been able to show that $p(u^*,v)$ is unique for the class of problems under consideration. However it will be unique and completely characterized by a variational inequality in the finite dimensional example given in the next section.♦

Remark 3.2. The characterization of the minimum given in (50) is different from the classical one given by MIGNOT and PUEL [1] which, if interpreted correctly, gives:
if $y^* = y(u^*)$ is the state corresponding to u^*, then

(51) $$\exists \bar{p} \in S(0),\ \forall \psi \in S(0),\ a(\psi, \bar{p}) \le (y^* - z_d, \psi)$$

and

(52) $$\bar{p} + u^* = 0.\ ♦$$

3.4. A simple finite dimensional example.

Let

(53) $$K = \{\varphi \in \mathbb{R} : \varphi \ge 0\} \quad , \quad E(u, \varphi) = \frac{1}{2}\varphi^2 - u\varphi$$

(54) $$F(u, \varphi) = \frac{1}{2}\{(\varphi + 1)^2 + u^2\},\ u \in \mathbb{R}$$

Given u, the solution $y = y(u)$ of the variational inequality

(55) $$y \ge 0,\ (y - u)(\varphi - y) \ge 0,\ \forall \varphi \ge 0$$

is given by

(56) $$y(u) = \begin{cases} u\ , & u \ge 0 \\ 0\ , & u < 0 \end{cases}$$

and the associated cost function by

(57) $$J(u) = F(u, y(u)) = \frac{1}{2}\begin{Bmatrix} (u+1)^2 + u^2, & u \geq 0 \\ 1 + u^2, & u < 0 \end{Bmatrix}.$$

The directional semi derivative at u in the direction v is given by

(58) $$dJ(u;v) = \begin{Bmatrix} (2u+1)v & , & u > 0 \\ \max\{v, 0\} & , & u = 0 \\ uv & , & u < 0 \end{Bmatrix}.$$

It is readily checked that J has a unique minimum at $u^* = 0$ where the function J is not differentiable.

Now we apply the results from the previous section to this example at $u = 0$ for $v \in \mathbb{R}$ and $t > 0$:

(59) $$y^t = \begin{Bmatrix} tv & , & v \geq 0 \\ 0 & , & v < 0 \end{Bmatrix}, \quad \nabla E(t, y^t)^{\perp} = \begin{Bmatrix} \mathbb{R} & , & v \geq 0 \\ \{0\} & , & v < 0 \end{Bmatrix}$$

(60) $$T_K(y^t) = \begin{Bmatrix} \mathbb{R} & , & v > 0 \\ \mathbb{R}^+ & , & v \leq 0 \end{Bmatrix}, \quad S(t) = \begin{Bmatrix} \mathbb{R}, & v > 0 \\ \mathbb{R}^+, & v = 0 \\ \{0\}, & v < 0 \end{Bmatrix}, \quad S(0) = \mathbb{R}^+$$

By solving the variational inequality

(61) $$p^t \in S(t), \quad (p^t, \psi - p^t) - (tv, \psi - p^t) \geq 0, \ \forall \psi \in S(t)$$

we obtain

(62) $$p^t = \begin{Bmatrix} -1 & , & v \geq 0 \\ 0 & , & v < 0 \end{Bmatrix}, \quad p^0 = -1.$$

Finally

(63) $$J(tv) = J(0) + \int_0^t (sv - p^s)v\,ds$$

and

(64) $$\lim_{\substack{t > 0 \\ t \to 0}} \frac{J(tv) - J(0)}{t} = -p^*(0, v)v,$$

where

(65) $$p^*(0, v) = \lim_{s > 0 \ s \to 0} p^s = \begin{Bmatrix} -1 & , & v \geq 0 \\ 0 & , & v < 0 \end{Bmatrix}.$$

Notice that for $v < 0$ $p^*(0, v) \neq p^0$ so that in general the limit points of $\{p^t\}$ as t goes to zero are ot the solution of (61) for $t = 0$ and that they do indeed depend on the direction v.

Finally at $u^* = 0$ the optimality condition (50) gives

(66) $$-p^*(0, v)v \geq 0, \ \forall v \in \mathbb{R}.$$

Remark 3.3. It is readily seen that $p^*(0, v)$ is the unique solution of the variational inequality

(61) $$p = p^*(0,v) \in S^*(0,v), \ (p, \psi - p) \geq 0, \forall \psi \in S^*(0,v),$$

where $S^*(0,v) = \mathbb{R}$, if $v > 0$, $\mathbb{R}^+$, if $v = 0$ and $\{0\}$ if $v < 0$. Moreover $S^*(0,v) = S(0)$ for $v \geq 0$ and $S^*(0,v) \neq S(0)$ for $v < 0$.♦

Remark 3.4. Again, if interpreted correctly, the condition (51) –(52) yields

(67) $$y^* = y(0) = 0$$

(68) $$\exists \bar{p} \in S(0) = \mathbb{R}^+, \forall \psi \in S(0) = \mathbb{R}^+, \psi \bar{p} \leq \psi \Rightarrow 0 \leq \bar{p} \leq 1$$

and

(69) $$\bar{p} + u^* = \bar{p} + 0 = 0 \Rightarrow \bar{p} = 0.$$

As indicated in Remark 3.2, our characterization (66) of the minimum is different from the one given by NIGNOT and PUEL [1]. But, up to a change in the sign of $\bar{p}$, the two conditions applied to this special example do not seem incompatible.♦

REFERENCES.

P. CANNARSA and H.M. SONER [1], On the singularities of the viscisity solutions to Hamilton-Jacobi-Bellman Equations, Indiana Univ. Math. J. 36 (1987), 501-524.

J. CEA [1], Conception optimale ou identification de formes: calcul rapide de la dérivée directionnelle de la fonction coût, Mathematical Modelling and Numerical Analysis(Modélisation mathématique et analyse numérique) 20 (1986), 371-402.

R. CORREA and A. SEEGER [1], Directional derivatives of a mimimax function, Nonlinear Analysis, Theory, Methods and Applications 9 (1985), 13-22.

M.C. DELFOUR and J.P. ZOLESIO [1], Dérivation d'un MinMax et application à la dérivation par rapport au contrôle d'une observation non-différentiable de l'état, C.R. Acad. Sc. Paris, t. 302, Sér. I, no. 16 (1986), 571-574.

[2], Shape Sensitivity Analysis via Min Max Differentiability, SIAM J. on Control and Optimization 26 (1988), 834-862.

[3], Differentiability of a Min Max and Application to Optimal Control and Design Problems, Part I, in "Control Problems for Systems Described as Partial Differential Equations and Applications", I. Lasiecka and R. Triggiani, eds., pp. 204-219, Springer-Verlag, New York, 1987.

[4], Differentiability of a Min Max and Application to Optimal Control and Design Problems, Part II, in "Control Problems for Systems Described as Partial Differential Equations and Applications", I. Lasiecka and R. Triggiani, eds., pp. 220-229, Springer-Verlag, New York, 1987.

[5], Further Developments in Shape Sensitivity Analysis via a Penalization Method, in "Boundary Control and Boundary Variations", J.P. Zolésio, ed., pp. 153-191, Springer- Verlag, Berlin, Heidelberg, New-York, Tokyo, 1988.

[6], Shape Sensitivity Analysis via a Penalization Method, Annali di Matematica Pura ed Applicata, Ser. 4, T. CLI (1988), pp. 179-212.

[7], Analyse des problèmes de forme par la dérivation des Min Max, Annales de l'Institut Henri-Poincaré, Série Analyse Non Lineaire,(AIHP, Nonlinear Analysis) to appear.

J. HADAMARD [1], Mémoire sur le problème d'analyse relatif à l'équilibre des plaques élastiques encastrées, in "Oeuvres de J. Hadamard, Vol II, pp. 515-641, C.N.R.S, Paris, 1968 (original reference: Mem. Sav. Etrang. 33 (1907), mémoire couronné par l'Académie des Sciences).

F. MIGNOT [1], Contrôle dans les inequations variationnellles elliptiques, J. Funct. Anal. 22 (1976), pp. 130-185.

F. MIGNOT and J.P. PUEL [1], Optimal Control in some Variational Inequalities, SIAM J. Control and Optimization 22 (1984), 466-476.

J. SOKOLOWSKI [1], Conical differentiability of projection on convex sets-an application to sensitivity analysis of Signorini variational inequaality, Technical Report, Institute of Mathematics of the University of Genova, 1981.

J. SOKOLOWSKI and J.P. ZOLESIO [1], Dérivée par rapport au domaine de la solution d'un problème unilatéral, C.R. Acad. Sci. Paris, Sér. I, 3301 (1985), pp. 103-106.

[2], Shape sensitivity analysis of unilateral problems, SIAM J. on Math. Anal., 18 (1987), pp.1416-1437.

J.P. ZOLESIO [1], Identification de domaines par déformation, Thèse de doctorat d'état, Université de Nice, France, 1979.

[2], The Material Derivative (or Speed) Method for Shape Optimization, in "Optimization of Distributed Parameter Structures", Vol. II, J. Céa and E.J. Haug, eds., pp. 1089-1151, Sijthoff and Nordhoff, Alphen aan den Rijn, 1981.

A NEWTON'S METHOD IN A DOMAIN OPTIMIZATION PROBLEM

Y. GOTO and N. FUJII

Department of Control Engineering, Osaka University, Toyonaka,

Osaka 560, JAPAN

ABSTRACT

This paper is concerned with a Newton's method for a kind of shape optimization problems. The first and the second variations of the object function are derived. These variations are discretized by introducing a set of boundary value problems in order to derive the second order numerical method. The boundary value problems are solved by the conventional finite element method. A simple numerical example is examined and shows the efficiency of the method.

1. INTRODUCTION

The domain optimization problem treated in this paper is a kind of shape optimization problems. It is an optimization problem in which an object function depending on a domain through the solution of a boundary value problem defined on the domain should be minimized (or maximized) with respect to the domain. Cea [1] enumerated various domain optimization problems. Zolesio [2] derived the first order necessary conditions for domain optimization problems with various boundary value problems as the constraint. One of the present authors [3] gave the first order necessary conditions for a domain optimization problem with a nonlinear boundary value problem. Pironneau gave the first order necessary conditions for the optimal shape of an obstacle in both Stokes flow [4] and Navier-Stokes [5] flow.

As to the second order necessary conditions, one [6] of the present author gave the second order necessary conditions to an optimization problem with a Dirichlet problem. The present authors and their colleague [7] showed the second order necessary conditions for an optimization problem with a Neumann problem assuming the differentiabilities .

A domain optimization problem has strong nonlinearity in general. Therefore, it is very difficult to obtain an analytic solution except a special case [8]. We can not help relying upon numerical methods. Pironneau [9] surveyed such numerical methods. Queau and Trompette [10] applied the finite element method to two-dimensional shape optimal design. Arumugam and Pironneau [11] studied numericaly the optimal shape of riblets. Kikuchi, Chung, Torigaki, and Taylor [12] developed their adaptive finite element methods to the shape optimization of elastic structures. Mota Soares and Choi used [13] the boundary element method to the optimal shape design of structures. Trompette, Marcelin and Lallemaud [14] studied minimal stress concentration design problem of axisymmetric structures. The book [15] edited by Mota Soares contains many papers on numerical shape optimization techniques.

These numerical techniques are based upon the first order variation of the object function considered. In this sense, these techniques are called first order ones; they, usually, adopt the steepest descent method with one-dimensional optimization. In this paper, we shall calculate the first and the second variations of the object function corresponding to a boundary variation caused by an arbitrary vector field defined only on the boundary. Discretizing these variations by the conventional finite element method, we shall propose a second order numerical technique (a Newton's method). We shall examine the efficiency of the second order technique compared with the conventional steepest descent method by a simple example.

2. PROBLEM STATEMENT, THE FIRST AND THE SECOND VARIATIONS OF THE SOLUTION AND THE OBJECT FUNCTION

In this section, we shall pose a typical domain optimizatin problem with a Neumann problem as a constraint. We shall characterize the first and the second variations of the solution to the boundary value problem assuming the existence of them. As for the case of Dirichlet problem, we can similarly obtain the results.

Let $g(x,u,q)$ be a smooth function of $(x,u,q) \in R^2 \times R \times R^2$. Let us introduce an object function $J(\Omega;u)$by

$$J(\Omega;u) = \int_\Omega g(x,u,\nabla u)\, dx, \tag{1}$$

where u is the solution of the following Neumann problem:

$$\Delta u(x) - k(x)u(x) = f(x) \qquad (x \in \Omega), \tag{2}$$

$$\frac{\partial u}{\partial n}(x) = \kappa(\text{const.}) \qquad (x \in \Gamma \equiv \partial\Omega). \tag{3}$$

The boundary value problem (2),(3) appears in, for example, the steady-state flow of the ideal fluid, the steady-state thermal conduction and so on.

Our problem is as follows:

<u>Problem</u> Find Ω which minimize $J(\Omega;u)$ above.

Our final goal is to find an efficient numerical method for our problem. To this end, let us define a boundary variation. Let us introduce an arclength s counterclockwise on the boundary Γ of Ω smooth enough. Let $\vec{\rho}(s)$ be a vector field defined on Γ and is assumed at least once differentiable with respect to s. Let ε be a small positive number. We place the vector $\varepsilon\vec{\rho}(s)$ at each point on Γ labeled by s(Fig. 1). If ε is

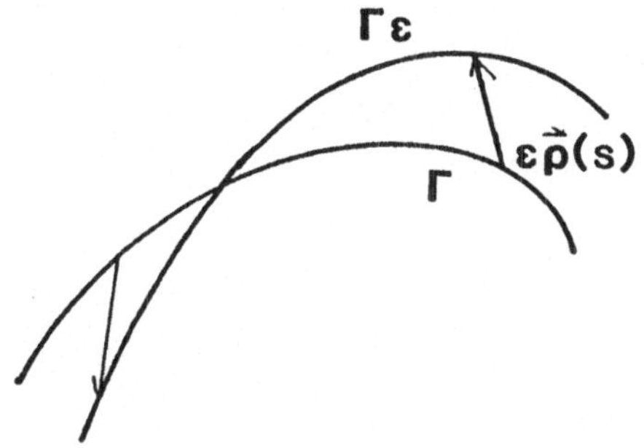

Fig. 1

sufficiently small, the end points of every $\varepsilon\vec{\rho}(s)$ form a continuous closed curve. Let Γ_ε be this curve; Ω_ε stands for the domain bounded by Γ_ε. For this newly defined domain Ω_ε , let us consider the following boundary value problem:

$$\Delta u_\varepsilon(x) - k(x)u_\varepsilon(x) = f(x) \qquad (x \in \Omega_\varepsilon), \qquad (4)$$

$$\frac{\partial u_\varepsilon}{\partial n}(x) = \kappa \qquad (x \in \Gamma_\varepsilon). \qquad (5)$$

Let us make a following assumption; i. e., we assume the differentiabilities.

Assumption 1

For the solution u of the boundary value problem (2), (3) and u_ε of (4), (5), we supose that there exist $\phi(x)$ and $\psi(x)$ defined on Ω such that the following relations hold for every $x \in \Omega \cap \Omega_\varepsilon$:

$$u_\varepsilon(x) - u(x) = \varepsilon\phi(x) + \varepsilon^2\psi(x) + o(\varepsilon^2); \qquad (6)$$

and the similar relations up to the third order derivatives. Here, $o(\varepsilon^2)$ denotes a quantity such that $\varepsilon^{-2}o(\varepsilon^2) \to 0$ as $\varepsilon \to 0$.

For Dirichlet problem, the existence of such ϕ and ψ are proved [6]. We shall call the above $\phi(x)$ and $\psi(x)$ the first and the second variations of the solution u, respectively. We have the following theorem on Assumption 1.

Theorem 1

The first and the second variation ϕ,ψ of the solution of (2), (3) are the solutions of the following boundary value problems, respectively:

$$\Delta\phi(x) - k(x)\phi(x) = 0 \qquad (x \in \Omega), \qquad (7)$$

$$\frac{\partial \phi}{\partial n}(x) = \frac{\partial u}{\partial s}(\vec{n}\cdot\frac{\partial \vec{\rho}}{\partial s}) - \vec{\rho}\cdot Du\cdot\vec{n} \qquad (x \in \Gamma); \qquad (8)$$

$$\Delta\psi(x) - k(x)\psi(x) = 0 \qquad (x \in \Omega), \qquad (9)$$

$$\frac{\partial \psi}{\partial n}(x) = \frac{\partial \phi}{\partial s}(\vec{n}\cdot\frac{\partial \vec{\rho}}{\partial s}) - \vec{\rho}\cdot D\phi\cdot\vec{n}$$

$$+ \frac{1}{2}\kappa(\vec{n}\cdot\frac{\partial \vec{\rho}}{\partial s})^2 - \frac{\partial u}{\partial s}(\vec{s}\cdot\frac{\partial \vec{\rho}}{\partial s})(\vec{n}\cdot\frac{\partial \vec{\rho}}{\partial s})$$

$$+ (\vec{\rho}\cdot Du\cdot\vec{s})(n\cdot\frac{\partial \vec{\rho}}{\partial s}) - \frac{1}{2}\vec{\rho}\cdot\vec{\rho}\cdot\vec{n}\cdot Tu \qquad (x \in \Gamma), \qquad (10)$$

where $\vec{n} = (n_1, n_2)$ denotes the unit outward normal vector of Γ, $\vec{s} = (s_1, s_2)$ stands for the unit tangent vector along Γ,

$$\vec{\rho}\cdot Du\cdot\vec{n} = \sum_{i,j=1}^{2} \rho_i n_j \frac{\partial^2 u}{\partial x_i \partial x_j},$$

and

$$\vec{\rho}\cdot\vec{\rho}\cdot\vec{n}\cdot Tu = \sum_{i,j,k=1}^{2} \rho_i \rho_j n_k \frac{\partial^3 u}{\partial x_i \partial x_j \partial x_k}.$$

The proof of this theorem is omitted.

We can derive the similar results for the case of Dirichlet problem:

$$\Delta u(x) - k(x)u(x) = f(x) \qquad (x \in \Omega), \qquad (11)$$

$$u(x) = k \text{ (const.)} \qquad (\quad). \qquad (12)$$

Theorem 2

The first and the second variation ϕ,ψ of the solution u to (11), (12) are the solutions of the following boundary value problems, respectively:

$$\Delta\phi(x) - k(x)\phi(x) = 0 \qquad (x \in \Omega), \qquad (13)$$

$$\phi(x) = -\vec{\rho}(x)\cdot\nabla u(x) \qquad (x \in \Gamma); \qquad (14)$$

$$\Delta\psi(x) - k(x)\psi(x) = 0 \qquad (x \in \Omega), \qquad (15)$$

$$\psi(x) = -\vec{\rho}(x)\cdot\nabla\phi(x) - \frac{1}{2}\vec{\rho}(x)\cdot Du\cdot\vec{\rho}(x) \qquad (x \in \Gamma). \qquad (16)$$

Now, we shall calculate the first and the second variations of the object function defined by (1). We can define an object function $J(\Omega_\varepsilon;u_\varepsilon)$ for Ω_ε by

$$J(\Omega_\varepsilon;u_\varepsilon) \equiv \int_{\Omega_\varepsilon} g(x,u_\varepsilon,\nabla u_\varepsilon)\,dx \tag{17}$$

Let us define the first and the second variations $\delta^{(1)}J$, $\delta^{(2)}J$ of $J(\Omega;u)$ by

$$J(\Omega_\varepsilon;u_\varepsilon) - J(\Omega;u) = \varepsilon\delta^{(1)}J + \varepsilon^2\delta^{(2)}J + o(\varepsilon^2) \tag{18}$$

where $o(\varepsilon^2)$ is a quantity such that $\varepsilon^{-2}o(\varepsilon^2) \to 0$ as $\varepsilon \to 0$. We shall call $\delta^{(1)}J$ the first, $\delta^{(2)}J$ the second variations of the object function $J(\Omega;u)$ corresponding to the boundary variation caused by the vector field $\vec{\rho}(s)$. We shall often call simply $\delta^{(1)}J$ the first, $\delta^{(2)}J$ the second variations of $J(\Omega;u)$, when the boundary variation is clearly realized in the context.

We shall use the notations $\nabla_q g$ and $\nabla_q{}^2 g$ defined by

$$\nabla_q g = \left(\frac{\partial g}{\partial q_1}, \frac{\partial g}{\partial q_2}\right),$$

$$\nabla_q{}^2 g = \begin{pmatrix} \dfrac{\partial^2 g}{\partial q_1{}^2} & \dfrac{\partial^2 g}{\partial q_1 \partial q_2} \\ \dfrac{\partial^2 g}{\partial q_2 \partial q_1} & \dfrac{\partial^2 g}{\partial q_2{}^2} \end{pmatrix}.$$

In order to simplify the expression for $\delta^{(1)}J$, we introdue an adjoint variable p as the solution of the following boundary value problem:

$$\Delta p(x) - k(x)p(x) = \frac{\partial g}{\partial u} - \mathrm{div}\,(\nabla_q g) \qquad (x \in \Omega), \tag{19}$$

$$\frac{\partial p}{\partial n}(x) = -\nabla_q g\cdot\vec{n} \qquad (x \in \Gamma). \tag{20}$$

Then we can obtain $\delta^{(1)}J$ and $\delta^{(2)}J$ as follows; the details of the derivations are omitted.

$$\delta^{(1)}J = \int_\Gamma [-p\{\frac{\partial u}{\partial s}(\vec{n}\cdot\frac{\partial\vec{\rho}}{\partial s}) - \vec{\rho}\cdot Du\cdot\vec{n}\} + g\vec{\rho}\cdot\vec{n}]\,d\Gamma \tag{21}$$

$$\delta^{(2)}J = -\int_\Gamma p[\frac{\partial\phi}{\partial s}(\vec{n}\cdot\frac{\partial\vec{\rho}}{\partial s}) + \vec{\rho}\cdot Du\cdot\vec{s}(\vec{n}\cdot\frac{\partial\vec{\rho}}{\partial s})$$

$$+ \frac{1}{2}\kappa(\vec{n}\cdot\frac{\partial\vec{\rho}}{\partial s})^2 - \frac{\partial u}{\partial s}(\vec{s}\cdot\frac{\partial\vec{\rho}}{\partial s})(\vec{n}\cdot\frac{\partial\vec{\rho}}{\partial s})$$

$$- \frac{1}{2}\vec{\rho}\cdot\vec{\rho}\cdot\vec{n}\cdot Tu + \vec{\rho}\cdot D\phi\cdot\vec{\rho}]\ d\Gamma$$

$$+ \frac{1}{2}\int_\Omega \frac{\partial^2 g}{\partial u^2}\phi^2\ dx + \frac{1}{2}\int_\Gamma g(\rho_1\frac{\partial\rho_2}{\partial s} - \rho_2\frac{\partial\rho_1}{\partial s})\ d\Gamma$$

$$+ \frac{1}{2}\int_\Gamma(\nabla g\cdot\vec{\rho})(\vec{\rho}\cdot\vec{n})\ d\Gamma + \int_\Gamma \frac{\partial g}{\partial u}\phi\vec{\rho}\cdot\vec{n}\ d\Gamma$$

$$+ \int_\Gamma(\nabla_q g\cdot\nabla\phi)(\vec{\rho}\cdot\vec{n})\ d\Gamma + \frac{1}{2}\int_\Omega \nabla\phi\cdot\nabla_q{}^2 g\cdot\nabla\phi\ dx$$

$$+ \int_\Omega \phi(\frac{\partial}{\partial u}\nabla_q g)\cdot\nabla\phi\ dx. \tag{22}$$

Note that ψ does not appear in these expressions because of the introduction of the adjoint variable p.

As for the case of the Dirichlet problem (11), (12), we can quite similarly obtain the first and the second variations of the object function. In this case, however, we introduce the adjoint variable p as the solution of the boundary value problem:

$$\Delta p(x) - k(x)p(x) = \frac{\partial g}{\partial u} - \mathrm{div}\ (\nabla_q g) \qquad (x \in \Omega), \tag{23}$$

$$p(x) = 0 \qquad (x \in \Gamma). \tag{24}$$

The results read

$$\delta^{(1)}J = \int_\Gamma (\vec{\rho}\cdot\vec{n})\{g - \frac{\partial u}{\partial n}(\frac{\partial p}{\partial n} + \nabla_q g\cdot\vec{n})\}\ d\Gamma, \tag{25}$$

$$\delta^{(2)}J = -\int_\Gamma (\nabla p\cdot\vec{n})(\vec{\rho}\cdot\nabla\phi + \tfrac{1}{2}\vec{\rho}\cdot Du\cdot\vec{\rho})\, d\Gamma$$

$$- \frac{1}{2}\int_\Gamma \frac{\partial g}{\partial u}(\nabla u\cdot\vec{\rho})(\vec{\rho}\cdot\vec{n})\, d\Gamma$$

$$+ \int_\Gamma \{(\nabla_q g\cdot\nabla\phi)(\vec{\rho}\cdot\vec{n}) - (\vec{\rho}\cdot\nabla\phi)(\nabla_q g\cdot\vec{n})\}\, d\Gamma$$

$$+ \int_\Gamma \{(\nabla_q g\cdot Du\cdot\vec{\rho})(\vec{\rho}\cdot\vec{n}) - (\vec{\rho}\cdot Du\cdot\vec{\rho})(\nabla_q g\cdot\vec{n})\}\, d\Gamma$$

$$+ \frac{1}{2}\int_\Gamma g(\rho_1\frac{\partial\rho_2}{\partial s} - \rho_2\frac{\partial\rho_1}{\partial s})\, d\Gamma + \frac{1}{2}\int_\Gamma (\nabla g\cdot\vec{\rho})(\vec{\rho}\cdot\vec{n})\, d\Gamma$$

$$+ \frac{1}{2}\int_\Omega \frac{\partial^2 g}{\partial u^2}\phi^2\, dx + \frac{1}{2}\int_\Omega \nabla\phi\cdot\nabla_q{}^2 g\cdot\nabla\phi\, dx$$

$$+ \int_\Omega \phi(\frac{\partial}{\partial u}\nabla_q g)\cdot\nabla\phi\, dx. \qquad (26)$$

3. DISCRETIZATION AND ALGORITHM

In this section, we shall discretize the first and the second variations in order to obtain a quadratic form of the variation of the object function. Based on the quadratic form thus obtained, we shall propose a second order numerical algorithm, a Newton's method.

A domain Ω is determined by its boundary Γ and Γ can only be described by an infinite number of parameters. We can not deal with an infinite number of parameters in numerical computations. Therefore, Γ hence Ω should be approximated by a finite number of parameters. We choose n points x^1, x^2, $\cdots$, x^n on the boundary Γ. We shall approximate Ω or Γ by the polygon formed by n points x^1, x^2, $\cdots$, x^n; the domain bounded by this polygon is expressed by Ω^h and Γ^h stands for the polygon. Let X be a 2n-dimensional vector defined by $X=(x^1,\cdots,x^n)$. Let us consider the boundary value problem induced from the original boundary value problem by the conventional finite element method. Let u^h be the solution of the approximating boundary value problem. Hence, the object function is approximated by $J(\Omega^h;u^h)$; i.e.,

$$J(\Omega;u) \simeq J(\Omega^h;u^h)$$

holds. In what follows we shall deal with $J(\Omega^h;u^h)$ instead of $J(\Omega;u)$. Domain Ω^h is completely determined by X and u^h is obtained as the solution of the boundary value problem on Ω^h. In this sense, $J(\Omega^h;u^h)$ is determined by X; hence, we can introduce a notation $J^h(X)$ by

$$J^h(X) \equiv J(\Omega^h;u^h). \qquad (27)$$

Thus, the original optimization problem is reduced to the approximating optimization problem:

$$\min_{X} J^h(X). \tag{28}$$

In order to solve this problem, we cannot help adopting an iteration procedure, since the problem has a strong nonlinearity. Let $\Omega^h{}_k$ and $\Gamma^h{}_k$ be the k-th domain and its boundary in the k-th iteration, respectively. To get (k+1)-th domain $\Omega^h{}_{k+1}$ we shift each point x^i on $\Gamma^h{}_k$ by $\vec{q}^i$; vectors $x^i + \vec{q}^i$ make a polygon $\Gamma^h{}_{k+1}$ which encloses $\Omega^h{}_{k+1}$ (Fig. 2). Hence, we we can write $\vec{\rho}(x)$ ($x \in \Gamma^h{}_k$) as

Fig. 2

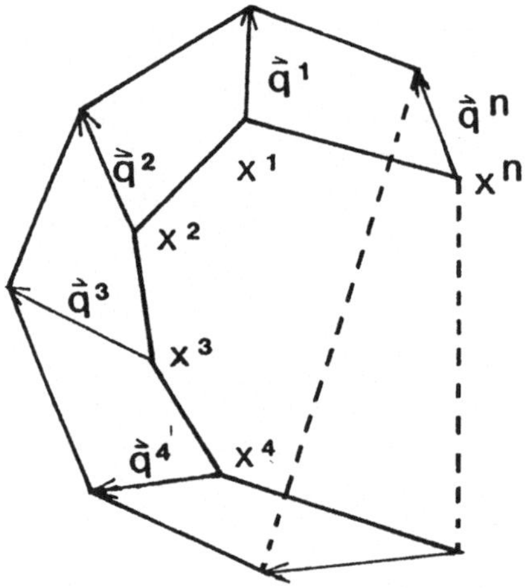

$$\vec{\rho}(x) = \sum_{i=1}^{n} q^i \vec{w}_i(x) \qquad (x \in \Gamma^h{}_k) \tag{29}$$

where $w_i(x)$ is a linear shape function such that (Fig. 3)

Fig. 3

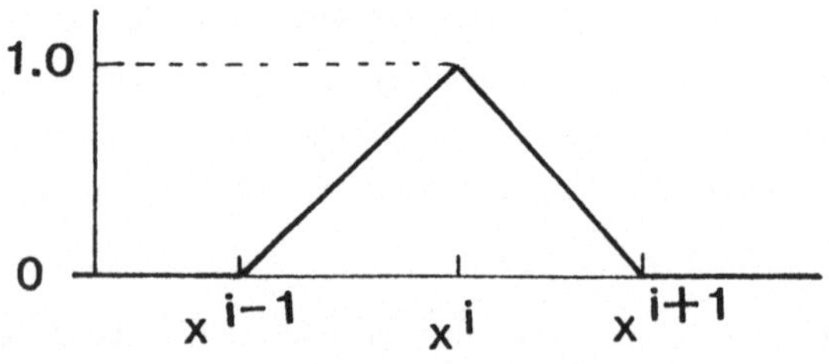

$$w_i(x^j) = \delta_{ij} \text{ (Kronecker's delta)} \tag{30}$$

Then the approximation $\delta^{(1)}J^h$ for $\delta^{(1)}J$ in the former section is written by

$$\delta^{(1)}J^h = \sum_{i=1}^{n} \vec{q}^i \cdot \int_{\Gamma^h_k} [-p\{\frac{\partial u}{\partial s}(\frac{\partial}{\partial s}w_i)\vec{n} - w_i Du\cdot\vec{n}\} + g w_i \vec{n}]\, d\Gamma \qquad (31)$$

If we introduce 2n-dimensional vectors q, d by

$$\vec{q} = (q^1{}_1, q^1{}_2 . q^2{}_1, q^2{}_2, \cdots, q^n{}_2), \qquad (32)$$

$$\vec{d} = (d^1{}_1, d^1{}_2, d^2{}_1, d^2{}_2, \cdots, d^n{}_2), \qquad (33)$$

where

$$d^i{}_j = \int_\Gamma [-p\{\frac{\partial u}{\partial s}(\frac{\partial}{\partial s}w_i)n_j - w_i(\frac{\partial^2 u}{\partial x_1 \partial x_j}n_1 + \frac{\partial^2 u}{\partial x_2 \partial x_j}n_2)\}$$

$$+ g w_i n_j]\, d\Gamma, \qquad (34)$$

then we can write as

$$\delta^{(1)}J^h = \vec{q}\cdot\vec{d}. \qquad (35)$$

As for $\delta^{(2)}J^h$, we first introduce a set of boundary value problems:

$$\Delta\phi^i{}_j(x) - k(x)\phi^i{}_j(x) = 0 \qquad (x \in \Omega^h{}_k), \qquad (36)$$

$$\frac{\partial\phi^i{}_j}{\partial n}(x) = \frac{\partial u}{\partial s}(\frac{\partial}{\partial s}w_i)n_j - w_i(\frac{\partial^2 u}{\partial x_1 \partial x_j}n_1 + \frac{\partial^2 u}{\partial x_2 \partial x_j}n_2) \quad (x \in \Gamma^h{}_k),$$

$$j = 1,2, \quad i = 1,\cdots,n. \qquad (37)$$

Let $\phi^i{}_j$ be the solution of (36), (37) by the finite element method. Then the approximation of the first variation ϕ, still denoted by ϕ, is given by

$$\phi(x) = \sum_{i=1}^{n} \sum_{j=1}^{2} q^i{}_j \phi^i{}_j \qquad (x \in \Omega^h{}_k) \qquad (38)$$

substituting (29) and (38) into (22), we find that the approximation $\delta^{(2)}J^h$ of $\delta^{(2)}J$ is given by

$$\delta^{(2)}J^h = \sum_{i=1}^{n} \sum_{j=1}^{2} \sum_{k=1}^{n} \sum_{\ell=1}^{2} q^i{}_j q^k{}_1 \vec{A}_{ijk\ell} = q^T A q, \qquad (39)$$

where $\vec{q}$ is given by (41) and A is given by

$$A = \begin{bmatrix} A_{1111} & A_{1112} & A_{1121} & \cdots\cdots & A_{11n1} & A_{11n2} \\ A_{1211} & A_{1212} & A_{1221} & \cdots\cdots\cdots & & A_{12n2} \\ \vdots & & & & & \\ A_{n211} & A_{n212} & \cdots\cdots & & A_{n2n1} & A_{n2n2} \end{bmatrix}, \tag{40}$$

$$\begin{aligned} A_{ijk\ell} = & - \int_\Gamma p[\frac{\partial \phi^i{}_j}{\partial s} n_\ell (\frac{\partial}{\partial s} w_k) + w_i (\frac{\partial^2 u}{\partial x_i \partial x_1} s_1 + \frac{\partial^2 u}{\partial x_i \partial x_2} s_2)(\frac{\partial}{\partial s} w_k n_\ell) \\ & + \frac{1}{2} \kappa n_\ell \frac{\partial w_k}{\partial s} n_j \frac{\partial w_i}{\partial s} - \frac{\partial u}{\partial s} s_j \frac{\partial w_i}{\partial s} n_\ell \frac{\partial w_k}{\partial s} \\ & - w_k (\frac{\partial^2 \phi^i{}_j}{\partial x_\ell \partial x_1} n_1 + \frac{\partial^2 \phi^i{}_j}{\partial x_\ell \partial x_2} n_2) \\ & - \frac{1}{2} w_k w_i (\frac{\partial^3 u}{\partial x_j \partial x_\ell \partial x_1} n_1 + \frac{\partial^3 u}{\partial x_j \partial x_\ell \partial x_2} n_2)] \, d\Gamma \\ & + \frac{1}{2}(1 - \delta_{j\ell}) \int_\Gamma g (w_i \frac{\partial w_k}{\partial s} - w_k \frac{\partial w_i}{\partial s}) \, d\Gamma \\ & + \frac{1}{2} \int_\Gamma \frac{\partial g}{\partial x_j} w_i w_k n_\ell \, d\Gamma + \int_\Gamma \frac{\partial g}{\partial u} \phi^i{}_j w_k n_\ell \, d\Gamma \\ & + \int_\Gamma (\nabla_q g \cdot \nabla \phi^i{}_j) w_k n_\ell \, d\Gamma + \frac{1}{2} \int_\Omega \frac{\partial^2 g}{\partial u^2} \phi^i{}_j \phi^k{}_\ell \, dx \\ & + \frac{1}{2} \int_\Omega \nabla \phi^i{}_j \cdot \nabla_q{}^2 g \cdot \nabla \phi^k{}_\ell \, dx + \int_\Omega \phi^i{}_j \frac{\partial}{\partial u} (\nabla_q g) \cdot \nabla \phi^k{}_\ell \, dx. \end{aligned} \tag{41}$$

Thus, we can obtain the following expression:

$$J^h(X + \vec{q}) \simeq J^h(X) + \vec{q} \cdot \vec{d}^T + \vec{q} \cdot A \cdot \vec{q}^T; \tag{42}$$

i.e., we obtain the quadratic form of $\vec{q}$. Hence, if A is positive definite, we can find $\vec{q}$ which minimizes the right side of (42) as

$$\vec{q} = - \frac{1}{2} \vec{d} A^{-1} \tag{43}$$

Thus, we obtain the following algorithm, a Newton's method, for minimization of $J(\Omega;u)$.

Algorithm

1. Guess the initial value $\chi^{(0)}$;
2. $k = 0$;
3. from (34) and (41) compute $\vec{d}^{(k)}$, $A^{(k)}$
4. compute $\vec{q}$ by $\vec{q} = -\frac{1}{2}\vec{d}^{(k)}(A^{(k)})^{-1}$;
5. $\chi^{(k+1)} = \chi^{(k)} + \vec{q}$
6. compute $J^h_{k+1} - J^h_k$, if it is small enough, stop, otherwise go to 7;
7. change k to k+1;
8. go to 3.

As to the case of the Dirichlet problem, we can obtain the same expression as (42) if we redefine $\vec{d}$ and A by

$$d^i_j = \int_{\Gamma h} w_i n_j [g + \frac{\partial u}{\partial n}(\frac{\partial p}{\partial n} - \nabla_q g\cdot\vec{n})]\, d\Gamma \tag{44}$$

$$\begin{aligned} A_{ijk\ell} = &-\frac{1}{2}\int_{\Gamma h}\frac{\partial g}{\partial u}(\frac{\partial u}{\partial x_j}w_i)(w_k n_\ell)\, d\Gamma \\ &+ \int_{\Gamma h}\{(\nabla_q g\cdot\nabla\phi^i{}_j)w_k n_\ell - w_k\frac{\partial\phi^i{}_j}{\partial x_\ell}\nabla_q g\cdot\vec{n}\}\, d\Gamma \\ &+ \frac{1}{2}\int_{\Gamma h}\{\nabla_q g\cdot\frac{\partial}{\partial x_j}\nabla u)w_i w_k n_\ell - w_i\frac{\partial^2 u}{\partial x_j\partial x_\ell}w_k\nabla_q g\cdot\vec{n}\}\, d\Gamma \\ &+ \frac{1}{2}(1-\delta_{ij})\int_{\Gamma h} g(w_i\frac{\partial w_k}{\partial s} - w_k\frac{\partial w_i}{\partial s})\, d\Gamma \\ &+ \frac{1}{2}\int_{\Gamma h}[\frac{\partial g}{\partial x_j}w_i w_k n_\ell + 2(\nabla p\cdot\vec{n})(w_i\frac{\partial\phi^k{}_\ell}{\partial x_j} + \frac{1}{2}w_i\frac{\partial^2 u}{\partial x_j\partial x_\ell}w_k)]\, d\Gamma \\ &+ \frac{1}{2}\int_{\Omega h}\{\frac{\partial^2 g}{\partial u^2}\phi^i{}_j\phi^k{}_\ell + \nabla\phi^i{}_j\cdot\nabla_q{}^2 g\cdot\nabla\phi^k{}_\ell\}\, dx \\ &+ \int_{\Omega h}\phi^i{}_j(\frac{\partial}{\partial u}\nabla_q g)\cdot\nabla\phi^k{}_\ell\, dx. \end{aligned} \tag{45}$$

Thus, we realize that the algorithm for the case of the Dirichlet problem is the same as above.

4. NUMERICAL EXAMPLE

In this section, we shall give a numerical example in order to illustrate the Newton's method developed in the previous sections. The optimization problem of this section is not a special case of the previous problems nor the generalization of them; it can be, however, trated in the quite similar way.

Nozzle problem. Let us consider a steady-state flow of an ideal incompressible fluid in a two dimensional nozzle illustrated in Fig. 4. We suppose that the flow is a

Fig. 4

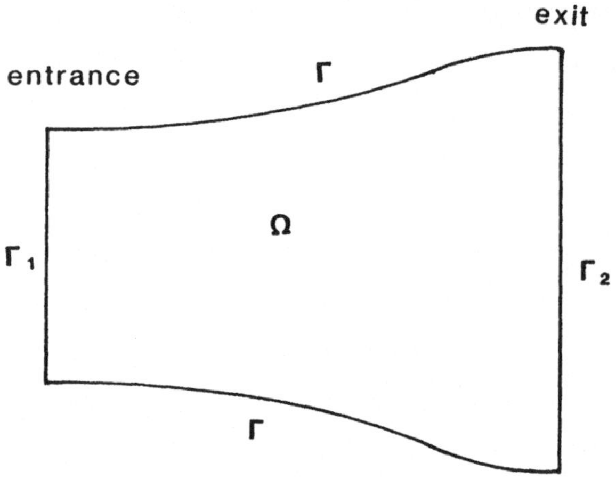

potential flow. In other words, the velocity $\vec{v}$ of the flow is expressed by the gradient of a potential u; more precisely, $\vec{v}$ is given by

$$\vec{v} = \nabla u \tag{46}$$

Our problem is to design the shape of the lateral boundary Γ so that the velocity field ∇u at the exit of the nozzle is as close as possible to the prescribed velocity field $\vec{v}_d$. More precisely, the problem is as follows:

Minimize $J(\Omega;u)$ defined by

$$J(\Omega;u) \equiv \int_{\Gamma_2} |\nabla u - \vec{v}_d|^2 \, d\Gamma \tag{47}$$

with respect to the shape of the lateral boundary Γ, where the velocity potential u is, as is well known (9), given as the solution of the boundary value problem:

$$\Delta u(x) = 0 \qquad (x \in \Omega), \tag{48}$$

$$u = u_d \qquad (x \in \Gamma_1 \cup \Gamma_2), \qquad (49)$$

$$\frac{\partial u}{\partial n} = 0 \qquad (x \in \Gamma), \qquad (50)$$

u_d being a prescribed scalar field on Γ_1 and Γ_2.

First and second variations of the object function. The first and the second variations of the solution to the boundary value problem above are not obtained directly from the previous expressions; they can, however, be obtained easily provided that we suppose Assumption 1 on the solution of (48)-(50). In fact, we can calculate as follows. Let Ω_ε be the domain obtained by the variation of the lateral boundary Γ, u_ε be the corresponding solution of (48)-(50) for Ω_ε and ϕ, ψ be the first and the second variation of the solution of (48)-(50). With these notations, we observe that

$$\begin{aligned} J(\Omega_\varepsilon;u_\varepsilon) - J(\Omega;u) &= \int_{\Gamma_2} |\nabla u_\varepsilon - \vec{v}_d|^2 \, d\Gamma - \int_{\Gamma_2} |\nabla u - \vec{v}_d|^2 \, d\Gamma \\ &= 2\varepsilon \int_{\Gamma_2} (\nabla u - \vec{v}_d)\cdot\nabla\phi \, d\Gamma \\ &\quad + 2\varepsilon^2 \int_{\Gamma_2} (\nabla u - \vec{v}_d)\cdot\nabla\psi \, d\Gamma + \varepsilon^2 \int_{\Gamma_2} |\nabla\phi|^2 \, d\Gamma \\ &\quad + o(\varepsilon^2) \end{aligned} \qquad (51)$$

Hence we can immediately see that

$$\delta^{(1)}J = 2\int_{\Gamma_2} (\nabla u - \vec{v}_d)\cdot\nabla\phi \, d\Gamma, \qquad (52)$$

$$\delta^{(2)}J = 2\int_{\Gamma_2} (\nabla u - \vec{v}_d)\cdot\nabla\psi \, d\Gamma + \int_{\Gamma_2} |\nabla\phi|^2 d\Gamma \qquad (53)$$

We shall not introduce the adjoint variable p, because the second variation ψ appears in the boundary integration and the introduction of p does not simplify the problem at all.

Discretization. We are dealing with the nozzle which is symmetry with respect to the horizontal center line; in what follows, we shall hence consider only the upper half of the nozzle. The center line becomes a new lateral boundary; we know from the symmetry that (50) holds on it.

We discretize the problem by the conventional finite element method. The number of the triangle elements is 80 and the number of the nodes is 55 in this discretization. We assume that only a part of the lateral boundary can move. Its nodes in the discretization are x^1, x^2, x^3, x^4 as illustrated in Fig. 5 (see also Fig.

Fig. 5

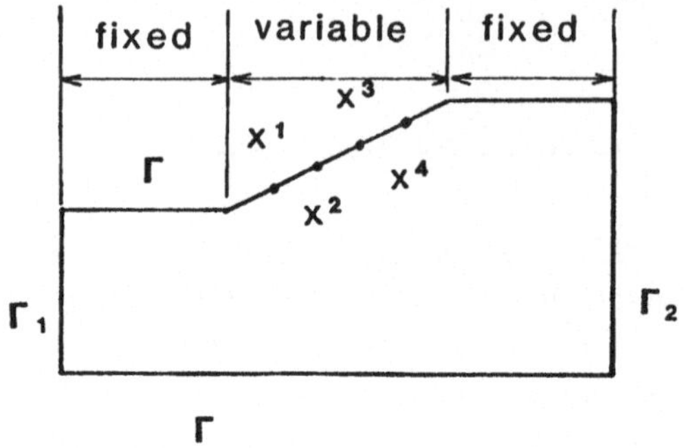

3). Thus, if we introduce shape functions w_i $(i = 1,2,\cdots,4)$ by (30), the variation $\rho(s)$ of the boundary is represented by

$$\rho(s) = \sum_{i=1}^{4} q^i w_i(s) \tag{54}$$

The coefficient q^i denotes the amount of motion of x^i along x_2-direction. Let us introduce $\phi^i(x)$ as the solutions of the boundary value problems

$$\Delta\phi^i(x) = 0 \qquad (x \in \Omega), \tag{55}$$

$$\phi^i(x) = 0 \qquad (x \in \Gamma_1 \cup \Gamma_2), \tag{56}$$

$$\frac{\partial\phi^i}{\partial n} = \frac{\partial u}{\partial s} n_2 \left(\frac{\partial}{\partial s} w_i\right) - w_i\left(\frac{\partial^2 u}{\partial x_1 \partial x_2} n_1 + \frac{\partial^2 u}{\partial x_2^2} n_2\right) \quad (x \in \Gamma), \tag{57}$$

and $\psi^{ij}(x)$ as the solutions of

$$\Delta\psi^{ij}(x) = 0 \qquad (x \in \Omega), \tag{58}$$

$$\psi^{ij}(x) = 0 \qquad (x \in \Gamma_1 \cup \Gamma_2), \tag{59}$$

$$\begin{aligned} \frac{\partial\psi^{ij}}{\partial n}(x) &= \frac{\partial\phi^i}{\partial s} n_2 \frac{\partial w_j}{\partial s} - w_i\left(\frac{\partial^2\phi^j}{\partial x_1 \partial x_2} n_1 + \frac{\partial^2\phi^j}{\partial x_2^2} n_2\right) \\ &- \frac{\partial u}{\partial s} \frac{\partial w_i}{\partial s} \frac{\partial w_j}{\partial s} n_1 n_2 + w_i \frac{\partial w_j}{\partial s} n_2\left(- \frac{\partial^2 u}{\partial x_1 \partial x_2} n_2 + \frac{\partial^2 u}{\partial x_2^2} n_1\right) \\ &- \frac{1}{2} w_i w_j\left(\frac{\partial^3 u}{\partial x_1 \partial x_2^2} n_1 + \frac{\partial^3 u}{\partial x_2^3} n_2\right) \quad (x \in \Gamma). \end{aligned} \tag{60}$$

Then, the first and the second variation $\phi(x)$, $\psi(x)$ of the solution of (48)-(50) are expressed by

$$\phi(x) = \sum_{i=1}^{4} q^i \phi^i(x), \tag{61}$$

$$\psi(x) = \sum_{i,j=1}^{4} q^i \psi^{ij}(x) q^j. \tag{62}$$

Substituting (61), (62) into (52), (53), we immediately obtain

$$\delta^{(1)}J = \sum_{i=1}^{4} 2q^i \int_{\Gamma_2} (\nabla u - \vec{v}_d)\cdot\nabla\phi^i \, d\Gamma, \tag{63}$$

$$\delta^{(2)}J = \sum_{i,j=1}^{4} q^i q^j \int_{\Gamma_2} \{2(\nabla u - \vec{v}_d)\cdot\nabla\psi^{ij} + \nabla\phi^i\cdot\nabla\phi^j\} \, d\Gamma. \tag{73}$$

Let us introduce vectors X and $\vec{q}$ by

$$X = (x^1{}_2,\ x^2{}_2,\ x^3{}_2,\ x^4{}_2),$$

$$\vec{q} = (q^1,\ q^2,\ q^3\ q^4).$$

Then we can write as

$$J^h(X+\vec{q}) \simeq -\vec{q}\cdot A\cdot\vec{q}^T + \vec{d}\cdot\vec{q}^T + J^h(X), \tag{65}$$

where

$$A = \{A_{ij}\}_{i,j=1,\cdots,4},$$

$$A_{ij} = 2\int_{\Gamma_2} \{2(\nabla u - \vec{v}_d)\cdot\nabla\psi^{ij} + \nabla\phi i\cdot\nabla\phi j\} \, d\Gamma, \tag{66}$$

and

$$\vec{d} = (d^1,\ d^2,\ d^3,\ d^4)$$

$$d^i = 2\int_{\Gamma_2} (\nabla u - \vec{v}_d)\cdot\nabla\phi^i \, d\Gamma, \quad i = 1,\cdots,4. \tag{67}$$

Now we have the following algorithm:

1. Guess the initial state $x^{(0)}$;
2. $k = 0$;
3. from (55)-(57) and (58)-(60), compute ϕ^i and ψ^{ij};
4. compute A and $\vec{d}$ from (66) and (67);
5. $x^{(k+1)} = x^{(k)} - \frac{1}{2}\ \vec{d}A^{-1}$

6. compute $J^h(x^{(k+1)})$, if it is smaller than a prescribed value ε, stop, otherwise go to 7;

7. change k to k+1;

8. go to 3.

Here we can use the following facts for the computations of the derivatives of the right hand sides of (57) and (60): We can easily obtain the following expressions on Γ;

$$n_1\frac{\partial^2 u}{\partial x_2 \partial x_1} = \frac{\partial^2 u}{\partial n^2} n_1{}^2 n_2 + \frac{\partial^2 u}{\partial s \partial n} n_1{}^3 + \frac{1}{R}\frac{\partial u}{\partial n} n_1{}^2 n_2$$

$$- \frac{\partial^2 u}{\partial n \partial s} n_1 n_2{}^2 - \frac{\partial^2 u}{\partial s^2} n_1{}^2 n_2 + \frac{1}{R}\frac{\partial u}{\partial s} n_1{}^3,$$

$$n_2\frac{\partial^2 u}{\partial x_2{}^2} = \frac{\partial^2 u}{\partial n^2} n_1 n_2{}^3 + \frac{\partial^2 u}{\partial s \partial n} n_1 n_2{}^2 - \frac{1}{R}\frac{\partial u}{\partial n} n_1{}^2 n_2$$

$$+ \frac{\partial^2 u}{\partial n \partial s} n_1 n_2{}^2 + \frac{\partial^2 u}{\partial s^2} n_1{}^2 n_2 + \frac{1}{R}\frac{\partial u}{\partial s} n_1 n_2{}^2,$$

where R denotes the radius of the curvature of Γ and defined to be positive when Γ is convex to Ω. Hence, we see that

$$n_1\frac{\partial^2 u}{\partial x_2 \partial x_1} + n_2\frac{\partial^2 u}{\partial x_2{}^2} = \frac{\partial^2 u}{\partial s^2} n_2 + \left(\frac{\partial^2 u}{\partial s \partial n} + \frac{1}{R}\frac{\partial u}{\partial s}\right) n_1$$

When we make use of the relations, which are valid on Γ,

$$\frac{\partial u}{\partial n} = 0, \quad \Delta u = \frac{\partial^2 u}{\partial s^2} + \frac{\partial^2 u}{\partial n^2} - \frac{1}{R}\frac{\partial u}{\partial n} = \frac{\partial^2 u}{\partial s^2} + \frac{\partial^2 u}{\partial n^2} = 0,$$

we can rewrite (57) as

$$\frac{\partial \phi^i}{\partial n} = \frac{\partial u}{\partial s}\frac{\partial w_i}{\partial s} n_2 - w^i\left(- \frac{\partial^2 u}{\partial s^2} n_2 + \frac{1}{R}\frac{\partial u}{\partial s} n_1\right). \tag{68}$$

Thus, the boundary condition on Γ for ϕ^i can be expressed only by the derivatives with respect to s. Similarly, we can replace (60) by

$$\frac{\partial\psi^{ij}}{\partial n} = w_i w_j \{ \frac{1}{R^2}\frac{\partial u}{\partial s} n_1 n_2 + \frac{1}{2R}\frac{\partial^2 u}{\partial s^2} + \frac{1}{2}\frac{\partial u}{\partial s}\frac{\partial}{\partial s}(\frac{1}{R})$$

$$+ \frac{\partial^3 u}{\partial s^3} n_1 n_2 \} - w_i \frac{\partial w_j}{\partial s}(\frac{\partial^2 u}{\partial s^2} n_1 n_2 + \frac{2}{R}\frac{\partial u}{\partial s} n_1{}^2)$$

$$- w_i \frac{\partial^2 w_j}{\partial s^2}\frac{\partial u}{\partial s} n_1 n_2 - \frac{\partial u}{\partial s}\frac{\partial w_i}{\partial s}\frac{\partial w_j}{\partial s} n_1 n_2$$

$$+ \frac{\partial w_i}{\partial s}\frac{\partial \phi^j}{\partial s} n_2 - \frac{1}{R} w_i \frac{\partial \phi^j}{\partial s} n_1 + w_i \frac{\partial^2 \phi^j}{\partial s^2} n_2, \qquad (69)$$

which contains only the derivatives with respect to s. These expressions make the computations much easier.

Results of computations. We show in Fig. 6 the final shape of the optimal nozzle

Fig. 6

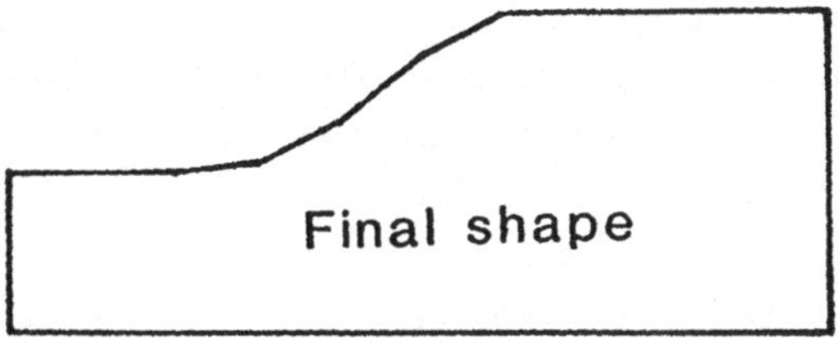

(upper half part). Since our optimization problem has a strong nonlinearity, it is difficult to estimate the cost of computation beforehand. We can not help estimating the efficiency of the computational method by real simulations. We computed the optimal shape also by means of the method of steepest descent. We show in Fig. 7 the

Fig. 7

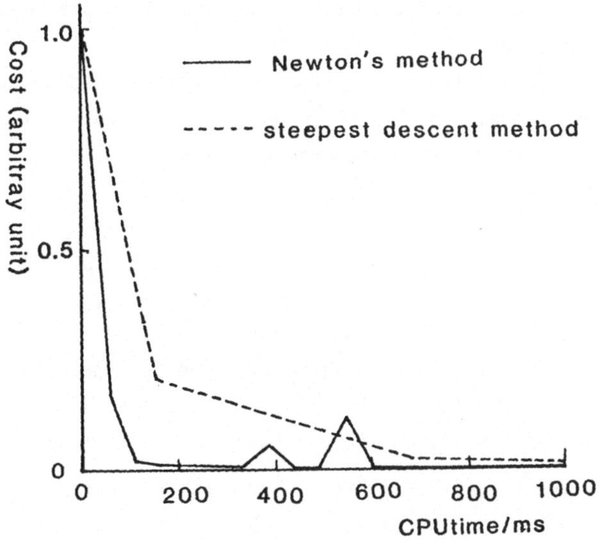

cost vs. cpu time needed for computation by the two methods. This illustration shows that we can compute by our Newton's method much faster than by the conventional steepest descent method. Indeed, our Newton's method works about ten times as fast as the steepest descent method. This is mainly because of the fact that the steepest descent method needs one-dimensional minimization and this optimization requires several to more than ten times of solutions of the boundary value problems. On the other hand, our Newton's method requires one time of solutions of the boundary value problems per one iteration.

Thus, our Newton's method appears to be very efficient to our nozzle problem.

5. CONCLUDING REMARKS

We gave the first and the second variations of the object function corresponding to the boundary variation caused by an arbitrary vector field defined only on the boundary for the cases of both Neumann and Dirichlet problems. Discretization of these variations is carried out by introducing the solutions of new boundary value problems and by the conventional finite element method. Thus, the quadratic forms for the differences of the object functions caused by the boundary variation are obtained for the cases of both Neumann and Dirichlet problems to derive a Newton's method for computation of the optimal domain.

A numerical example is examined by both our Newton's method and the conventional steepest descent method. The results show the efficiency of the Newton's method compared with the conventional steepest descent method. By our other few experiments, however, the Newton's method is not necessarily always efficient. In particular, when the matrix of the quadratic form is not positive definite from the beginning of the iterations, the Newton's method has not tendency toward efficiency.

It seems to be important to find out the adequate numerical method for each optimization problem.

The authors would like to express their thanks to Mr. T. Masanao, who was an under graduate student, for his cooperation and comments. They also thank Professor Y. Sakawa of Osaka University for his encouragements.

REFERENCES

[1] J. Cea, 'Problems shape optimal design', in E. J. Haug and J. Cea (eds.), Optimization of Distributed Parameter Structures, vol.2, Sijthoff and Noordhoff, Alphen aan den Rijn, Holland, 1005-1048(1981).

[2] J. -P. Zolesio, 'The material derivative (or speed) method for shape optimization', ibd., 1152-1194(1981).

[3] N. Fujii, 'Necessary conditions for a domain optimization problem in elliptic boundary value problems', SIAM J. Control and Optimization 24, 346-360(1986).

[4] O. Pironneau, 'On optimum problems in Stokes flow', J. Fluid Mech. 59, 117-128(1973)

[5] O. Pironneau, 'On optimum design in fluid mechnics', J. Fluid Mech. 64, 97-110(1974).

[6] N. Fujii, 'Second variation and its application in a domain optimization problem', Proceedings of the 4th IFAC Symposium on Control of Distributed Parameter Systems, Pergamon, 431-436(1986).

[7] Y. Goto, N. Fujii and Y. Muramatsu, 'Second order necessary optimality conditions for domain optimization problems with a Neumann problem', Proceedings of the 13th IFIP Conference on System Modelling and Optimization, in press(1988)

[8] G. Polya, 'Torsional rigidity, principal frequency, elctrostatic capacity and symmetrization', Quarterly Appl. Math. 6, 267-277(1948).

[9] O. Pironneau, Optimal Shape Design for Elliptic Systems, Springer-Verlag, New York, 1984.

[10] J. P. Queau and Ph. Trompette, 'Two-dimensional shape optimal design by the finite element method', Int. J. Num. Meth. Eng. 15, 1603-1612(1980).

[11] G. Arumugam and O. Pironneau, 'On the feasibility of riblets for airplane', Proceedings of the 13th IFIP Conference on System Modelling and Optimization, in press(1988).

[12] N. Kikuchi, K. Y. Chung T. Torigaki and J. E. Taylor, 'Adaptive finite element method for shape optimization of linear elastic structures', in J. A. Bennet and M. E. Botkin (eds.), The Optimal Shape, Plenum, New York, 139-169(1986).

[13] C. A. Mota Soares and K. K. Choi, 'Boundary elements in shape optimal design of structures', ibd., 199-231(1986).

[14] Ph. Trompette, J. L. Marcelin and C. Lallemaud, 'Optimal shape design of axisymmetric structures', ibd., 283-295(1986).

[15] C. A. Mota Soares (ed.), Computer Aided Optimal Design: Structural and Mechanical Systems, Springer-Verlag, Berlin, 1987.

OPTIMAL CONTROL OF HEMIVARIATIONAL INEQUALITIES

J. Haslinger
KFK, MFF, Charles University
Prague, Czechoslovakia

P.D. Panagiotopoulos
Dept.of Civil Engineering, Aristotle University
Thessaloniki, Greece

Abstract

Here the following optimal control problem is considered:

minimize $J(y,u)$

where (y,u) are related by a *hemivariational inequality* written equivalently as the inclusion

$$A(u)y + \bar{\partial}\Phi(y) \ni f + Bu,$$

where the control $u \in U_{ad}$, $A(u) \in \mathcal{L}(V,V')$ and $\Phi: V \to \mathbb{R}^1$ is a *locally Lipschitz continuous functional* and $\bar{\partial}$ is the *generalized gradient of F.H. Clarke*. Applications in mechanics are presented.

1. Introduction

In the present paper we study the optimal control of mechanical problems involving nonconvex superpotentials. These problems have as variational formulations hemivariational inequalities, which express the principle of virtual work or power in inequality form.

The notion of superpotential was introduced in mechanics by J.J. Moreau [1] for convex but generally nondifferentiable energy functionals. The convex superpotential leads to variational inequalities and to corresponding convex minimization problems. Relations between mechanical quantities expressed in terms of convex superpotentials are generally monotone. There is, however, a large class of mechanical laws between stresses and strains (or strain rates etc.) between boundary displacements (or velocities) and reactions, or generally between fluxes and forces, which are nonmonotone and possibly multivalued, and therefore they cannot be derived from a convex superpotential via subdifferentiation. These laws are of phenomenological nature, and both the lack of monotonicity, as well as the multivaluedness, have been observed in reliable experiments. In order to study mathematically the aforementioned laws the second author has introduced [2] the nonconvex superpotentials through the notion of generalized gradient of F.H. Clarke [3]. In this case the corresponding variational expressions are no longer variational inequalities but hemivariational inequalities, which give rise to substationarity problems. For a study of the hemivariational inequalities the reader is referred to [4],[5],[6]. It should be noted here, that in engineering sciences the lack of monotonicity is the rule in most realistic problems,whereas the monotonicity constitutes a first but important approximation to the physical problem. We mention here as examples the nonmonotone holonomic laws of concrete, rocks, plastic and granular media, and the sawtooth stress-strain diagrams of reinforced concrete in tension (Scanlon's diagram) of composite materials, and of rock interfaces; also the delamination effect in composite beams and plates and generally in adhesive joints. The sawtooth form of the laws describes the cracking or pull-out of the fibres, the local stick-slip phenomena, and the debonding between adjacent laminae. All these mechanical problems lead to hemivariational inequalities. Analogous formulations in terms of nonconvex superpotentials and hemivariational inequalities, but of incremental nature, can be obtained in nonholonomic problems, e.g. in loading-unloading processes.

With respect to all the aforementioned mechanical problems the optimal control problem has a great importance, not only from the theoretical point of view, because it leads to a new type of mathematical problem, but also from the practical point of view, because it permits a better understanding of the behaviour of the corresponding physical problem. For instance, let us consider a parameter identification problem for a rock mass having interfaces, whose behaviour is described by nonmonotone multivalued interface laws. It is well-known that the form of these laws cannot be determined accurately. However, if we have measurements of some variables, i.e. of some displacements, we can introduce a fictitious interface law given by a set of control parameters and then we can minimize the discrepancy between the measured values of the displacements and the predicted ones. The same happens if the rock mass is completely inhomogeneous and we want to determine one or possibly some fictitious moduli of elasticity, which will minimize the same descrepancy as before. (Along these lines a different approach to the homogeneisation problem can be obtained.) In the general optimal control problem, the cost functional may have a more general meaning. It can represent for example, the weight of a structure, its safety, the deviation from a wished state, or combinations of the above.

The resulting problem is of highly nonclassical nature, because, instead of state equations we have state hemivariational inequalities. We recall here that optimal control problems governed by variational inequalities have been already studied (cf. e.g. [7]-[14]). If, however, the state variables are connected with the control variables through a hemivariational inequality we cannot use monotonicity arguments and the development of a mathematical theory based on compactness arguments is necessary. This is attempted in the present paper.

As a model problem we consider the optimal control of a hemivariational inequality arising in the plate theory (cf. e.g. [15]). Let us consider a plate in adhesive contact with a rigid support. The plate is built-in at its boundary, i.e. both the rotations and the deflexions of the plate are zero. The debonding of the plate is described by a nonmonotone sawtooth law. Then the mathematical problem reads: Find $y \in H_0^2(\Omega)$ such that

$$a(y,y^*-y) + \Phi^o(y,y^*-y) \geq (f,y^*-y) \quad \forall y^* \in H_0^2(\Omega). \tag{1.1}$$

Here $a(.,.)$ denotes the bilinear form expressing the energy of the elastic plate, f are the external forces, Φ is the nonconvex superpotential of the adhesive joints and $\Phi^o(.,.)$ is the directional derivative in the sense of F.H. Clarke [3]. In order to formulate the corresponding optimal control problem we introduce a control variable u via an operator B, i.e. we replace f by $f + Bu$, assuming that the elastic energy of the plate depends also on u, and we look for u and y which minimize an appropriately defined cost functional and satisfy the control-state hemivariational inequality (1.1).

2. Mathematical setting of the problem

Let $b \in L^\infty_{loc}(\mathbb{R}^1)$. For any $\delta > 0$ we define

$$\underline{b}_\delta(\xi) = \operatorname*{ess\,inf}_{|t-\xi| \leq \delta} b(t), \quad \bar{b}_\delta(\xi) = \operatorname*{ess\,sup}_{|t-\xi| \leq \delta} b(t)$$

$$\underline{b}(\xi) = \lim_{\delta \to 0_+} \underline{b}_\delta(\xi), \quad \bar{b}(\xi) = \lim_{\delta \to 0_+} \bar{b}_\delta(\xi).$$

Let $\hat{b}(\xi) = [\underline{b}(\xi), \bar{b}(\xi)]$ be a multivalued function in $\mathbb{R}^1$ and

$$j(\xi) = \int_0^\xi b(t)\,dt$$

a locally Lipschitz continuous function. It is known [16], that the generalized derivative of j at ξ satisfies

$$\bar{\partial} j(\xi) \subseteq \hat{b}(\xi)$$

and if $b(\xi\pm)$ exists at any $\xi \in \mathbb{R}^1$ then $\partial j(\xi) = \hat{b}(\xi)$.

Let $\Omega \subset \mathbb{R}^2$ be a bounded domain with Lipschitz boundary $\partial\Omega$ and $\Omega' \subset\subset \Omega$. Define $\Phi: V = H_0^2(\Omega) \to \mathbb{R}^1$ by

$$\Phi(y) = \int_{\Omega'} j(y(x))dx = \int_{\Omega'} \int_0^{y(x)} b(t)dt \ .$$

It is easy to see that Φ is locally Lipschitz continuous in V. *Next we shall assume* that for any bounded measurable selection $f(\xi) \in \partial j(\xi)$ the following growth condition holds:

$$\exists p \geq 2,\ c_1, c_2 = \text{const.} > 0: \quad |f(\xi)| \leq c_1 + c_2 |\xi|^{p-1} \ . \tag{2.1}$$

Then Φ is locally Lipschitz continuous in $L^p(\Omega)$ and

$$\bar{\partial}\Phi(y(x)) \subseteq \hat{b}(y(x)) \quad \text{a.e. in } \Omega' \ .$$

Let us assume the hemivariational inequality of the following type:

$$(\mathcal{P}(u)) \begin{cases} \text{Find } y = y(u) \in V \quad \text{such that} \\ A(u)y + \bar{\partial}\Phi(y) \ni f + Bu \ . \end{cases}$$

Here we assume that the convex variable $u \in U_{ad}$, where U_{ad} is a non-empty, *closed, convex* and *compact* subset of a Hilbert space $U \subset V'$ (dual to V), $B \in \mathcal{L}(U, V')$ (i.e. a linear, continuous mapping), $f \in V'$ a given fixed element. We assume that for any $u \in U_{ad}$, $A(u) \in \mathcal{L}(V, V')$ is generated by a bilinear, continuous form

$$\langle A(u)y, v \rangle = a_u(y, v) \quad \forall y, v \in V \ ,$$

which satisfies:

$$\exists \alpha = \text{const.} > 0: \quad a_u(y,y) \geq \alpha \|y\|^2 \quad \forall y \in V \quad \forall u \in U_{ad}; \tag{2.2}$$

$$a_u(y,v) = a_u(v,y) \quad \forall y, v \in V, \quad \forall u \in U_{ad} \ ; \tag{2.3}$$

$$u_n \to u \text{ in } U \Rightarrow A(u_n) \to A(u) \text{ in } \mathcal{L}(V, V') \ . \tag{2.4}$$

Definition 1.2 A function $y=y(u)\in V$ *will be called* a solution of $(\mathcal{P}(u))$ if there exists $\chi\in L^2(\Omega')$ for some $q\in(1,\infty]$ such that

$$\begin{cases} \langle A(u)y,v\rangle + \int_{\Omega'}\chi v\,dx = \langle f+Bu,v\rangle & \forall v\in V, \\ \chi(x)\in\hat{b}(y(x)) & \text{a.e. in } \Omega' . \end{cases}$$

Finally, let $J: V\times U\to \mathbb{R}^1$ be a cost functional, satisfying

J is coercive in u, uniformly with respect to y, i.e.

$$\forall k>0 \quad \exists r>0 \quad \forall_{u\in U_{ad}} \|u\|\geq r \quad \forall y\in V: J(y,u)\geq k; \tag{2.5}$$

$$\left.\begin{array}{ll} y_n \rightharpoonup y \text{ (weakly) in } V \\ u_n\to u \quad \text{in } U \end{array}\right\} \Rightarrow \lim_{n\to\infty} J(y_n,u_n) = J(y,u). \tag{2.6}$$

The optimal control problem of hemivariational inequality $(\mathcal{P}(u))$ is stated as follows:

$$(\mathbb{P})\quad \begin{cases} \text{Find } \bar{u}\in U_{ad} \text{ such that} \\ J(y(\bar{u}),\bar{u}) \leq J(y(u),u) \quad \forall u\in U_{ad}, \end{cases}$$

where $y(u)\in V$ and $u\in U_{ad}$ are related by $(\mathcal{P}(u))$.

In order to prove the existence of a solution of $(\mathbb{P})$ we shall need the following assumption concerning b (and guaranteeing the existence of at least one solution of $(\mathcal{P}(u))$:

$\exists\bar{\xi}>0$ such that

$$\operatorname*{ess\,sup}_{\xi\in(-\infty,-\bar{\xi})} b(\xi) \leq 0 \leq \operatorname*{ess\,inf}_{\xi\in(\bar{\xi},+\infty)} b(\xi) . \tag{2.7}$$

3. Approximation of $(\mathbb{P})$

Let $V_h\subset V$, $U_h\subset U$, $h\in(0,1)$ be finite-dimensional subspaces of V, U, respectively and U^h_{ad} a closed, convex subset of U_h, not necessarily contained in U_{ad}. Instead of $(\mathcal{P}(u))$ we assume its discretized form $(u_h\in U^h_{ad})$:

$$(\mathcal{P}(u_h))_h \quad \begin{cases} \text{Find } y_h = y_h(u_h)\in V_h \quad \text{such that} \\ \langle A(u_h),v_h\rangle + \int_{\Omega'}\chi_h v_h\,dx = \langle f+Bu_h,v_h\rangle \quad \forall v_h\in V_h, \\ \chi_h\in L^q(\Omega') \quad \text{for some } q\in(1,\infty] \text{ and} \\ \chi_h(x)\in\hat{b}(y_h(x)) \quad \text{a.e. in } \Omega' . \end{cases}$$

Problem (TP) is now replaced by

$$(TP)_h \qquad \begin{array}{l} \text{Find } \bar{u}_h \in U_{ad}^h \text{ such that} \\ J(y_h(\bar{u}_h),(\bar{u}_h) \quad J(y_h(u_h),u_h) \quad \forall u_h \in U_{ad}^h , \end{array}$$

with $y_h(u_h)$ being the solution of $(\mathcal{P}(u_h))_h$. In this Section, we establish the existence of at least one solution of $(TP)_h$.

Before doing it, we prove that $(\mathcal{P}(u_a))_R$ has at least one solution for any $u_h \in U_{ad}^h$. To this end we suppose that (2.2) is satisfied not only for $u \in U_{ad}$, but also for $u \in U_{ad}^h$.

Let $\omega \in \mathcal{D}(\mathbb{R}^1)$ be a smoothing kernel,

$$\int_{-\infty}^{+\infty} \omega(\xi)d\xi = 1, \quad \omega_\varepsilon(\xi) = \omega(\frac{\xi}{\varepsilon}) \text{ and}$$

$$b_\varepsilon(\xi) = \int_{-\infty}^{+\infty} \omega_\varepsilon(t)b(\xi-t)dt = b*\omega_\varepsilon \qquad (\varepsilon>0).$$

<u>Lemma 1.3</u> For any $u_h \in U_{ad}^h$ there exists at least one solution of

$$\left. \begin{array}{l} \text{Find } y_{h\varepsilon} = y_{h\varepsilon}(u_h) \in V_h \text{ such that} \\ \langle A(u_h)y_{h\varepsilon},v_h\rangle + \int_\Omega b_\varepsilon(y_{h\varepsilon})v_h dx = \langle f + Bu_h,v_h\rangle \quad \forall v_h \in V_h . \end{array} \right\} \tag{3.1}$$

<u>Proof</u> Because of (2.7), there exist positive numbers ρ_1,ρ_2 such that

$$\left. \begin{array}{ll} |b_\varepsilon(\xi)| \leq \rho_2 & \forall |\xi| \leq \rho_1 \\ b_\varepsilon(\xi) \geq 0 & \forall \xi \geq \rho_1 \\ b_\varepsilon(\xi) \leq 0 & \forall \xi \leq -\rho_1 \end{array} \right\} \tag{3.2}$$

Define a mapping $T:V_h \to V_h$ by means of

$$\langle Tz_h,\cdot\rangle \equiv \langle A(u_h)z_h,\cdot\rangle + \int_{\Omega'} b_\varepsilon(z_h)\cdot dx - \langle f + Bu_h,\cdot\rangle ,$$

we see that

$$\langle Tz_h,z_h\rangle \geq \alpha ||z_h||^2 - \rho_1\rho_2 \text{ meas } \Omega' - c||z_h||,$$

making use of (3.1) and the fact that $A(\cdot)$ is V-elliptic for $u \in U_{ad} \cup U_{ad}^h$. Thus then exists $r > 0$ such that

$$\langle Tz_h,z_h\rangle > 0 \quad \forall ||z_h|| = r .$$

Applying Brouwer's fixed point theorem we arrive at the assertion of lemma q.e.d.

<u>Theorem 1.3</u> For any $u_h \in U_{ad}^h$ there exists at least one solution $y_h \in V_h$ of $(\mathcal{P}(u_h))_h$.

<u>Proof</u> Let $y_{h\varepsilon} \in V_h$ be a solution of (3.1) and let $\varepsilon \to 0_+$. It is easily seen that $\{y_{h\varepsilon}\}$ is bounded independently of $\varepsilon > 0$. Without loss of generality we may assume that

$$y_{h\varepsilon} \to y_h \in V_h, \quad \varepsilon \to 0_+ . \tag{3.3}$$

At the same time

$$|b_\varepsilon(y_{h\varepsilon})| \leq c \quad \text{a.e. in } \Omega' . \tag{3.4}$$

There exists a function $\chi_h \in L^q(\Omega')$, $q \in (1,\infty)$ such that

$$b_\varepsilon(y_{h\varepsilon}) \rightharpoonup \chi_h \quad \text{in } L^2(\Omega'), \quad \varepsilon \to 0_+ . \tag{3.5}$$

Passing to the limit with $\varepsilon \to 0_+$ in (3.1) we obtain

$$\langle A(u_h) y_h, v_h \rangle + \int_{\Omega'} \chi_h v_h dx = \langle f + Bu_h, v_h \rangle \qquad \forall v_h \in V_h .$$

It remains to show that $\chi_h \in \hat{b}(y_h)$ a.e. in Ω'.

Indeed, starting from the inequality

$$\operatorname*{ess\,inf}_{|t| \leq \varepsilon} b(\xi - t) \leq b_\varepsilon(\xi) \leq \operatorname*{ess\,sup}_{|t| \leq \varepsilon} b(\xi - t),$$

we obtain

$$\operatorname*{ess\,inf}_{|\xi - y_{h\varepsilon}| \leq \varepsilon} b(\xi) \leq b_\varepsilon(y_{h\varepsilon}) \leq \operatorname*{ess\,sup}_{|\xi - y_{h\varepsilon}| \leq \varepsilon} b(\xi).$$

Let $\rho > 0$ be given and $0 < \varepsilon \leq \frac{\rho}{2}$. Then

$$\operatorname*{ess\,inf}_{|\xi - y_{h\varepsilon}| \leq \frac{\rho}{2}} b(\xi) \leq b_\varepsilon(y_{h\varepsilon}) \leq \operatorname*{ess\,sup}_{|\xi - y_{h\varepsilon}| \leq \frac{\rho}{2}} b(\xi)$$

and

$$\underline{b}_\rho(y_h) = \operatorname*{ess\,inf}_{|\xi - y_h| \leq \rho} b(\xi) \leq b_\varepsilon(y_{h\varepsilon}) \leq \operatorname*{ess\,sup}_{|\xi - y_h| \leq \rho} b(\xi) = \overline{b}_\rho(y_h)$$

because of (3.3).

Multiplying the previous inequality by a non-negative function $\varphi \in L^2(\Omega')$ and using (3.5) we have

$$\int_{\Omega'} \underline{b}_\rho(y_h) \varphi dx \leq \int_{\Omega'} \chi_h \varphi dx \leq \int_{\Omega'} \overline{b}_\rho(y_h) \varphi dx . \tag{3.6}$$

Finally, setting $\rho\to 0_+$ in (3.6)we arrive at

$$\int_{\Omega'} \underline{b}(y_h)\varphi dx \le \int_{\Omega'} \chi_h \varphi dx \le \int_{\Omega'} \overline{b}(y_h)\varphi dx \quad ,$$

which is equivalent to

$$\chi_h \in \hat{b}(y_h) \quad \text{a.e. in } \Omega' \quad \text{q.e.d.}$$

The main result of this Section is

Theorem 2.3 There exists at least one solution of $(\mathbb{P})_h$.

Proof Set

$$q = \inf_{u_h \in U_{ad}^h} J(y_h, u_h)$$

and let y_h^n, u_h^n be a minimizing sequence, i.e.

$$q = \lim_{n\to\infty} J(y_h^n, u_h^n) \ . \tag{3.7}$$

Due to (2.5), $\{u_h^n\}$ and at the same time $\{y_h^n\}$ is bounded. Without loss of generality we may, again assume that the whole sequences $\{y_h^n\}$, $\{u_h^n\}$ converge

$$y_h^n \to \overline{y}_h$$

$$u_h^n \to \overline{u}_h \ , \quad n \to \infty \ .$$

Let us recall that y_h^n, u_h^n are related by

$$\langle A(u_h^n) y_h^n, v_h\rangle + \int_{\Omega'} \chi_h^n v_h dx = \langle f + Bu_h^n, v_h\rangle \qquad \forall v_h \in V_h ,$$

$$\chi_h^n \in \hat{b}(y_h^n) \quad \text{a.e. in } \Omega' \ . \tag{3.8}$$

Due to

$$\underline{b}(y_h^n) \le \chi_h^n \le \overline{b}(y_h^n) \qquad \text{a.e. in } \Omega' \ ,$$

there exists $\chi_h \in L^\infty(\Omega')$ such that

$$\chi_h^n \to \chi_h, \quad n \to \infty \text{ in } L^2(\Omega') \qquad \forall q \in (1,\infty) .$$

Passing to the limit with $n\to\infty$ in (3.8) and using (2.4) we arrive at

$$\langle A(\overline{u}_h)\overline{y}_h, v_h\rangle + \int_{\Omega'} \chi_h v_h dx = \langle f + B\overline{u}_h, v_h\rangle \qquad \forall v_h \in V_h \ .$$

Let us prove now that

$\chi_h \in \hat{b}(\bar{y}_h)$ a.e. in Ω' .

To this end we use the fact that $\underline{b}, \bar{b}$ is lower, upper semicontinuous function, respectively.
Let $\varphi \in L^2(\Omega')$ be a non-negative function. Then

$$\int_{\Omega'} \chi_h \varphi dx = \lim_{n\to\infty} \int_{\Omega'} \chi_h^n \varphi dx \leq \limsup_{n\to\infty} \int_{\Omega'} \bar{b}(y_h^n)\varphi dx \leq$$

$$\leq \int_{\Omega'} \limsup_{n\to\infty} \bar{b}(y_h^n)\varphi dx \leq \int_{\Omega'} \bar{b}(\bar{y}_h)\varphi dx \ . \tag{3.9}$$

On the other hand

$$\lim_{n\to\infty} \int_{\Omega'} \chi_h^n \varphi dx \geq \liminf_{n\to\infty} \int_{\Omega'} \underline{b}(y_h^n)\varphi dx \geq \int_{\Omega'} \liminf_{n\to\infty} \underline{b}(y_h^n)\varphi dx \geq$$

$$\geq \int_{\Omega'} \underline{b}(\bar{y}_h)\varphi dx \ . \tag{3.10}$$

Relations (3.9) and (3.10) yield:

$\chi_h \in \hat{b}(\bar{y}_h)$ a.e. in Ω' ,

i.e. $\bar{y}_h$ is a solution of $(\mathcal{P}(\bar{u}_h))_h$. Combining (2.6) with (3.7) we conclude that

$q = J(\bar{y}_h, \bar{u}_h)$ q.e.d.

4. Existence of a solution of $(\mathbb{P})$

Repeating exactly the same approach as in the previous section and using the fact that U_{ad} is compact we arrive at

Theorem 1.4 There exists at least one solution of $(\mathbb{P})$.

5. Relation between $(\mathbb{P})$ and $(\mathbb{P})_h$

Let $h>0$ be the parameter of discretization, tending to zero. With any $h>0$ we associate finite dimensional spaces $V_h \subset V$, $U_h \subset U$. Let $U_{ad}^h \subset U_h$ be a closed convex subset of U_h. We assume that systems of $\{V_h\}$, $\{U_{ad}^h\}$ satisfy the following assumptions:

$$\forall v \in V \quad \exists v_h \in V_h: \quad v_h \to v, \quad h \to 0_+ \ ; \tag{5.1}$$

$$\forall u \in U_{ad} \ \exists u_h \in U_{ad}^h: \quad u_h \to u, \quad h \to 0_+ \ ; \tag{5.2}$$

$u_h \in U_{ad}^h$, $u_h \to u$, $h \to 0_+ \Rightarrow u \in U_{ad}$; (5.3)

system $\{U_{ad}^h\}$ is compact in the following sense: (5.4)

for any $\{u_h\}$, $u_h \in U_{ad}^h$ there exist a subsequence $\{u_{h'}\} \subset \{u_h\}$ and an element $u \in U_{ad}$ such that

$$u_{h'} \to u, \quad h' \to 0_+ .$$

Moreover we suppose that (2.5) is satisfied for all $u \in \overline{U_{ad} \cup \bigcup U_{ad}^h}$ (if $U_{ad}^h \subset U_{ad}$ $\forall h \in (0,1)$, this is automatically satisfied by (2.5)).

<u>Lemma 1.5</u> Let $u_h \in U_{ad}^h$ be such that

$$u_h \to u, \quad h \to 0_+$$

and let $y_h = y_h(u_h)$ be solutions of $(\mathcal{P}(u_h))_h$. Then there exists a subsequence of $\{y_h\}$ (still denoted by $\{y_h\}$) such that

$$y_h \rightharpoonup y, \quad h \to 0_+ \tag{5.5}$$

and $y = y(u) \in V$ is solution of $(\mathcal{P}(u))$.

<u>Proof</u> $\{u_h\}$ is bounded, thus $\{y_h\}$ is bounded as well. There exists a function $y \in V$ such that (5.5) holds. Let $v \in V$ be an arbitrary function. In accordance with (5.1) there exists a sequence $\{\bar{v}_h\}$, $\bar{v}_h \in V_h$ such that

$$\bar{v}_h \to v, \quad h \to 0_+ .$$

It is easily seen that

$$\langle A(u_h)y_h, \bar{v}_h \rangle \to \langle A(u)y, v \rangle ,$$

$$\langle f + Bu_h, \bar{v}_h \rangle \to \langle f + Bu, v \rangle , \quad h \to 0_+ .$$

It remains to study the behaviour of

$$\int_{\Omega'} \chi_h \bar{v}_h ds , \quad h \to 0_+ ,$$

where $\chi_h \in \hat{b}(y_h)$ a.e. in Ω'.

As χ_h is bounded in $L^2(\Omega')$ $\forall q \in (1,\infty)$, there exists a function $\chi \in L^2(\Omega')$ such that

$$\chi_h \rightharpoonup \chi, \; h \to 0_+ .$$

The proof that $\chi \in \hat{b}(y)$ a.e. in Ω' proceeds exactly in the same way as in Theorem 2.3. Here we use the fact that $y_h \rightharpoonup y$ in $H^2(\Omega) \Rightarrow y_h \to y$ uniformly in $\bar{\Omega}$. q.e.d.

Let $u \in U_{ad}$ and denote by

$$X(u) = \{y \in V \mid \quad y \text{ solves } (\mathcal{P}(u))\}$$

and by

$$\tilde{X}(u) = \{y \in V \mid \quad y_h \rightharpoonup y,\ h \to 0_+,\ y_h \in V_h \text{ solves } (\mathcal{P}(u_h))_h,$$
$$u_h \to u,\ h \to 0_+\} .$$

From lemma 1.5 it follows that $\tilde{X}(u) \subseteq X(u)$. $\tilde{X}(u)$ contains those solutions of $(\mathcal{P}(u))$, which can be attained as limits of Galerkin approximations - solutions of $(\mathcal{P}(u_h))$.

The main result of this Section is

<u>Theorem 1.5</u> Let $(\bar{y}_h, \bar{u}_h)$ be an optimal pair of $(\mathbb{P})_h$. Then there exists a subsequence of $\{(\bar{y}_h, \bar{u}_h)\}$ (still denoted by the same symbol) such that

$$\bar{y}_h \rightharpoonup \bar{y} \tag{5.6}$$

$$\bar{u}_h \to \bar{u} \in U_{ad}, \quad h \to 0_+ \tag{5.7}$$

where $\bar{y} \in \tilde{X}(u)$ and

$$J(\bar{y}, \bar{u}) \leq J(y, u) \qquad \forall u \in U_{ad}, \quad \forall y \in X(u). \tag{5.8}$$

<u>Proof</u> Sequences $\{\bar{y}_h\}, \{\bar{u}_h\}$ are bounded. (5.7) is a consequence of (5.3) and (5.4). (5.6) immediately follows from lemma 1.5. Let us prove (5.8). Let $u \in U_{ad}$ be a fixed element. Then there exists a sequence $\tilde{u}_h \in U^h_{ad}$ such that (see 5.2)

$$\tilde{u}_h \to u \ , \quad h \to 0_+$$

and at the same time

$$\tilde{y}_h \rightharpoonup \tilde{y} \ ,$$

where $\tilde{y}_h = y_h(\tilde{u}_h)$ is a solution of $(\mathcal{P}(\tilde{u}_h))$. Because of lemma 1.5, $\tilde{y} \in X(u)$. From the definition of $(\mathbb{P})_h$ one has

$$J(\bar{y}_h, \bar{u}_h) \leq J(\tilde{y}_h, \tilde{u}_h) \ .$$

Passing to the limit with $h \to 0_+$ and using (2.6) we arrive at the assertion of Theorem. q.e.d.

References

[1] J.J. Moreau: La notion de sur-potential et les liaisons unilatérales en élastostatique. C.R. Acad. Sc. Paris 267A (1968) 954-957

[2] P.D. Panagiotopoulos: Non-convex superpotentials in the sense of F.H. Clarke and applications. Mech.Res.Comm.8 (1981) 335-340

[3] F.H. Clarke: Nonsmooth analysis and optimization. J. Wiley, New York 1984

[4] P.D. Panagiotopoulos: Inequality Problems in Mechanics and Applications. Convex and Non-convex Energy Functions. Birkhäuser Verlag, Basel,Boston,Stuttgart 1985 (Russian transl. MIR Publ. Moscow 1987)

[5] J.J. Moreau, P.D. Panagiotopoulos, G. Strang: Topics in Nonsmooth Mechanics. Birkhäuser Verlag, Boston,Basel 1988

[6] J.J. Moreau, P.D. Panagiotopoulos (eds.): Nonsmooth Mechanics and Applications. CISM Lect. Notes Vol.302, Springer Verlag, Wien, N.York 1988

[7] J.P. Yvon: Etude de quelques problèmes de contrôle pour des systèmes distribués. Thèse de Doctorat d'Etat, Université Paris VI, 1973

[8] J.L. Lions: Optimal control of systems governed by partial differential equations. Springer-Verlag 1971

[9] P.D. Panagiotopoulos: Optimal Control in the Unilateral Thin Plate Theory. Archives of Mechanics 29(1977) 25-39

[10] F. Mignot: Contrôle dans les inéquations variationnelles elliptiques. J.Funct. Anal., 22(1976) 130-185

[11] F. Mignot, J.P. Puel: Optimal control in some variational inequalities. SIAM Journal on Control and Optimization 22(1984) 466-476

[12] Shuzhong Shi: Optimal control of Strongly Monotone Variational Inequalities. SIAM Journal on Control and Optimization 26(1988) 274-290

[13] J. Haslinger, P. Neittaanmäki, T. Tiihonen: Shape Optimization on Contact Problems Based on Penalization of the State Inequality. Aplikace Matematiky 31 (1986) 1-88

[14] J. Haslinger, P. Neittaanmäki: On Optimal Shape Design of Systems Governed by Mixed Dirichlet-Signorini Boundary Value Problems. Math. Meth. in the Appl. Sci. 8(1986) 157-181

[15] G. Stavroulakis, P.D. Panagiotopoulos: Laminated orthotropic plates under Subdifferential Boundary Conditions. A Variational-Hemivariational Inequality Approach. ZAMM 68(1988) 213-224

[16] K.C. Chang: Variational methods for non-differentiable functionals and their applications to partial differential equations. J. Math. Anal. Appl. 80(1981) 102-129

EXACT CONTROLLABILITY IN SHORT TIME

Vilmos Komornik

Eötvös University, Department of Analysis, Budapest, Múzeum krt. 6-8.,H-1088, Hungary

Abstract.

The Hilbert Uniqueness Method, introduced by J.-L. Lions, allows one to obtain explicit estimates for the minimal time of exact controllability. We present here a constructive method to improve these estimates. It is based on a slightly more precise form of the usual a priori estimates in the HUM method and on an estimation method introduced recently (for different purposes) by A. Haraux.

1. INTRODUCTION.

In J.-L. Lions (1) a new method was introduced for the study of exact controllability of evolution systems: the so-called Hilbert Uniqueness Method (HUM). It is based on the construction of new Hilbert spaces corresponding to different uniqueness theorems. In general the results involve a "sufficiently large" time. The proofs are based on *a priori* estimates of the type introduced by L. F. Ho (1). These a priori estimates allow one to derive explicit upper estimates of the minimal time of exact controllability, cf. J.-L. Lions (2), (3).

Let us recall the main types of these estimates.

1/ In general it is easy to derive explicit time estimates from the usual a priori estimates by *direct*, *elementary* methods. Sometimes the results are satisfactory. In most cases, however, the results are far from the optimal ones. Several applications of this method are given in J.-L. Lions (2), (3) and J. E. Lagnese and J.-L. Lions (1).

2/ In order to improve the results obtained by the first way, an indirect method was proposed by P.-L. Lions. It is based on a compactness argument and on a suitable unique continuation theorem. This method led to sharp time estimates in several cases, cf. J.-L. Lions (2), (3), E. Zuazua (2). On the other hand this method is technically more involved than the former one and in many situations the appropriate unique continuation theorems are not available. However, in several cases this is the only existing method.

3/ In V. Komornik (1), (2) a different, constructive method was introduced. This method, based on Green's formula, led in some special cases to the sharp time estimates, replacing the just mentioned compactness arguments by much simpler proofs. It was applied later in other situations, too, cf. J.-L. Lions (4), J. E. Lagnese and J.-L. Lions (1), E. Zuazua (3). However, this method did not apply in many cases.

4/ In V. Komornik (3) another constructive method was introduced which leads to sharp time estimates in most cases. It is based on a strenghtened form of the a priori estimates used already in the first method above, and on a powerful estimation method introduced recently by A. Haraux (1).

The purpose of this note is to outline this last method. First we present it by giving a simpler proof for a result of J.-L. Lions and E. Zuazua concerning the exact controllability of a Petrowsky system. In the second part we apply this method to answer a question raised by J.-L. Lions (2) concerning a simultaneous exact controllability problem. For further applications of this method we refer to V. Komornik (4), (5).

2. THE MODEL PROBLEM.

Let G be a non-empty, bounded, open, connected set in R^n ($n = 1,2,\ldots$) having a boundary B of class C^4, and let T be a positive number. Let us consider the system

$$(1)\quad \left| \begin{array}{l} y'' + \Delta^2 y = 0 \quad \text{in} \quad G \times (0,T), \\ \\ y(0) = y^0 \quad \text{and} \quad y'(0) = y^1, \\ \\ y = 0 \quad \text{and} \quad \partial y/\partial \nu = v \quad \text{on} \quad B \times (0,T). \end{array} \right.$$

Following J.-L. Lions (2), (3) we say that the system (1) is exactly controllable at time T if for every initial data $y^0 \in L^2(G)$ and $y^1 \in H^{-2}(G)$ there exists a corresponding control $v \in L^2(B \times (0,T))$ such that the solution y of the system (1) satisfies the final conditions $y(T) = y'(T) = 0$.

Let us consider the homogeneous system

$$(2)\quad \left| \begin{array}{l} z'' + \Delta^2 z = 0 \quad \text{in} \quad G \times (0,T), \\ \\ z(0) = z^0 \quad \text{and} \quad z'(0) = z^1, \\ \\ z = \partial z/\partial \nu = 0 \quad \text{on} \quad B \times (0,T). \end{array} \right.$$

It is well-known that the energy of the solution defined by

$$(3)\qquad E = 2^{-1}\int_G |z'(t)|^2 + |\Delta z(t)|^2 \, dx$$

is in fact independent of $t \in [0,T]$.

Assume that there exists a positive constant $c = c(T)$ such that for every initial data $z^0 \in H^2_0(G)$ and $z^1 \in L^2(G)$ the following inequality holds true:

$$(4) \qquad \int_0^T \int_G |\Delta z|^2 \, dB \, dt \geq cE.$$

Then it follows by applying the Hilbert Uniqueness Method (cf. J.-L. Lions (2), (3)) that the system (1) is exactly controllable at time T.

The first positive result is due to J.-L. Lions. Let us put

$$(5) \qquad R_0 = \inf_{y \in R^n} \sup_{x \in G} |x-y|;$$

fix a point y where this infimum is achieved and set

$$(6) \qquad m(x) = x-y, \; x \in R^n.$$

Using these notations he proved the following estimate:

$$(7) \qquad \left| \begin{array}{l} R_0 \int_0^T \int_B |\Delta z|^2 \, dB \, dt \geq 4TE + 2 \left[\int_G z'(t) m \cdot \nabla z(t) dx\right]_0^T \\ + (n-2) \left[\int_G z'(t) z(t) \, dx\right]_0^T \end{array} \right.$$

($\cdot$ denotes the scalar product in R^n).

Denoting by μ_1, λ_1 the first eigenvalues of the eigenvalue problems

$$(8) \qquad z \in H^2_0(G), \; \int_G \Delta z \, \Delta u dx = \mu \int_G \nabla z \cdot \nabla u \, dx, \; \forall u \in H^2_0(G)$$

and

$$(9) \qquad z \in H^2_0(G), \; \int_G \Delta z \, \Delta u dx = \lambda \int_G zu \, dx, \; \forall u \in H^2_0(G),$$

he deduced from (7) the inequality

$$(10)\qquad \int_0^T\int_B |\Delta z|^2\, dB\, dt \geqslant R_0^{-1}(4T - 4R_0\mu_1^{-1/2} - 2|n-2|\lambda_1^{-1/2})E.$$

Hence the estimate (4) and the exact controllability follows if

$$(11)\qquad T > R_0\mu_1^{-1/2} + 2^{-1}|n-2|\lambda_1^{-1/2}.$$

This result was slightly improved in V. Komornik (1). It was shown that if we keep together the two integrals on the right side of (7), then by an application of Green's formula we may obtain instead of (10) the stronger inequality

$$(12)\qquad \int_0^T\int_B |\Delta z|^2\, dB\, dt \geqslant R_0^{-1}(4T - 4R_0\mu_1^{-1/2})E$$

whence the estimate (4) and the exact controllability follows under the weaker hypothesis

$$(13)\qquad T > R_0\mu_1^{-1/2}.$$

However, as E. Zuazua (2) proved later, in fact the system (1) is exactly controllable for <u>every</u> $T > 0$. His proof is based on the estimates (7) and he applies non-trivial unique continuation and compactness arguments.

We are going to give a simpler and constructive proof for this last result. We begin with an important observation: the assumption (11) can be weakened considerably if we restrict our attention to initial data (z^0, z^1) belonging to a suitable <u>finite-codimensional</u> subspace of $H^2_0(G)\times L^2(G)$.

Indeed, let us denote by $\lambda_1 < \lambda_2 < \ldots$ the increasing sequence of the different eigenvalues of the eigenvalue problem (9) and let us denote by Z_1, $Z_2, \ldots$ the corresponding sequence of finite-dimensional eigenspaces. Fix a positive integer k arbitrarily and assume that the initial data z^0, z^1 are both orthogonal in $L^2(G)$ to the eigenspaces Z_j for all $j < k$. Then the usual proof of the inequality (10) leads to the following stronger estimate:

$$(14) \qquad \int_0^T \int_B |\Delta z|^2 \, dB \, dt \geqslant R_0^{-1}(4T - 4R_0 \lambda_k^{-1/4} - 2|n-2| \lambda_k^{-1/2})E$$

(we used also the relations $\mu_k \geqslant \lambda_k^{1/2}$).

Let us recall finally that the reverse inequality of (4) is always valid (cf. J.-L. Lions (2), (3)): for every $T > 0$ there exists a positive constant C such that for every initial data $z^0 \in H^2_0(G)$ and $z^1 \in L^2(G)$ we have

$$(15) \qquad \int_0^T \int_B |\Delta z|^2 \, dB \, dt \leqslant CE.$$

As we shall see in the next section, the estimates (14) and (15) are already sufficient for the proof of the exact controllability of the system (1) in arbitrarily small time.

3. AN ABSTRACT NORM EQUIVALENCE THEOREM AND ITS APPLICATION FOR THE MODEL PROBLEM.

Let H be a real Hilbert space, A a positive, selfadjoint, inversible operator in H and put $V = D(A^{1/2})$. Given $z^0 \in V$ and $z^1 \in H$ arbitrarily, consider the solution $z \in C(R;V) \cap C^1(R;H)$ of the evolution problem

$$(16) \qquad \left| \begin{array}{l} z'' + Az = 0 \quad \text{in} \quad R, \\ \\ z(0) = z^0 \quad \text{and} \quad z'(0) = z^1. \end{array} \right.$$

We define the energy of z by

$$E = 2^{-1}(\|A^{1/2}z^0\|_H^2 + \|z^1\|_H^2).$$

Assume that A^{-1} is compact and let $0 < \lambda_1 < \lambda_2 < \ldots$ be the sequence of the different eigenvalues of A. Let Z_k be the eigenspace of A associated to λ_k and set

$$Z = \oplus_{k \geq 1} Z_k$$

(vector space direct sum).

Assume that there exists a sequence $T_1 < T_2 < \ldots$ of positive numbers such that the semi norm

$$\left(\int_0^{T_k} |p(z(t),z'(t))|^2 \, dt\right)^{1/2}$$

is equivalent to the energy norm $(E(z^0,z^1))^{1/2}$ on the subspace defined by $z^0, z^1 \in \oplus_{j \geq k} Z_j$.

Then, as it is proved in V. Komornik (5), for every positive number T such that

$$T > \inf \{T_k : k \geq 1\},$$

the semi norm

$$\left(\int_0^{T} |p(z(t),z'(t))|^2 \, dt\right)^{1/2}$$

is in fact a norm which is equivalent to the energy norm $(E(z^0,z^1))^{1/2}$ on the whole space $Z \times Z$. □

Let us apply this theorem for the above exact controllability problem. Putting $H = L^2(G)$, $A = \Delta^2$, $V = H^2_0(G)$,

$$p(z^0, z^1) = \left(\int_B |\Delta z^0|^2 \, dB\right)^{1/2}$$

and

$$T_k = R_0 \lambda_k^{-1/4} + 2^{-1}|n-2| \lambda_k^{-1/2} + 1/k,$$

it follows from the estimates (14) and (15) that the hypotheses of the theorem are satisfied. Hence the quadratic form

$$\int_0^T \int_B |\Delta z|^2 \, dB \, dt$$

is equivalent to the energy E on the whole of $Z \times Z$, for all $T > 0$. In particular the estimate (4) holds true for every $T > 0$ and thus the system (1) is exactly controllable for every $T > 0$. □

The proof of the above norm equivalence theorem is based on a recent method of A. Haraux. Let we are given a discrete (separated) sequence ω_k of real numbers and a positive number T. We say that the system of exponential functions

$$e^{i\omega_1 t}, \; e^{i\omega_2 t}, \ldots$$

is a Riesz basis in $L^2(0,T)$ if there exist two positive constants c_1 and c_2 such that

$$c_1 \sum_{k \geq 1} |z_k|^2 \leq \int_0^T \left|\sum_{k \geq 1} z_k e^{i\omega_k t}\right|^2 dt \leq c_2 \sum_{k \geq 1} |z_k|^2$$

for every sequence of real numbers z_k belonging to l_2. (According to this definition the linear hull of the system is not necessarily dense in $L^2(0,T)$.)

It is well-known from the theory of non-harmonic Fourier series that if

$$e^{i\omega_1 t},\ e^{i\omega_2 t},\ldots$$

is a Riesz basis in $L^2(0,T)$ and if ω_0 is an arbitrary real number, different from $\omega_1, \omega_2,\ldots,$ then the system

$$e^{i\omega_0 t},\ e^{i\omega_1 t},\ldots$$

is a Riesz basis in $L^2(0,S)$ for every $S > T$. Recently, in connection with some interior control problems a constructive and elegant proof was given for this result by A. Haraux (1).

Now observe that by the method of Fourier the solutions of the evolution problem (16) may be written in the form

$$z(t) = \sum_{k\geqslant 1} z_k\, e^{i\omega_k t}$$

where the numbers ω_k depend only on the spectrum of A and where the coefficients z_k belong to the corresponding eigenspaces Z_k of A. Furthermore the energy of z is then given by the formula

$$E = \sum_{k\geqslant 1} \|z_k\|_V^2.$$

Therefore the estimates of the above theorem may be regarded as a Riesz basis property for vectorial coefficients. The method introduced by A. Haraux is then easily adapted to prove this result. For the details we refer to V. Komornik (5).

4. A SIMULTANEOUS CONTROL PROBLEM.

Given a point $y \in R^n$ arbitrarily, define

$$B_+ = \{ x \in B : (x-y)\cdot\nu(x) > 0 \}$$

and

$$B_- = \{ x \in B : (x-y)\cdot\nu(x) \leq 0 \} .$$

Let us choose y such that

$$B_- \neq 0.$$

Let us consider simultaneously the following two systems, governed by the wave equation:

$$(17)\quad \left| \begin{array}{l} y_i'' - \Delta y_i = 0 \quad \text{in} \quad G \times (0,T),\ i = 1,\ 2, \\ y_i(0) = y_i^0 \quad \text{and} \quad y_i'(0) = y_i^1,\ i = 1,\ 2, \\ y_i = 0 \quad \text{on} \quad B_- \times (0,T),\ i = 1,\ 2, \\ y_1 = v \quad \text{and} \quad \partial y_2/\partial\nu = v' \quad \text{on} \quad B_+ \times (0,T). \end{array} \right.$$

The problem is to drive the two systems simultaneously to rest at time T, using the <u>same</u> control function v.

We introduce the Hilbert space

$$V = \{ z \in H^1(G) : z = 0 \quad \text{on} \quad B_- \}$$

and we put

$$R_0 = \sup_{x \in R^n} |x - y| .$$

The purpose of this section is to prove the following theorem:

THEOREM. If $T > 4R_0$, then for every y_i^0, y_i^1 such that

$$y_i^0 \in L^2(G),\ i = 1,2, \quad y_1^1 \in H^{-1}(G) \quad \text{and} \quad y_2^1 \in V',$$

there exists a control function $v \in L^2(B_+ \times (0,T))$ such that the solution y_i of (17) satisfies the final conditions $y_i(T) = y_i'(T) = 0$, $i = 1, 2$. □

REMARK. This result answers a question raised in J.-L. Lions (2), Remark 6.4. He proved the exact controllability of this system under a stronger assumption on T. □

PROOF. Consider the homogeneous system

$$(18) \quad \left| \begin{array}{l} z_i'' - \Delta z_i = 0 \quad \text{in} \quad G \times (0,T),\ i = 1, 2, \\ z_i(0) = z_i^0 \quad \text{and} \quad z_i'(0) = z_i^1,\ i = 1, 2, \\ z_i = 0 \quad \text{on} \quad B_- \times (0,T),\ i = 1, 2, \\ z_1 = \partial z_2/\partial \nu = 0 \quad \text{on} \quad B_+ \times (0,T). \end{array} \right.$$

Defining the energy of the solution of this system by

$$E = 2^{-1} \int_G |\nabla z_1^0|^2 + |\nabla z_2^0|^2 + |z_1^1|^2 + |z_2^1|^2 \, dx,$$

the following estimates were established in J.-L. Lions (2):

Firstly, applying his Hilbert Uniqueness Method he proved that the system (17) is exactly controllable at time T if there exist positive constants c_1, c_2 such that for every initial data $z_1^0 \in H^1_0(G)$, $z_2^0 \in V$, $z_1^1 \in L^2(G)$ and $z_2^1 \in L^2(G)$ the solution of the system (18) satisfies the inequalities

$$(19) \qquad c_1 E \leq \int_{B_+ \times (0,T)} |\partial z_1/\partial \nu + z_2'|^2 \, dB \, dt \leq c_2 E.$$

Secondly, by a multiplier method he proved for every $T > 0$ the following a priori estimates:

$$(20) \qquad \left| \begin{aligned} & 2R_0^{-1}(T - 4R_0 - (n-1)\mu_1^{-1/2})E \\ & \leq \int_{B_+ \times (0,T)} |\partial z_1/\partial \nu + z_2'|^2 \, dB \, dt \leq c_3(1 + T)E, \end{aligned} \right.$$

where μ_1 denotes the first eigenvalue of $-\Delta$ in V and c_3 is a suitable positive constant, depending only on the geometry of G.

Now let us denote by $\mu_1 < \mu_2 < \ldots$ the eigenvalues of the operator $-\Delta$ in V and let X_1, $X_2, \ldots$ be the corresponding eigenspaces. By the multiplier method we obtain that if k is a positive integer and if z_2^0, z_2^1 are both orthogonal in $L^2(G)$ to the eigenspaces $X_1, \ldots, X_{k-1}$, then instead of (20) the following stronger inequalities hold true:

$$(21) \qquad \left| \begin{aligned} & 2R_0^{-1}(T - 4R_0 - (n-1)\mu_k^{-1/2})E \\ & \leq \int_{B_+ \times (0,T)} |\partial z_1/\partial \nu + z_2'|^2 \, dB \, dt \leq c_3(1 + T)E. \end{aligned} \right.$$

Now we are going to apply the general norm equivalence theorem mentioned in the preceeding section. Set $H = L^2(G) \times L^2(G)$, $D(A) = \{(z_1, z_2) \in H^1_0(G) \times V \colon (-\Delta z_1, -\Delta z_2) \in L^2(G) \times L^2(G)\}$ and $A(z_1, z_2) = (-\Delta z_1, -\Delta z_2)$. Then in view of the estimates (21) the general theorem applies with

$$T_k = 4R_0 + (n-1)\mu_k^{-1/2} + 1/k.$$

Hence the estimates (19) follow and this proves our theorem. □

The author is grateful to A. Haraux and J.-L. Lions for the fruitful discussions. The results of this note were obtained while the author was visiting the Université de Savoie, Chambéry, France, as invited professor.

REFERENCES

A. Haraux (1), Quelques propriétés des séries lacunaires utiles dans l'étude des vibrations élastiques, Res. Notes in Math., H. Brézis and J.-L. Lions editors, Nonlinear Partial Differential Equations and Their Applications, Collège de France seminar, 1987-88, Pitman, to appear.

L. F. Ho (1), Observabilité frontière de l'équation des ondes, C. R. Acad. Sci. Paris Sér. I Math. 302(1986), 443-446.

V. Komornik (1), Contrôlabilité exacte en un temps minimal, C. R. Acad. Sci. Paris Sér. I Math. 304(1987), 223-225.

V. Komornik (2), Exact controllability in short time for the wave equation, Ann. Inst. H. Poincaré, Anal. Nonlinéaire, to appear.

V. Komornik (3), Une méthode générale pour la contrôlabilité exacte en temps minimal, C. R. Acad. Sci. Paris Sér. I Math., to appear.

V. Komornik (4), Contrôlabilité exacte en temps minimal de quelques modèles de plaques, C. R. Acad. Sci. Paris Sér. I Math., to appear.

V. Komornik (5), A general method of exact controllability in short time and applications, to appear.

J. E. Lagnese and J.-L. Lions (1), Modelling, Analysis amd Control of Thin Plates, Masson, Collection R. M. A., Paris, 1988.

J.-L. Lions (1), Contrôlabilité exacte des systèmes distribués, C. R. Acad. Sci. Paris Sér. I Math. 302(1986), 471-475.

J.-L. Lions (2), Exact controllability, stabilization and perturbations for distributed systems, SIAM Rev. 30(1988), 1-68.

J.-L. Lions (3), Contrôlabilité exacte des systèmes distribués, I-II, Masson, Collection R. M. A., Paris, 1988.

J.-L. Lions (4), Contrôlabilité exacte et perturbations singulières (II). La méthode de dualité, to appear.

E. Zuazua (1), Contrôlabilité exacte d'une modèle de plaques en un temps arbitrairement petit, C. R. Acad. Sci. Paris Sér. I Math. 304(1986), 173-176.

E. Zuazua (2), Contrôlabilité exacte de quelques modèles de plaques en un temps arbitrairement petit, Appendix in J.-L. Lions (3).

E. Zuazua (3), Some remarks on the boundary stabilizability of the wave equation, in Control of boundaries and stabilization, J. Simon editor, Lecture Notes in Control and Information Sciences, Springer-Verlag, 1989.

UNIFORM BOUNDARY STABILIZATION OF THERMOELASTIC PLATES

J. E. Lagnese
Department of Mathematics
Georgetown University
Washington, DC 20057 USA

1. ORIENTATION

Let Ω be a bounded, open, connected set in $\mathbb{R}^2$ having a Lipschitz boundary Γ. We assume that $\Gamma = \Gamma_0 \cup \Gamma_1$ where Γ_0 and Γ_1 are relatively open, disjoint subsets of Γ with $\Gamma_1 \neq \emptyset$. We denote by $\nu = (\nu_1, \nu_2)$ the unit normal vector to Γ (when it exists) pointing out of Ω, and set $\tau = (-\nu_2, \nu_1)$, a unit positively oriented tangent vector to Γ. Let $T > 0$, and set

$$Q = \Omega \times (0,T), \quad \Sigma = \Gamma \times (0,T), \quad \Sigma_0 = \Gamma_0 \times (0,T), \quad \Sigma_1 = \Gamma_1 \times (0,T).$$

We consider the following boundary value problem which describes the small vibrations of a thin, homogeneous, isotropic, thermoelastic plate of uniform thickness h:

$$\begin{cases} \rho h w'' - \frac{\rho h^3}{12} \Delta w'' + D[\Delta^2 w + \frac{1+\mu}{2} \Delta \vartheta] = f_3, \\ \frac{1}{\kappa} \vartheta' - \Delta \vartheta - \alpha \eta \Delta w' + \frac{12}{h^2}(1 + \frac{h\lambda_1}{2})\vartheta = \frac{\alpha}{\lambda_0} p + \frac{6\alpha\lambda_1}{h^2}(\tau_2 - \tau_1) \quad \text{in } Q, \end{cases}$$

$$w = \frac{\partial w}{\partial \nu} = 0 \quad \text{on } \Sigma_0,$$

$$\begin{cases} D[\Delta w + (1-\mu)B_1 w + \frac{1+\mu}{2} \vartheta] = -M_\tau, \\ D[\frac{\partial \Delta w}{\partial \nu} + (1-\mu)\frac{\partial B_2 w}{\partial \tau} + \frac{1+\mu}{2}\frac{\partial \vartheta}{\partial \nu}] - \frac{\rho h^3}{12}\frac{\partial w''}{\partial \nu} = -\frac{\partial}{\partial \tau} M_\nu - g_3 \quad \text{on } \Sigma_1, \end{cases}$$

$$\frac{\partial \vartheta}{\partial \nu} = -\lambda_2(\vartheta - \tilde{\vartheta}) \quad \text{on } \Sigma,$$

where $' = \partial/\partial t$ and where

$$B_1 w = 2\nu_1\nu_2 \frac{\partial^2 w}{\partial x \partial y} - \nu_1^2 \frac{\partial^2 w}{\partial y^2} - \nu_2^2 \frac{\partial^2 w}{\partial x^2},$$

$$B_2 w = (\nu_1^2 - \nu_2^2)\frac{\partial^2 w}{\partial x \partial y} + \nu_1\nu_2(\frac{\partial^2 w}{\partial y^2} - \frac{\partial^2 w}{\partial x^2}).$$

In the above system, w denotes vertical deflection of the plate and ϑ the deviation of the temperature from a reference temperature at which the plate is free of thermal stress. The various parameters which appear have the following meanings:

ρ: mass density;
D: flexural rigidity;
μ: Poisson's ratio;
f_3: vertical loading on the faces of the plate;

M_τ, M_ν: bending and twising moments, respectively;

g_3: shear force in the vertical direction;

α, η, κ, λ_0, λ_1, λ_2: various *positive* thermal parameters;

p: heat sources within the plate;

$\tau_2 - \tau_1$: temperature differential of the external medium across the two faces of the plate;

$\bar{\vartheta}$: temperature of the external medium at the edge of the plate, measured from the reference temperature.

The reader is referred to [3, Chapter I] for a heuristic derivation of the model.

In what follows we shall assume

$$f_3 = 0, \quad p = 0, \quad \lambda_1(\tau_2 - \tau_1) = 0. \tag{1.1}$$

Assumptions (1.1) mean, respectively, that there are no heat sources within the plate, and that the heat flux through either face of the plate vanishes. We make the change $t \to t\sqrt{D/\rho h}$ in the time scale and also change notation in order to bring the system to the form

$$\begin{cases} w'' - \gamma\Delta w'' + \Delta^2 w + \alpha\Delta\vartheta = 0, \qquad \gamma = h^2/12, \\ \beta\vartheta' - \eta\Delta\vartheta + \sigma\vartheta - \alpha\Delta w' = 0 \qquad \text{in } Q \end{cases} \tag{1.2}$$

where α, β, σ and η are *positive* constants,

$$w = \frac{\partial w}{\partial \nu} = 0 \qquad \text{on } \Sigma_0, \tag{1.3}$$

$$\begin{cases} \Delta w + (1-\mu)B_1 w + \alpha\vartheta = v_1, \\ \dfrac{\partial \Delta w}{\partial \nu} + (1-\mu)\dfrac{\partial B_2 w}{\partial \tau} - \gamma\dfrac{\partial w''}{\partial \nu} + \alpha\dfrac{\partial \vartheta}{\partial \nu} = v_2 \qquad \text{on } \Sigma_1, \end{cases} \tag{1.4}$$

$$\frac{\partial \vartheta}{\partial \nu} = -\lambda(\vartheta - \bar{\vartheta}) \qquad \text{on } \Sigma, \ \lambda \geq 0. \tag{1.5}$$

The initial conditions for (1.2)–(1.5) are

$$w(0) = w^0, \quad w'(0) = w^1, \quad \vartheta(0) = \vartheta^0 \qquad \text{in } \Omega. \tag{1.6}$$

The functions v_1 and v_2 in (1.4) are the *control variables*. The purpose of this paper is show that *linear feedback laws*

$$v_1 = l_1(w,w'), \quad v_2 = l_2(w,w')$$

can be found which *uniformly stabilize* the system (1.1)–(1.5) in the sense that

$$E(t) \leq Ce^{-\omega t}E(0), \qquad t \geq 0,$$

for some *positive* constant ω, where $E(t)$ is the *thermoelastic energy* defined below. We note that v_1 is a bending moment while v_2 is (for example) vertical shear force. Note also that ϑ *cannot* be a control variable since it is a measure of the temperature of the external medium along the edge of the plate and as such is not a quantity which can be expected to be at our disposal. However, it is evident that some restrictions on $\tilde{\vartheta}$ will be required if we expect to be able to stabilize the motion of the plate through feedbacks in the moments and shear force on Γ. These restrictions will be discussed in the next section.

1.1. The Energy Functional. To arrive at an appropriate energy functional for the system (1.1)–(1.5), we multiply the two equations (1.2) by w' and ϑ, respectively, add the products and integrate the sum over Ω. We obtain

$$(w'' - \gamma\Delta w'' + \Delta^2 w, w') + \beta(\vartheta', \vartheta) - \eta(\Delta\vartheta, \vartheta) + \sigma(\vartheta, \vartheta) \\ + \alpha[(\Delta\vartheta, w') - (\vartheta, \Delta w')] = 0 \tag{1.7}$$

where (f,g) denotes the $L^2(\Omega)$ scalar product $\int_\Omega f(X)g(X)dX$ or $\int_\Omega f(X)\cdot g(X)dX$ depending on whether f, g are real or vector valued.

We have

$$(w'' - \gamma\Delta w'', w') = \frac{1}{2}\frac{\partial}{\partial t}[\|w'\|^2 + \gamma\|\nabla w'\|^2] - \gamma\int_{\Gamma_1} w'\frac{\partial w''}{\partial\nu}\, d\Gamma \tag{1.8}$$

where $\|\cdot\| = (\cdot,\cdot)$. To transform the term $(\Delta^2 w, w')$ we use the following *Green's formula* (see [1], p. 206):

$$(\Delta^2\varphi, \hat{\varphi}) = a(\varphi;\hat{\varphi}) + \int_\Gamma \{[\frac{\partial\Delta\varphi}{\partial\nu} + (1-\mu)\frac{\partial B_2\varphi}{\partial\tau}]\hat{\varphi} - [\Delta\varphi + (1-\mu)B_1\varphi]\frac{\partial\hat{\varphi}}{\partial\nu}\}d\Gamma. \tag{1.9}$$

where $a(\varphi;\hat{\varphi})$ denotes the bilinear form

$$a(\varphi;\hat{\varphi}) = \int_\Omega [\frac{\partial^2\varphi}{\partial x^2}\frac{\partial^2\hat{\varphi}}{\partial x^2} + \frac{\partial^2\varphi}{\partial y^2}\frac{\partial^2\hat{\varphi}}{\partial y^2} + \mu(\frac{\partial^2\varphi}{\partial x^2}\frac{\partial^2\hat{\varphi}}{\partial y^2} + \frac{\partial^2\varphi}{\partial y^2}\frac{\partial^2\hat{\varphi}}{\partial x^2}) + 2(1-\mu)\frac{\partial^2\varphi}{\partial x\partial y}\frac{\partial^2\hat{\varphi}}{\partial x\partial y}]dX. \tag{1.10}$$

It follows from (1.8), (1.9) and the boundary conditions (1.3), (1.4) that

$$(w'' - \gamma\Delta w'' + \Delta^2 w, w') = \frac{1}{2}\frac{\partial}{\partial t}[\|w'\|^2 + \gamma\|\nabla w'\|^2 + a(w)] \\ + \int_{\Gamma_1} [(v_2 - \alpha\frac{\partial\vartheta}{\partial\nu})w' - (v_1 - \alpha\vartheta)\frac{\partial w'}{\partial\nu}]d\Gamma. \tag{1.11}$$

The last term in (1.7) is

$$\alpha[(\Delta\vartheta, w') - (\vartheta, \Delta w')] = \alpha\int_{\Gamma_1} (w'\frac{\partial\vartheta}{\partial\nu} - \vartheta\frac{\partial w'}{\partial\nu})d\Gamma. \tag{1.12}$$

From (1.11) and (1.12) we have

$$(w'' - \gamma\Delta w'' + \Delta^2 w, w') + \alpha[(\Delta\vartheta, w') - (\vartheta, \Delta w')] \\ = \frac{1}{2}\frac{\partial}{\partial t}[\|w'(t)\|^2 + \gamma\|\nabla w'(t)\|^2 + a(w(t))] + \int_{\Gamma_1} (v_2 w' - v_1\frac{\partial w'}{\partial\nu})d\Gamma. \tag{1.13}$$

In addition, from (1.5) we have

$$(\Delta\vartheta,\vartheta) = -\|\nabla\vartheta\|^2 - \lambda\int_\Gamma \vartheta(\vartheta - \tilde{\vartheta})d\Gamma. \tag{1.14}$$

Substitution of (1.13) and (1.14) into (1.7) allows (1.7) to be rewritten

$$E'(t) = -\eta\|\nabla\vartheta\|^2 - \sigma\|\vartheta\|^2 - \eta\lambda\int_\Gamma \vartheta(\vartheta - \tilde{\vartheta})d\Gamma - \int_{\Gamma_1}(v_2 w' - v_1\frac{\partial w'}{\partial\nu})d\Gamma \tag{1.15}$$

where the terms on the right are evaluated at time t and where

$$E(t) = \frac{1}{2}[\|w'(t)\|^2 + \gamma\|\nabla w'(t)\|^2 + a(w(t)) + \beta\|\vartheta(t)\|^2] \tag{1.16}$$

is what we shall call the *thermoelastic energy*. We note that $\frac{1}{2}[\|w'\|^2 + \gamma\|\nabla w'\|^2]$ is the *kinetic energy* and a(w(t)) the *elastic strain energy* in Kirchhoff elastic plate theory.

The first three terms on the right side of (1.15) are the contributions of thermal effects and forces to energy dissipation, while the last term represents the effects of mechanical forces and moments on Γ_1. In order to obtain a uniform asymptotic decay rate for (1.16) we shall first of all require that the system be *strongly thermally dissipative*, that is

$$-\eta\|\nabla\vartheta\|^2 - \sigma\|\vartheta\|^2 - \eta\lambda\int_\Gamma \vartheta(\vartheta - \tilde{\vartheta})d\Gamma \le -k(\|\vartheta\|^2 + \|\nabla\vartheta\|^2) \tag{1.17}$$

for some *positive* constant k. (1.17) is a restriction on the thermal parameters and/or the temperature of the surrounding medium. It will be satisfied if, for example, $\lambda = 0$ (insulated edge condition); or if $\tilde{\vartheta}$ *sgn* $\vartheta \le c|\vartheta|$ on Γ, where c satisfies

$$\frac{(c-1)\lambda\eta}{2}\int_\Gamma \varphi^2 d\Gamma \le \sigma\|\varphi\|^2 + \eta\|\nabla\varphi\|^2, \quad \forall\varphi\in H^1(\Omega).$$

The latter inequality will be satisfied for $|c-1|\lambda$ sufficiently small, depending on σ/η and on the diameter of Ω.

Next, we set

$$v_1 = -\mathscr{F}_1(w'), \quad v_2 = -\mathscr{F}_2(w') \tag{1.18}$$

where $\mathscr{F}_1$, $\mathscr{F}_2$ are chosen in such a way that the system is mechanically dissipative, i.e.,

$$\int_{\Gamma_1}(v_2 w' - v_1\frac{\partial w'}{\partial\nu})d\Gamma = \int_{\Gamma_1} F\{\frac{\partial w'}{\partial x},\frac{\partial w'}{\partial y},w'\}\cdot\{\frac{\partial w'}{\partial x},\frac{\partial w'}{\partial y},w'\}d\Gamma \tag{1.19}$$

where the "gain matrix" F is a given symmetric 3×3 matrix $[f_{ij}]$ of $L^\infty(\Gamma_1)$ functions which is *positive definite* on Γ_1. The last term in (1.15) then produces mechanical dissipation in the system. It may be verified that (1.19) holds if $\mathscr{F}_1$ and $\mathscr{F}_2$ are defined by

$$\begin{aligned}\mathscr{F}_1(\varphi) &= (\nu_1^2 f_{11} + \nu_2^2 f_{22} + 2\nu_1\nu_2 f_{12})\frac{\partial\varphi}{\partial\nu} \\ &\quad + [\nu_1\nu_2(f_{22} - f_{11}) + (\nu_1^2 - \nu_2^2)f_{12}]\frac{\partial\varphi}{\partial\tau} + (\nu_1 f_{13} + \nu_2 f_{23})\varphi,\end{aligned} \tag{1.20}$$

$$\mathcal{F}_2(\varphi) = -f_{33}\varphi - (\nu_1 f_{13} + \nu_2 f_{23})\frac{\partial\varphi}{\partial\nu}$$
$$+ \frac{\partial}{\partial\tau}\{[\nu_1\nu_2(f_{22} - f_{11}) + (\nu_1^2 - \nu_2^2)f_{12}]\frac{\partial\varphi}{\partial\nu} + (\nu_1^2 f_{22} + \nu_2^2 f_{11} - 2\nu_1\nu_2 f_{12})\frac{\partial\varphi}{\partial\tau}\}. \quad (1.21)$$

In particular, if F is chosen to be diagonal with equal diagonal elements $f_{ii} = f_0$, then (1.20), (1.21) simplify to

$$\mathcal{F}_1(\varphi) = f_0\frac{\partial\varphi}{\partial\nu},$$

$$\mathcal{F}_2(\varphi) = -f_0\varphi + \frac{\partial}{\partial\tau}(f_0\frac{\partial\varphi}{\partial\tau}).$$

Remark 1.1. J. Prüss and Ph. Clément [5] have recently obtained some very interesting results regarding asymptotic stability of thermoelastic systems with strong thermal damping. They have shown that when the dimension $n = 1$, thermal dissipation *alone* will induce a *uniform decay rate* in the thermoelastic energy, while if $n > 1$, *strong stability* will hold for a class of domains and for certain conservative boundary conditions on the elastic components. (The latter result is probably valid in general.) Their work suggests that the system (1.2)–(1.5) is strongly stable even in the absence of dissipative boundary feedback, although such a result has not been proved and will not be pursued here. On the other hand, it is highly unlikely that thermal dissipation alone is sufficient to induce a *uniform* decay rate for the uncontrolled system (1.2)–(1.5) (i.e., $v_1 = v_2 = 0$) since such uniform asymptotic stability appears not to hold for thermoelastic systems in dimension higher than one.

2. ASYMPTOTIC BEHAVIOR OF SOLUTIONS

2.1. Well–posedness of the Problem. We consider the system (1.2)–(1.6) with v_1, v_2 given by the feedback laws (1.18) with $\mathcal{F}_1$, $\mathcal{F}_2$ defined by (1.20), (1.21). We introduce the spaces

$$H^1_{\Gamma_0}(\Omega) = \{\varphi | \ \varphi \in H^1(\Omega), \ \varphi = 0 \ \text{ on } \Gamma_0\},$$

$$H^2_{\Gamma_0}(\Omega) = \{\varphi | \ \varphi \in H^2(\Omega), \ \varphi = \frac{\partial\varphi}{\partial\nu} = 0 \ \text{ on } \Gamma_0\},$$

and we assume

$$w^0 \in H^2_{\Gamma_0}(\Omega), \quad w^1 \in H^1_{\Gamma_0}(\Omega), \quad \vartheta^0 \in L^2(\Omega), \quad \tilde{\vartheta} \in L^2(\Gamma \times (0,T)). \quad (2.1)$$

A *solution* to the problem is, by definition, a pair $\{w, \vartheta\}$ such that, for each $T > 0$,

$$w \in L^\infty(0,T;H^2_{\Gamma_0}(\Omega)), \quad w' \in L^\infty(0,T;H^1_{\Gamma_0}(\Omega)), \quad \vartheta \in L^\infty(0,T;L^2(\Omega)) \cap L^2(0,T;H^1(\Omega)), \quad (2.2)$$

$$\frac{d}{dt}[c(w';\hat{w}) + b(w;\hat{w}) + \beta(\vartheta,\hat{\vartheta}) - \alpha(\Delta w,\hat{\vartheta})] + a(w;\hat{w}) + \alpha(\vartheta,\Delta\hat{w}) + \eta(\nabla\vartheta,\nabla\hat{\vartheta})$$
$$+ \eta\lambda\int_\Gamma \vartheta\hat{\vartheta}d\Gamma + \sigma(\vartheta,\hat{\vartheta}) = \eta\lambda\int_\Gamma \tilde{\vartheta}\hat{\vartheta}d\Gamma, \quad \forall\hat{w} \in H^2_{\Gamma_0}(\Omega), \ \forall\hat{\vartheta} \in H^1(\Omega), \quad (2.3)$$

where the form $a(\cdot;\cdot)$ is defined by (1.10) and where $b(\cdot;\cdot)$ and $c(\cdot;\cdot)$ are bilinear forms defined by

$$c(\varphi;\hat{\varphi}) = \int_\Omega (\varphi\hat{\varphi} + \gamma\nabla\varphi\cdot\nabla\hat{\varphi})dX, \quad b(\varphi;\hat{\varphi}) = \int_{\Gamma_1} F\{\frac{\partial\varphi}{\partial x},\frac{\partial\varphi}{\partial y},\varphi\}\cdot\{\frac{\partial\hat{\varphi}}{\partial x},\frac{\partial\hat{\varphi}}{\partial y},\hat{\varphi}\}d\Gamma.$$

In the first term in (2.3) the time derivative is interpreted in the sense of distributions on (0,T). Definition (2.3) is justified because of (1.9) and (1.19), and it is satisfied by classical solutions of (1.2)–(1.5).

A key point is that $a(\varphi;\hat{\varphi})$ *is strictly coercive on* $H^2_{\Gamma_0}(\Omega)$. This fact is known as *Korn's Lemma* (see e.g., [1,2]). On the basis of this fact and the trivial observation that $c(\varphi;\hat{\varphi})$ is strictly coercive on $H^1_{\Gamma_0}(\Omega)$ and $b(\varphi;\varphi) \geq 0$ on $H^2_{\Gamma_0}(\Omega)$, equation (2.3) may be solved by a Galerkin approximation method, for example. In fact, formally setting $\hat{w} = w'$ and $\hat{\vartheta} = \vartheta$ in (2.3) and integrating on (0,t) gives

$$c(w'(t)) + \int_0^t b(w')ds + a(w(t)) + \beta\|\vartheta(t)\|^2 + \int_0^t\int_\Omega (\eta|\nabla\vartheta|^2 + \sigma\vartheta^2)dXds$$

$$+ \eta\lambda\int_0^t\int_\Gamma \vartheta^2 d\Gamma ds = c(w^1) + a(w^0) + \beta\|\vartheta^0\|^2 + \eta\lambda\int_0^t\int_\Gamma \vartheta\tilde{\vartheta}d\Gamma ds$$

$$\leq c(w^1) + a(w^0) + \beta\|\vartheta^0\|^2 + \frac{\eta\lambda}{2}\int_0^t\int_\Gamma \vartheta^2 d\Gamma ds + \frac{\eta\lambda}{2}\int_0^t\int_\Gamma \tilde{\vartheta}^2 d\Gamma ds. \tag{2.4}$$

where $c(\varphi) = c(\varphi;\varphi)$ and $b(\varphi) = b(\varphi;\varphi)$. From (2.4) follows that

$$\|w'\|^2_{L^\infty(0,T;H^1_{\Gamma_0}(\Omega))} + \|w\|^2_{L^\infty(0,T;H^2_{\Gamma_0}(\Omega))} + \|\vartheta\|^2_{L^\infty(0,T;L^2(\Omega))} + \|\vartheta\|^2_{L^2(0,T;H^1(\Omega))}$$

$$\leq K[\|w^1\|_{H^1_{\Gamma_0}(\Omega)} + \|w^0\|_{H^2_{\Gamma_0}(\Omega)} + \|\vartheta^0\|_{L^2(\Omega)} + \|\tilde{\vartheta}\|_{L^2(\Gamma\times(0,T))}]. \tag{2.5}$$

On the basis of the estimate (2.5), the existence of a unique solution of (2.2), (2.3) satisfying the initial conditions (1.6) is standard (for example, [1, Chapter 3]).

2.2. An Energy Identity. Let $\{w,\vartheta\}$ by a solution of (1.2)–(1.5). We shall assume that this solution is *sufficiently smooth* to justify the computations which follow. In particular, this will require that w have at least $H^{7/2+\delta}(\Omega)$ regularity and ϑ at least $H^{3/2+\delta}(\Omega)$ regularity in the spatial variables for some $\delta > 0$.

Let $\{x_0,y_0\}$ be a fixed but arbitrary point on $\mathbb{R}^2$, and define the vector field m(x,y) in $\mathbb{R}^2$ by

$$m(x,y) = \{x - x_0, y - y_0\}.$$

For any measureable set $\Lambda\subset\Gamma$ we introduce the notation

$$a_\Lambda(w) = \int_\Lambda |m\cdot\nu|[w^2_{xx} + w^2_{yy} + 2\mu w_{xx}w_{yy} + 2(1-\mu)w^2_{xy}]d\Gamma.$$

Let us note that since $w = \partial w/\partial\nu = 0$ on Γ_0 one has

$$w_{xx}^2 + w_{yy}^2 + 2\mu w_{xx}w_{yy} + 2(1-\mu)w_{xy}^2 = (\Delta w)^2 \quad \text{on } \Gamma_0.$$

For reasons of notation *only*, we assume $m\cdot\nu \geq 0$ on Γ_1 for the remainder of this section. However, this assumption will be *essential* in the derivation of *a priori* estimates given in the next section.

The main result of this section is the following energy identity.

LEMMA. *For every* $T > 0$,

$$\begin{aligned}
&Y_1 - \tfrac{1}{2}Y_2 + Y_3 + \int_0^T E(t)dt + \int_0^T\int_\Omega w'^2 dXdt + \frac{\beta\eta}{2\sigma}\int_0^T\int_\Omega |\nabla\vartheta|^2 dXdt \\
&= \alpha\int_0^T\int_\Omega \nabla\vartheta\cdot[\nabla(m\cdot\nabla w) - \tfrac{1}{2}\nabla w - \frac{\beta}{2\sigma}\nabla w']dXdt - \int_0^T b(w';m\cdot\nabla w)dt \\
&+ \tfrac{1}{2}\int_0^T\int_{\Gamma_0} m\cdot\nu(\Delta w)^2 d\Gamma dt + \tfrac{1}{2}\int_0^T\int_{\Gamma_1} m\cdot\nu(w'^2 + \gamma|\nabla w'|^2)d\Gamma dt - \tfrac{1}{2}\int_0^T a_{\Gamma_1}(w)dt \\
&- \alpha\int_0^T\int_{\Gamma_1} \vartheta\frac{\partial}{\partial\nu}(m\cdot\nabla w - \tfrac{1}{2}w - \frac{\beta}{2\sigma}w')d\Gamma dt - \frac{\beta\eta\lambda}{2\sigma}\int_0^T\int_\Gamma \vartheta(\vartheta - \tilde{\vartheta})d\Gamma dt,
\end{aligned} \tag{2.6}$$

where

$$Y_1 = c(w'(t);m\cdot\nabla w(t))\,\Big|_{t=0}^{T},$$

$$Y_2 = [c(w'(t);w(t)) + \tfrac{1}{2}b(w(t))]\,\Big|_{t=0}^{T},$$

$$Y_3 = \frac{\beta^2}{4\sigma}\|\vartheta(t)\|^2\,\Big|_{t=0}^{T}.$$

Proof: From the first equation in (1.2),

$$\int_0^T\int_\Omega m\cdot\nabla w(w'' - \gamma\Delta w'' + \Delta^2 w + \alpha\Delta\vartheta)dXdt = 0. \tag{2.7}$$

Let us calculate the left side of (2.7). We start with

$$\begin{aligned}
\int_0^T\int_\Omega m\cdot\nabla w(w'' - \gamma\Delta w'')dXdt &= Y_1 - \int_0^T\int_\Omega w'(m\cdot\nabla w')dXdt \\
&- \gamma\int_0^T\int_\Omega \nabla w'\cdot\nabla(m\cdot\nabla w')dXdt - \gamma\int_0^T\int_{\Gamma_1}(m\cdot\nabla w)\frac{\partial w''}{\partial\nu}d\Gamma dt.
\end{aligned} \tag{2.8}$$

We have

$$\begin{aligned}
\int_0^T\int_\Omega w'(m\cdot\nabla w')dXdt &= \int_0^T\int_\Omega[\tfrac{1}{2}\,\mathrm{div}(m|w'|^2) - w'^2]dXdt \\
&= \tfrac{1}{2}\int_0^T\int_{\Gamma_1}(m\cdot\nu)w'^2 d\Gamma dt - \int_0^T\int_\Omega w'^2 dXdt,
\end{aligned} \tag{2.9}$$

$$\gamma\int_0^T\int_\Omega \nabla w'\cdot\nabla(m\cdot\nabla w')dXdt = \frac{\gamma}{2}\int_0^T\int_\Omega \operatorname{div}(m|\nabla w'|^2)dXdt = \frac{\gamma}{2}\int_0^T\int_{\Gamma_1} m\cdot\nu|\nabla w'|^2 d\Gamma dt. \tag{2.10}$$

Substitution of (2.9) and (2.10) into (2.8) yields

$$\int_0^T\int_\Omega m\cdot\nabla w(w'' - \gamma\Delta w'')dXdt = Y_1 + \int_0^T\int_\Omega w'^2 dXdt$$

$$-\frac{1}{2}\int_0^T\int_{\Gamma_1} m\cdot\nu(w'^2 + \gamma|\nabla w'|^2)d\Gamma dt - \gamma\int_0^T\int_{\Gamma_1}(m\cdot\nabla w)\frac{\partial w''}{\partial\nu}d\Gamma dt. \tag{2.11}$$

Next, with the aid of (1.9) we calculate in (2.8) the term

$$\int_0^T\int_\Omega(m\cdot\nabla w)\Delta^2 w dXdt = \int_0^T a(w;m\cdot\nabla w)dt + \int_0^T\int_{\Gamma_1}\{[\frac{\partial\Delta w}{\partial\nu} + (1-\mu)\frac{\partial B_2 w}{\partial\tau}](m\cdot\nabla w)$$

$$-[\Delta w + (1-\mu)B_1 w]\frac{\partial}{\partial\nu}(m\cdot\nabla w)\}d\Gamma dt - \int_0^T\int_{\Gamma_0}\Delta w\frac{\partial}{\partial\nu}(m\cdot\nabla w)d\Gamma dt \tag{2.12}$$

since $B_1 w = 0$ on Γ_0. (One also has $B_2 w = 0$ on Γ_0.) We have (see, e.g., [3])

$$a(w;m\cdot\nabla w) = a(w)$$

$$+\frac{1}{2}\int_0^T\int_\Gamma m\cdot\nu\left[\left[\frac{\partial^2 w}{\partial x^2}\right]^2 + \left[\frac{\partial^2 w}{\partial y^2}\right]^2 + 2\mu\left[\frac{\partial^2 w}{\partial x^2}\right]\left[\frac{\partial^2 w}{\partial y^2}\right] + 2(\tfrac{1}{2}\mu)\left[\frac{\partial^2 w}{\partial x\partial y}\right]^2\right]d\Gamma dt$$

$$= \frac{1}{2}\int_0^T a_{\Gamma_1}(w)dt + \frac{1}{2}\int_0^T\int_{\Gamma_0} m\cdot\nu(\Delta w)^2 d\Gamma dt. \tag{2.13}$$

Since $\frac{\partial}{\partial\nu}(m\cdot\nabla w) = (m\cdot\nu)\Delta w$ on Γ_0, we obtain from (2.12), (2.13) and the boundary conditions (1.4) (with v_1, v_2 defined by (1.18))

$$\int_0^T\int_\Omega(m\cdot\nabla w)\Delta^2 w dXdt = \int_0^T a(w)dt + \frac{1}{2}\int_0^T a_{\Gamma_1}(w)dt - \frac{1}{2}\int_0^T\int_{\Gamma_0} m\cdot\nu(\Delta w)^2 d\Gamma dt$$

$$+\int_0^T\int_{\Gamma_1}\{[\gamma\frac{\partial w''}{\partial\nu} - \alpha\frac{\partial\vartheta}{\partial\nu} - \mathscr{F}_2(w')]m\cdot\nabla w + [\alpha\vartheta + \mathscr{F}_1(w')]\frac{\partial(m\cdot\nabla w)}{\partial\nu}\}d\Gamma dt. \tag{2.14}$$

We also have

$$\alpha\int_0^T\int_\Omega(m\cdot\nabla w)\Delta\vartheta dXdt = -\alpha\int_0^T\int_\Omega\nabla\vartheta\cdot\nabla(m\cdot\nabla w)dXdt + \alpha\int_0^T\int_{\Gamma_1}(m\cdot\nabla w)\frac{\partial\vartheta}{\partial\nu}d\Gamma dt. \tag{2.15}$$

Substitution of (2.11), (2.14) and (2.15) into (2.7) yields

$$Y_1 + \int_0^T\int_\Omega w'^2 dXdt + \int_0^T a(w)dt - \alpha\int_0^T\int_\Omega\nabla\vartheta\cdot\nabla(m\cdot\nabla w)dXdt$$

$$= \int_0^T \int_{\Gamma_1} [\mathscr{F}_2(w')(m\cdot\nabla w) - \mathscr{F}_1(w')\frac{\partial(m\cdot\nabla w)}{\partial\nu}]d\Gamma dt + \frac{1}{2}\int_0^T \int_{\Gamma_0} m\cdot\nu(\Delta w)^2 d\Gamma dt$$

$$-\frac{1}{2}\int_0^T a_{\Gamma_1}(w)dt + \frac{1}{2}\int_0^T \int_{\Gamma_1} m\cdot\nu(w'^2 + \gamma|\nabla w'|^2)d\Gamma dt$$

$$= -\int_0^T b(w';m\cdot\nabla w)dt + \frac{1}{2}\int_0^T \int_{\Gamma_0} m\cdot\nu(\Delta w)^2 d\Gamma dt$$

$$-\frac{1}{2}\int_0^T a_{\Gamma_1}(w)dt + \frac{1}{2}\int_0^T \int_{\Gamma_1} m\cdot\nu(w'^2 + \gamma|\nabla w'|^2)d\Gamma dt. \tag{2.16}$$

Next, we use

$$0 = \int_0^T \int_\Omega w(w'' - \gamma\Delta w'' + \Delta^2 w + \alpha\Delta\vartheta)dXdt$$

$$= c(w'(t);w(t))\Big|_{t=0}^T - \int_0^T \int_\Omega (w'^2 + \gamma|\nabla w'|^2)dXdt + \int_0^T a(w)dt$$

$$-\alpha\int_0^T \int_\Omega \nabla\vartheta\cdot\nabla w dXdt + \int_0^T \int_{\Gamma_1} \{[\mathscr{F}_1(w') + \alpha\vartheta]\frac{\partial w}{\partial\nu} - w\mathscr{F}_2(w')\}d\Gamma dt$$

$$= Y_2 - \int_0^T \int_\Omega (w'^2 + \gamma|\nabla w'|^2)dXdt + \int_0^T a(w)dt$$

$$-\alpha\int_0^T \int_\Omega \nabla\vartheta\cdot\nabla w dXdt + \alpha\int_0^T \int_{\Gamma_1} \vartheta\frac{\partial w}{\partial\nu}d\Gamma dt. \tag{2.17}$$

From (2.16), (2.17) we obtain

$$Y_1 - \frac{1}{2}Y_2 + \frac{1}{2}\int_0^T \int_\Omega (w'^2 + \gamma|\nabla w'|^2)dXdt + \frac{1}{2}\int_0^T a(w)dt + \int_0^T \int_\Omega w'^2 dXdt$$

$$= \alpha\int_0^T \int_\Omega \nabla\vartheta\cdot[\nabla(m\cdot\nabla w) - \frac{1}{2}\nabla w]dXdt - \int_0^T b(w';m\cdot\nabla w)d\Gamma dt$$

$$+\frac{1}{2}\int_0^T \int_{\Gamma_0} m\cdot\nu(\Delta w)^2 d\Gamma dt + \frac{1}{2}\int_0^T \int_{\Gamma_1} m\cdot\nu(w'^2 + \gamma|\nabla w'|^2)d\Gamma dt$$

$$-\frac{1}{2}\int_0^T a_{\Gamma_1}(w)dt - \alpha\int_0^T \int_{\Gamma_1} \vartheta\frac{\partial}{\partial\nu}(m\cdot\nabla w - \frac{1}{2}w)d\Gamma dt. \tag{2.18}$$

We next employ the identity

$$0 = \int_0^T \int_\Omega \vartheta(\beta\vartheta' - \eta\Delta\vartheta + \sigma\vartheta - \alpha\Delta w')dXdt$$

$$= \frac{\beta}{2}\|\vartheta(t)\|^2\Big|_{t=0}^T + \int_0^T \int_\Omega [\eta|\nabla\vartheta|^2 + \sigma\vartheta^2]dXdt$$

$$+ \alpha\int_0^T \int_\Omega \nabla\vartheta\cdot\nabla w' dXdt - \alpha\int_0^T \int_{\Gamma_1} \vartheta\frac{\partial w'}{\partial\nu}d\Gamma dt + \eta\lambda\int_0^T \int_\Gamma \vartheta(\vartheta - \tilde{\vartheta})d\Gamma dt. \tag{2.19}$$

If we multiply (2.19) by $\beta/2\sigma$ and add the product to (2.18) we obtain the identity (2.6).

2.3. A Priori Estimates. In order to obtain uniform asymptotic energy estimates we shall make the following geometric assumptions:

$$m\cdot\nu \le 0 \quad \text{on } \Gamma_0, \qquad m\cdot\nu \ge \rho_0 > 0 \quad \text{on } \Gamma_1. \tag{2.20}$$

We also suppose that

$$\Gamma_0 \neq \emptyset, \tag{2.21}$$

and that the gain matrix F is uniformly positive definite on Γ_1. Because of (2.20), we may write

$$\begin{cases} F = (m\cdot\nu)G, \quad G = [g_{ij}], \quad g_{ij}\in L^\infty(\Gamma_1), \\ g_0|\xi|^2 \le G\xi\cdot\xi \le G_0|\xi|^2, \quad \forall\xi\in\mathbb{R}^3, \quad 0 < g_0 \le G_0, \end{cases} \tag{2.22}$$

where g_0 and G_0 are constants. Then $b(\varphi;\hat{\varphi})$ has the form

$$b(\varphi;\varphi) = \int_{\Gamma_1} (m\cdot\nu)G\{\varphi_x,\varphi_y,\varphi\}\cdot\{\hat{\varphi}_x,\hat{\varphi}_y,\hat{\varphi}\}d\Gamma$$

and satisfies

$$g_0\int_{\Gamma_1} m\cdot\nu(|\nabla\varphi|^2 + \varphi^2)d\Gamma \le b(\varphi) \le G_0\int_{\Gamma_1} m\cdot\nu(|\nabla\varphi|^2 + \varphi^2)d\Gamma. \tag{2.23}$$

Finally, we shall require that the system (1.2)–(1.5) be *strongly thermally dissipative* in the sense of (1.17). Under the above assumptions, a uniform decay rate will be obtained for *regular* solutions of (1.2)–(1.5).

THEOREM. *There is a constant $\omega > 0$ such that if $\{w,\vartheta\}$ is any sufficiently regular solution of* (1.2)–(1.5), *then*

$$\int_t^\infty [E(s) + E_\Gamma(s)]ds \le \frac{1}{\omega}E(t), \qquad t \ge 0, \tag{2.24}$$

where

$$E_\Gamma(t) = \frac{1}{2}\left[\int_\Gamma (w'^2 + \gamma|\nabla w'|^2)d\Gamma + a_\Gamma(w)\right].$$

As a consequence of (2.24) *one has*

$$\int_t^\infty E(s)ds \le e^{-\omega t}\int_0^\infty E(s)ds, \qquad t \ge 0, \tag{2.25}$$

$$E(t) \le e\cdot e^{-\omega t}E(0), \qquad t \ge 0. \tag{2.26}$$

Remark 2.1. The requirement that the constant $\gamma > 0$ in (1.2) is *essential* in the proof which follows.

Remark 2.2. An *explicit* formula for the constant ω will be given.

Proof of (2.24). The proof is based on the identity (2.6). We introduce

$$\rho(t) = c(w'(t);m\cdot\nabla w(t)) - \tfrac{1}{2}[c(w'(t);w(t)) + \tfrac{1}{2}b(w(t))] + \frac{\beta^2}{4\sigma}\|\vartheta(t)\|^2. \tag{2.27}$$

Then

$$Y_1 - \tfrac{1}{2}Y_2 + Y_3 = \rho(T) - \rho(0).$$

Using (2.21) and Poincaré's inequality we have

$$|\rho(t)| \leq CE(t) \tag{2.28}$$

for an appropriate constant C.

For $\epsilon > 0$ define

$$F_\epsilon(t) = E(t) + \epsilon\rho(t). \tag{2.29}$$

It will be proved that for $\epsilon > 0$ sufficiently small,

$$F_\epsilon'(t) \leq -\tfrac{\epsilon}{2}[E(t) + E_\Gamma(t)]. \tag{2.30}$$

The extimate (2.24) is then obtained from (2.28) and (2.30) via the following argument. Let $\beta > 0$, multiply (2.30) by $e^{-\beta t}$ and integrate from t to ∞. After an integration by parts we obtain

$$\beta\int_t^\infty e^{-\beta s}F_\epsilon(s)ds + \frac{\epsilon}{2}\int_t^\infty e^{-\beta s}[E(s) + E_\Gamma(s)]ds \leq e^{-\beta t}F_\epsilon(t). \tag{2.31}$$

From (2.28) and (2.29) we have

$$(1-\epsilon C)E(t) \leq F_\epsilon(t) \leq (1+\epsilon C)E(t), \tag{2.32}$$

and therefore $F_\epsilon(t) \geq 0$ of $\epsilon C \leq 1$. For such ϵ we may drop the first term in (2.31). Letting $\beta \to 0$ in what remains and again using (2.32), we obtain

$$\int_t^\infty [E(s) + E_\Gamma(s)]ds \leq \frac{2(1+\epsilon C)}{\epsilon} E(t),$$

which is (2.24) with $\omega = \frac{\epsilon}{2(1+\epsilon C)}$.

The estimate (2.25) follows from (2.24) and a Gronwall's Lemma type argument. To obtain (2.26), we note that since E(t) is nonincreasing, for every $\tau > 0$

$$\tau E(t+\tau) \le \int_t^{t+\tau} E(s)ds \le e^{-\omega t}\int_0^{\infty} E(s)ds \le \frac{e^{-\omega t}}{\omega} E(0),$$

and therefore

$$E(t+\tau) \le \frac{e^{\omega\tau}}{\omega\tau} e^{-\omega(t+\tau)}E(0), \qquad t \ge 0, \quad \tau > 0.$$

The first factor on the right achieves its minimum on $\tau > 0$ at $\tau = 1/\omega$, so that

$$E(t+1/\omega) \le e\cdot e^{-\omega(t+1/\omega)}E(0), \qquad t \ge 0.$$

Proof of (2.30). To prove (2.30), we first estimate $\rho'(t)$ which is calculated by differentiating (2.6) in T. Writing t in place of T we obtain

$$\begin{aligned}\rho'(t) = &- E(t) - \|w'\|^2 - \frac{\beta\eta}{2\sigma}\|\nabla\vartheta\|^2 + \alpha(\nabla\vartheta,\nabla(m\cdot\nabla w) - \frac{1}{2}\nabla w - \frac{\beta}{2\sigma}\nabla w')\\ &- b(w';m\cdot\nabla w) + \frac{1}{2}\int_{\Gamma_0} m\cdot\nu(\Delta w)^2 d\Gamma + \frac{1}{2}\int_{\Gamma_1} m\cdot\nu(w'^2 + \gamma|\nabla w'|^2)d\Gamma\\ &-\frac{1}{2}a_{\Gamma_1}(w) - \alpha\int_{\Gamma_1}\vartheta\frac{\partial}{\partial\nu}(m\cdot\nabla w - \frac{1}{2}w - \frac{\beta}{2\sigma}w')d\Gamma - \frac{\beta\eta\lambda}{2\sigma}\int_{\Gamma}\vartheta(\vartheta-\tilde{\vartheta})d\Gamma,\end{aligned} \tag{2.33}$$

where the terms on the right are evaluated at t. We proceed to estimate various of these terms.

Let $\delta > 0$ be arbitrary. We have

$$\begin{aligned}|\alpha(\nabla\vartheta,\nabla(m\cdot\nabla w) - \frac{1}{2}\nabla w - \frac{\beta}{2\sigma}\nabla w')| &\le \alpha\|\nabla\vartheta\|\|\nabla(m\cdot\nabla w) - \frac{1}{2}\nabla w - \frac{\beta}{2\sigma}\nabla w'\|\\ &\le \alpha C_1\|\nabla\vartheta\|[a(w) + \gamma\|\nabla w'\|^2]^{\frac{1}{2}} \le \delta[a(w) + \gamma\|\nabla w'\|^2] + C_\delta\|\nabla\vartheta\|^2.\end{aligned} \tag{2.34}$$

The term $b(w';m\cdot\nabla w)$ is estimated as follows: for any $\zeta > 0$,

$$|b(w';m\cdot\nabla w)| \le \frac{1}{2\zeta} b(w') + \frac{\zeta}{2} b(m\cdot\nabla w). \tag{2.35}$$

Let us estimate the last term in (2.35). Utilizing (2.23) we have

$$\begin{aligned}b(m\cdot\nabla w) &= \int_{\Gamma_1} m\cdot\nu G\{(m\cdot\nabla w)_x,(m\cdot\nabla w)_y,m\cdot\nabla w\}\{(m\cdot\nabla w)_x,(m\cdot\nabla w)_y,m\cdot\nabla w\}d\Gamma\\ &\le G_0\int_{\Gamma_1} m\cdot\nu[|m\cdot\nabla w|^2 + |\nabla(m\cdot\nabla w)|^2]d\Gamma\\ &\le C\int_{\Gamma_1} m\cdot\nu[|\nabla w|^2 + \sum_{\alpha=2}|D^\alpha w|^2]d\Gamma \le C_1 a(w) + C_2 a_{\Gamma_1}(w).\end{aligned}$$

(The last estimate uses $0 < \mu < 1$. In fact, $0 < \mu \le 1/2$ in physical situation.) Inserting this last estimate into (2.35) yields

$$\begin{aligned}|b(w';m\cdot\nabla w)| &\le \frac{1}{2\zeta} b(w') + \frac{\zeta}{2}[C_1 a(w) + C_2 a_{\Gamma_1}(w)]\\ &\le \delta[a(w) + a_{\Gamma_1}(w)] + C_\delta b(w')\end{aligned} \tag{2.36}$$

for any $\delta > 0$.

The next to last term on the right side of (2.33) may be estimated in the following way, keeping in mind assumption (2.20):

$$|\alpha \int_{\Gamma_1} \vartheta \frac{\partial}{\partial \nu}(m \cdot \nabla w - \frac{1}{2} w - \frac{\beta}{2\sigma} w')d\Gamma|$$

$$\leq \frac{\alpha}{\rho_0}[\int_\Gamma \vartheta^2 d\Gamma]^{\frac{1}{2}}[\int_{\Gamma_1} m \cdot \nu |\frac{\partial}{\partial \nu}(m \cdot \nabla w - \frac{1}{2} w - \frac{\beta}{2\sigma} w')|^2 d\Gamma]^{\frac{1}{2}}$$

$$\leq \delta[a_{\Gamma_1}(w) + \gamma \int_{\Gamma_1} m \cdot \nu |\nabla w'|^2 d\Gamma + a(w)] + C_\delta(\|\vartheta\|^2 + \|\nabla \vartheta\|^2). \tag{2.37}$$

The last term on the right side of (2.33) is bounded from above with the aid of assumption (1.17), and one obtains

$$-\frac{\beta\eta\lambda}{2\sigma} \int_\Gamma (\vartheta(\vartheta - \tilde{\vartheta})d\Gamma \leq C_1(\|\vartheta\|^2 + \|\nabla \vartheta\|^2) \tag{2.38}$$

for some constant C_1.

If the inequalities (2.34), (2.36)–(2.38) are substituted for the corresponding terms in (2.33), an estimate for $\rho'(t)$ is obtained that has the following structure:

$$\rho'(t) \leq -\frac{3}{2}\|w'\|^2 - \frac{\gamma}{2}(1-2\delta)\|\nabla w'\|^2 - (\frac{1}{2} - 3\delta)a(w) - \frac{\beta}{2}\|\vartheta\|^2$$

$$+\frac{1}{2}\int_{\Gamma_0} m \cdot \nu(\Delta w)^2 d\Gamma + \frac{1}{2}\int_{\Gamma_1}(m \cdot \nu)w'^2 d\Gamma + \frac{\gamma}{2}(1-2\delta)\int_{\Gamma_1} m \cdot \nu |\nabla w'|^2 d\Gamma$$

$$-(\frac{1}{2} - 2\delta)a_{\Gamma_1}(w) + C_\delta[b(w') + \|\vartheta\|^2 + \|\nabla \vartheta\|^2]. \tag{2.39}$$

We choose $\delta = 1/12$ in (2.39) and obtain an estimate of the form

$$\rho'(t) \leq -\frac{1}{2} E(t) - \frac{1}{4} a_\Gamma(w) + C_1[b(w') + \|\vartheta\|^2 + \|\nabla \vartheta\|^2] \tag{2.40}$$

for some constant C_1.

The proof of (2.30) may now be completed as follows. We have

$$F_\epsilon'(t) = E'(t) + \epsilon\rho'(t). \tag{2.41}$$

From (1.15), (1.17) and (1.19),

$$E'(t) = -\eta\|\nabla \vartheta\|^2 - \sigma\|\vartheta\|^2 - \eta\lambda \int_\Gamma \vartheta(\vartheta - \tilde{\vartheta})d\Gamma - b(w') \leq -k(\|\vartheta\|^2 + \|\nabla \vartheta\|^2) - b(w'). \tag{2.42}$$

If (2.40) and (2.42) are inserted into (2.41), the result is

$$F_\epsilon'(t) \leq -\frac{\epsilon}{2} E(t) - \frac{\epsilon}{4} a_\Gamma(w) - (1-\epsilon C_1)b(w') - (k-\epsilon C_1)(\|\vartheta\|^2 + \|\nabla \vartheta\|^2)$$

$$\le -\frac{\epsilon}{2}E(t) - \frac{\epsilon}{4}a_\Gamma(w) - (k-\epsilon C_1)(\|\vartheta\|^2 + \|\nabla\vartheta\|^2)$$
$$- g_0(1-\epsilon C_1)\int_{\Gamma_1} m\cdot\nu(w'^2 + \gamma|\nabla w'|^2)d\Gamma \tag{2.43}$$

provided $\gamma = h^2/12 \le 1$ (as we may assume). The estimate (2.30) follows by choosing $\epsilon > 0$ so that

$$k - \epsilon C_1 \ge 0, \quad g_0(1 - \epsilon C_1) \ge \frac{\epsilon}{4}.$$

REFERENCES

[1] Duvaut, G. and J.–L. Lions, *Inequalities in Mechanics and Physics*, Springer–Verlag, Berlin, 1976.

[2] Gobert, J., *Une inequation fundamentale de la theorie de l'elasticite*, Bull. Soc. Royale Sci. Liege, 31 (1962), pp. 182–191.

[3] Lagnese, J. E. and J.–L. Lions, *Modelling, Analysis and Control of Thin Plates*, Recherches en Mathématiques Appliquées, Vol. 6, Masson, Paris, 1988.

[4] Lions, J.–L. and E. Magenes, *Problemes aux Limites Non–Homogeneous et Applications*, Vol. 1, Dunod, Paris, 1968.

[5] Prüss, J. and Ph. Clément, *Asymptotic stability of the equations of thermoelasitity*, to appear.

ASYMPTOTIC BEHAVIOR OF THE SOLUTIONS OF THE KIRKHOFF PLATE WITH NONLINEAR DISSIPATION IN THE BENDING MOMENT

I. Lasiecka
Department of Applied Mathematics
University of Virginia
Charlottesville, Virginia 22903, U.S.A.

1. Introduction

We study the question of asymptotic stability of the solutions to the classical equation of a thin, isotropic, homogeneous plate with nonlinear dissipation occurring on a portion of the edge of the plate. We assume, that the energy is dissipated by the application of appropriate nonlinear bending moments.

Let Ω be a bounded, open domain in R^2 with a boundary Γ consisting of two portions: Γ_0 and Γ_1. Let $\gamma(u)$ be a monotone increasing graph such that $\gamma(0) = 0$. Consider the following model:

$$\left.\begin{array}{l}
\text{(i) } u_{tt}(t,x) + \Delta^2 u(t,x) = 0 \quad \text{in } Q \equiv \Omega \times (0,\infty); \\
\text{(ii) } u(t{=}0,x) = u_0(x); \; u_t(t{=}0) = u_1(x), \quad \text{in } \Omega; \\
\text{(iii) } u(t,x) = 0; \quad x \in \Gamma_0; \; t > 0; \\
\text{(iv) } \Delta u(t,x) + (1-\mu) B_1 u(t,x) - \gamma(\frac{\partial}{\partial \eta} u_t(t,x)) \ni 0; \; x \in \Gamma_1; \\
\text{(v) } \frac{\partial}{\partial \eta} \Delta u(t,x) + (1-\mu) B_2 u(t,x) = 0, \; x \in \Gamma_1.
\end{array}\right\} \tag{1.1}$$

Here $0 < \mu < 1$ and the boundary operators B_1 and B_2 are given by

$$B_1 u = -\frac{\partial^2}{\partial \tau^2} u - k \frac{\partial}{\partial \eta} u,$$

$$B_2 u = \frac{\partial^2}{\partial \tau^2} \frac{\partial}{\partial \eta} u - 2 \frac{\partial}{\partial \tau} k \frac{\partial u}{\partial \tau},$$

where $\vec{\eta} = (n_1, n_2)$ is the outward normal to Γ, $\vec{\tau} = (\tau_1, \tau_2) = (-n_2, n_1)$ is the tangent to Γ and k stands for the curvature.

The problem of stability of the solutions of the equation (1.1) but with the dissipation occurring through the shear forces has been studied recently in the literature ([L.1], [L.2], [L-L]). In fact, in [L.1] (see also [L-L]) the model (1.1), ((i), (ii)) with the boundary conditions

$$\left.\begin{array}{l}\text{(i) } u(t,x) = \dfrac{\partial u}{\partial \eta}(t,x) = 0, \quad x \in \Gamma_0; \\ \text{(ii) } \Delta u\,(t,x) + (1-\mu)\, B_1\, u(t,x) = 0; \quad x \in \Gamma_1, \\ \text{(iii) } \dfrac{\partial}{\partial \eta}\, \Delta\, u(t,x) + (1-\mu)\, B_2\, u(t,x) + u_t(t,x) = 0: \ x \in \Gamma_1, \end{array}\right\} \tag{1.2}$$

has been considered. Under certain (rather natural) geometric conditions imposed on Ω, it was shown in [L.1], that the solution to (1.1) ((i), (ii)) and (1.2) decay exponentially to zero in the topology of $H^2(\Omega) \times L_2(\Omega)$.

In [L.2], the nonlinear model has been discussed where the boundary conditions in (1.2) (iii) are replaced by

$$\frac{\partial}{\partial \eta}\, \Delta u(t,x) + (1-\mu)\, B_2 u(t,x) + \gamma\,(u_t(t,x)) \ni 0 \tag{1.3}$$

with $\gamma\,(u)$ monotone graph such that $0 \in \gamma\,(0)$. Under the same geometric conditions as in [L.1], strong stability of the solutions of (1.1) (i,ii)-(1.3) has been established in [L.2].

Later, in [L.3]], it was shown that the uniform decay rate holds for the solution of the nonlinear model (1.3) but with

$$\gamma(u) = f(u)$$

where f is a single valued, monotone increasing function, f (0) = 0 and it satisfies certain growth conditions at the origin and at the infinity.

The main goal of this paper is to study the solutions with dissipation in the bending moment. Here, even in the linear case, the open question is whether or not we can stabilize system (1.1) by applying the dissipation <u>only</u> through the moments? This problem appears much more difficult than the one with the dissipation in shear forces. In fact, one of the main difficulties is a lack of an appropriate Holmgren type uniqueness theorem for the linear homogeneous on the boundary problem. In the present paper we shall attempt to provide a positive answer to the question above, at least for certain geometric configurations. Our main result is formulated below:

<u>Theorem</u>

I. For any initial data

$$u_0 \in \tilde{H}_2(\Omega); \quad u_1 \in L_2(\Omega)$$

where

$$\tilde{H}^2(\Omega) \equiv \{u \in H^2(\Omega);\ u|_{\tilde{\Gamma}_0} = 0\},$$

there exists the unique solution of (1.1); $(u,u_t) \in C[0\ \infty;\ \tilde{H}_2(\Omega) \times L_2(\Omega)]$.

II. Assume

$$\vec{h} \cdot \vec{\eta} \le 0 \text{ on } \Gamma_0;\ \ \vec{h} \cdot \vec{\eta} > 0 \text{ on} \Gamma_1 \tag{H-1}$$

where $\vec{h} \equiv x - x_0;\ x_0 \in R^2;$

$$k\vec{h}\,\vec{\eta} \le \frac{3-2\mu}{13+6\mu} \quad \text{on } \Gamma_1. \tag{H-2}$$

Then the solutions $(u, u_t) \to 0$ when $t \to \infty$ in the strong topology of $\tilde{H}_2(\Omega) \times L_2(\Omega)$.

Remark 1

Geometric condition (H1) is standard (see [L.13]). Instead the condition (H-2) requires that the curvature on Γ_1 be "small".

Remark 2

Stabilization via bending moment requires that the dissipation be applied on the entire boundary Γ. This is in contrast with the case when the control acts as a shear force, in which case it is enough that the dissipation occurs only on a portion of the boundary.

2. Proof of Theorem.1

In order to prove the Theorem, we find convenient to represent the solution of (1.1) in the semigroup form. To this end we introduce the following operators:

$$\begin{aligned} &A: L_2(\Omega) \to L_2(\Omega) \quad \text{given by} \\ &Au \equiv \Delta^2 u; \; u \in D(A), \\ &D(A) \equiv \{u \in L_2(\Omega);\; \Delta^2 u \in L_2(\Gamma);\; u|\Gamma_0 = 0; \\ &\Delta u + (1-\mu) B_1 u = 0 \text{ on } \Gamma,\; \frac{\partial}{\partial\eta}\Delta u + (1-\mu) B_2 u = 0 \text{ on } \Gamma_1\}. \end{aligned} \tag{2.1}$$

It can be easily shown that A is selfadjoint, positive on $L_2(\Omega)$.

$$\begin{aligned} &G: L_2(\Gamma) \to L_2(\Gamma) \quad \text{defined by} \\ &Gg \equiv v \;\text{ iff }\; \Delta^2 v = 0 \;\text{ and} \\ &v|_{\Gamma_0} = 0;\; \frac{\partial}{\partial\eta}\Delta v + (1-\mu) B_2 v = 0 \quad \text{on } \Gamma_1; \\ &\Delta v + (1-\mu) B_1 v = g. \end{aligned} \tag{2.2}$$

$$\begin{aligned} &B: L_2(\Gamma) \to \mathcal{D}(A)'; \\ &Bg \equiv \begin{bmatrix} 0 \\ AGg \end{bmatrix}. \end{aligned} \tag{2.3}$$

With the above notation we can rewrite (1.1) in an abstract form as:

$$\left.\begin{aligned} &u_{tt} + Au + B\,\partial\Phi\, B^* u_t \ni 0; \\ &u(0) = u_0;\; u_t(0) = u_1; \\ &(u_0, u_1) \in H \equiv \tilde{H}^2(\Omega) \times L_2(\Gamma); \end{aligned}\right\} \tag{2.4}$$

where $\langle B^* u, v\rangle_{L_2(\Gamma)} = (u, Bv)_H$ for $v \in L_2(\Gamma)$; $u \in D(A)$ and $\partial\Phi$ is the subgradient of the convex function Φ given by

$$\Phi(u) \equiv \begin{cases} \int_\Gamma j(g) & \text{if } j(g) \in L_1(\Gamma) \\ \infty & \text{otherwise} \end{cases}$$

with $\partial j(g) \equiv \gamma(g)$.

One can show ([L.2]) that in our situation $B\,\partial\,\Phi\,B^* = \partial\,(\Phi\,B^*)$, hence $B\,\partial\,\Phi\,B^*$ is maximal monotone on H. Thus, by using the results on nonlinear semigroups, it is straightforward to prove the existence and the uniqueness of the solutions (u, u_t) to (1.1).

For the proof of strong stability, we shall use la Salle's Invariance Principle. In fact, by using Sobolev's Imbeddings, one can show ([L.2]) that the resolvent R (1, C) is compact on H where

$$C \equiv \begin{bmatrix} 0 & I \\ A & B\partial\Phi B^* \end{bmatrix}.$$

Hence, by the Invariance Principle, for any initial data $(u_0, u_1) \in H$ the solution $e^{Ct}\begin{bmatrix} u_0 \\ u_1 \end{bmatrix} \to \omega\,(u_0, u_1)$, $t \to \infty$ where $\omega\,(u_0, u_1)$ is the ω-limit set — a compact subset of a ball with a radius equal to $\|\,(u_0, u_1)\,\|$. The main task, which remains, is to characterize the ω-limit set. One can show (see [L.2]) that if $(z_0, z_1) \in \omega$ then $z(t) \equiv e^{Ct}\begin{bmatrix} z_0 \\ z_1 \end{bmatrix}$ satisfies:

$$\left.\begin{aligned} & z_{tt} + \Delta^2 z = 0;\ \ z(t{=}0) = z_0;\ \ z_t(t{=}0) = z_1; \\ & z|_{\Gamma_0} = 0; \\ & \Delta z + (1-\mu)\,B_1 z = 0 \quad \text{on}\,\Gamma; \\ & \frac{\partial}{\partial\eta}\,\Delta z + (1-\mu)\,B_2 z = 0 \quad \text{on}\ \Gamma_1; \\ & \frac{\partial}{\partial\eta}\,z_t|_\Gamma = 0,\ \ \text{on}\ \Gamma. \end{aligned}\right\} \tag{2.5}$$

Since strong stability is equivalent showing that ω-limit set consists of only "zero" element, it is enough to show that z_t in (2.5) is equal to zero. In fact, this would automatically imply that $z = A^{-1} z_{tt} = 0$, hence $z \equiv 0$ and $z_0 = z_1 = 0$. On the other hand, to show that $z_t = 0$ it suffices to prove the following "uniqueness" result.

<u>Lemma:</u>

Let ϕ be the solution of

$$u_{tt} + \Delta^2\phi = 0,$$

$$\phi|_{\Gamma_0} = \Delta\phi|_{\Gamma_0} = 0;$$

$$\left.\begin{aligned} \Delta\phi + (1-\mu)\,B_1\,\phi = 0 \\ \frac{\partial}{\partial\eta}\,\Delta\phi + (1-\mu)\,B_2\phi = 0 \end{aligned}\right\} \quad \text{on}\ \Gamma_1$$

$$\frac{\partial}{\partial\eta}\phi = 0 \quad \text{on } \Gamma.$$

If (H-1)and (H-2) hold then $\phi \equiv 0$.

Remark

Notice that the result stated in the Lemma does not follow from the standard Holmgren's type Theorems. In fact, to apply the classical Holmgren's Theorem, one needs four boundary conditions to vanish on a portion of the boundary (we have only three).

3. Proof of Lemma

The proof of the Lemma is rather technical and we shall provide only the sketch (see [L.2]). The basic idea is to use a suitable combination of two multipliers for the equation (2.6). They are: $h \cdot \nabla \Delta\phi$ and $h \cdot \nabla \phi_\tau$ with an appropriate extension of $\vec{\tau}$ into the interior of the domain Ω. In fact, multiplying equation (2.6) by $h \cdot \nabla \Delta \phi$, integrating by parts and taking advantage of the boundary conditions, rather lengthy computations give

Proposition 1

$$-\int_Q [|\nabla \Delta \phi|^2 + |\nabla \phi_t|^2] dQ + (1/2 - \mu) \int_{\Sigma_1} |\phi_{t\tau}|^2 \vec{n}\,\vec{h}$$

$$+ 1/2 \int_{\Sigma_0} |\frac{\partial}{\partial\eta} \Delta \phi|^2 \vec{h}\,\vec{n} - \frac{(1-\mu)^2}{2} \int_{\Sigma_1} |\phi_{\tau\tau\tau}|^2 \vec{h}\,\vec{n}$$

$$- \frac{(1-\mu)}{2} \int_{\Sigma_1} |\phi_t|^2 k(1-khn) + (1-\mu)^2 \int_{\Sigma_1} (\phi_{\tau\tau})^2 k(3khn+1)$$

$$+ (\phi_t, h \nabla \Delta \phi)_\Omega \Big|_{0}^{T} - (\phi_t, \Delta \phi)_\Omega \Big|_0^T.$$

Now, for the simplicity of the exposition we assume that the curvature k is constant on each portion of the boundary Γ_0 and Γ_1. Let $\psi \equiv \phi_\tau$. Since

$$\Delta \psi_\tau - \frac{\partial}{\partial\tau} \Delta \psi = [-2n_{2x}\, \phi_{xx} + 2n_{1y}\, \phi_{yy} + 2(n_{1x} - n_{2y})\, \phi_{xy} - \Delta n_2\, \phi_x + \Delta n_1\, \phi_y], \quad \text{and}$$

$\Delta n_1, \Delta n_2, n_{2x}, n_{1y}, n_{1x} - n_{2y}$ are all equal to zero on Γ, we can extend vector $\vec{n}$

in the interior of Ω in such a way that $\Delta \phi_\tau - \frac{\partial}{\partial\tau} \Delta \phi = 0$ in Ω. Thus ψ satisfies the following equation.

$$\psi_{tt} + \Delta^2 \psi = 0$$

$$\psi|_{\Gamma_0} = \frac{\partial \psi}{\partial \eta}|_{\gamma_0} = \Delta \psi|_{\Gamma_0} = 0;$$

$$\frac{\partial}{\partial \eta} \psi = -k\psi; \quad \text{on } \Gamma_1$$

$$\Delta \psi = (1-\mu)\psi_{\tau\tau}; \quad \text{on } \Gamma_1$$

$$\frac{\partial}{\partial \eta} \Delta \psi = (1-\mu) k \psi_{\tau\tau}; \quad \text{on } \Gamma_1.$$

We next multiply both sides of equation (3.1) by $h \nabla \psi$, we integrate by parts and we use the boundary conditions. This gives

$$\int_\Sigma \psi_t^2 \, h\vec{n} = (1-\mu^2) \int_{\Sigma_1} \psi_{\tau\tau}^2 \, \vec{h}\vec{n} + 4 \int_Q (\Delta\psi)^2$$

$$+ (1-\mu) \int_{\Sigma_1} |\psi_\tau|^2 (-9k + 7k^2 hn) - 2(1-\mu) \int_{\Sigma_1} |\psi|^2 \cdot k^3 (1-khn) + O(E_0) \tag{3.2}$$

where $O(E_0)$ denotes the term bounded by $|\psi(0)|_{H^2(\Omega)}$ and $|\psi_t(0)|_{L_2(\Omega)}$. If we multiply (3.1) by ψ and we integrate by parts we obtain:

$$\int_Q [|\psi_t|^2 - |\Delta\psi|^2] = -2(1-\mu) \int_{\Sigma_1} k|\psi_\tau|^2 + O(E_0). \tag{3.3}$$

From (3.2) and (3.3) we obtain

__Proposition 2__

$$\int_{\Sigma_1} |\phi_{t\tau}|^2 hn = (1-\mu^2) \int_{\Sigma_1} |\phi_{\tau\tau\tau}|^2 \vec{h}\vec{n} + 2 \int_Q (\Delta \phi_\tau|^2 + 2 \int_Q |\phi_{t\tau}|^2 + (1-\mu) \int_{\Sigma_1} |\phi_{\tau\tau}|^2 k [-5 +$$

$$-2(1-\mu) \int_{\Sigma_1} |\phi_\tau|^2 k^3 (1-khn) = E_1 \tag{3.4}$$

where

$$E_1 = O(|\phi_0|^2_{H^3(\Omega)} + |\phi_1|^2_{H^1(\Omega)}) = O(|\phi_0|^2_{\mathcal{D}(A^{3/4})} + |\phi_1|^2_{D(A^{1/4})}).$$

with $\phi_0 = \phi(0)$; $\phi_1 = \phi_t(0)$. Let $0 < \alpha < 1$ (to be determined later).

We multiply (3.4) by α and we add to the result of Proposition 1. This yields

$$(\alpha-1) \int_Q [|\nabla \Delta \phi|^2 + |\nabla \phi_t|^2]$$

$$-\alpha[\int_Q [|\nabla \Delta \phi|^2 + |\nabla \phi_t|^2 - |\Delta \phi_\tau|^2 - |\phi_{t\tau}|^2]$$

$$+ 1/2 \int_{\Sigma_0} |\frac{\partial}{\partial \eta} \Delta \phi|^2 nh + (1/2 - \eta - \frac{\alpha}{2}) \int_{\Sigma_1} |\phi_{t\tau}|^2 hn$$

$$+ [(1-\mu^2) \frac{\alpha}{2} - \frac{(1-\mu)^2}{2}] \int_{\Sigma_1} |\phi_{\tau\tau\tau}|^2 hn$$

$$+ (1-\mu) \int_{\Sigma_1} k |\phi_{\tau\tau}|^2 [(1-\mu)(3khn+1) + \frac{\alpha}{2}(7khn-5)] \quad (1-\mu)\alpha \int_{\Sigma_1} |\phi_\tau|^2 k^3(1-khn)$$

$$\frac{-(1-\mu)}{2} \int_{\Sigma_1} |\phi_t|^2 k(1-khn) = O(E_1) \tag{3.5}$$

Now we select α such that

$$\alpha = \frac{1-\mu}{1+\mu}. \tag{3.6}$$

With the above choice of α, by our assumption (H-2) we have

$$\left.\begin{aligned} &1/2 - \mu - \frac{\alpha}{2} \leq 0 \\ &(1+\mu)\alpha - (1-\mu) \leq 0 \\ &(1-\mu)(3\,khn+1) + \frac{\alpha}{2}(7\,khn-5) \leq 0 \\ &1-khn \leq 0. \end{aligned}\right\} \tag{3.7}$$

From (3.5) and (3.7)

$$\frac{2\mu}{1+\mu} \int_Q [|\nabla \Delta \phi|^2 + |\nabla \phi_t|^2] \leq O(E_1) \tag{3.8}$$

Next we prove

<u>Proposition 3</u>

Let ϕ satisfy the boundary conditions as in (3.1)

$$\text{(i)} \quad |\nabla \Delta \phi|_{L_2(\Omega)} \sim |\phi|_{D(A^{3/4})} \quad |\nabla \phi_t|_{L_2(\Omega)} \sim |\phi_t|_{D(A^{1/4})} \tag{3.9}$$

(3.9 (ii)) simply follows from the general result in [G.1]. Thus it is enough to prove that

$$|\phi|_{D(A^{3/4})} \leq C\, |\nabla \Delta \phi|_{L_2(\Omega)} \tag{3.10}$$

(inequality in opposite direction follows again from [G.1]). Indeed

$$|\Delta \phi|_{H^1(\Omega)} \leq C |\nabla \Delta \phi|_{L_2(\Omega)}.$$

Thus in particular

$$|\Delta \phi|_{H^{1/2}(\Gamma)} \leq C |\nabla \Delta \phi|_{L_2(\Omega)}.$$

We shall prove that

$$|\Delta^2 \phi|_{(\tilde{H}^1(\Omega))'} = |A \phi|_{(\tilde{H}^1(\Omega))'} = |A \phi|_{(D(A^{1/4}))'} \leq C |\nabla \Delta \phi|_{L_2(\Omega)}.$$

where by [G.1] $\tilde{H}^1(\Omega) = \{y \in H^1(\Omega); y|_{\gamma_0} = 0\} = D(A^{1/4})$ and $(\tilde{H}^1(\Omega))'$ denotes the duality with respect to $L_2(\Omega)$ inner product. In fact, let $z \in \tilde{H}^1(\Omega)$. Then

$$(\Delta^2 \phi, z)_{L_2(\Omega)} = \langle \frac{\partial}{\partial \eta} \Delta \phi, z \rangle_{L_2(\Gamma_1)} - (\nabla \Delta \phi, \nabla z)_{L_2(\Omega)}.$$

Since

$$\frac{\partial}{\partial \eta} \delta \phi = 2(1-\mu) \frac{\partial}{\partial \tau} (k \phi_\tau) \text{ on } \Gamma_1 \text{ and } \Delta \phi = (1-\mu) \phi_{\tau\tau} \in H^{1/2}(\Gamma_1),$$

we obtain

$$\langle \frac{\partial}{\partial \eta} \Delta \phi, z \rangle_{L_2(\Gamma_1)} \leq C |\Delta \phi|_{H^{1/2}(\Gamma_1)} |z|_{\tilde{H}_1(\Omega)}.$$

Hence $\Delta^2 \phi$ with $\nabla \Delta \phi \in L_2(\Omega)$ defines a linear bounded functional on $(\tilde{H}^1(\Omega)' = (D(A^{1/4}))'$. Thus

$$|\Delta^2 \phi|_{(D(A^{1/y}))'} \leq C |\nabla \Delta \phi|_{L_2(\Omega)}$$

and consequently

$$|A^{3/4} \phi|_{L_2(\Omega)} \leq C |\nabla \Delta \phi|_{L_2(\Omega)}$$

as desired. ■

Since (3.1) is conservative in the topology at $D(A^{3/4}) \times D(A^{1/4})$, we obtain

$$|\phi(t)|^2_{D(A^{3/4})} + |\phi_t(t)|^2_{D(A^{1/4})} = |\phi_0|^2_{D(A^{3/4})} + |\phi_1|^2_{D(A^{1/4})}$$

Recalling (see [G.1]) that

$$|\phi|_{H^3(\Omega)} \leq C |\phi|_{D(A^{3/4})}$$
$$|\phi|_{H^1(\Omega)} \leq C |\phi|_{D(A^{1/4})},$$

from (3.8) and the result of the Proposition 3 we obtain

$$\frac{2\mu}{1+\mu} T [|\phi_0|^2_{D(A^{3/4})} + |\phi_1|^2_{D(A^{1/4})}] \leq C [|\phi_0|^2_{D(A^{3/4})} + |\phi_1|^2_{D(A^{1/4})}].$$

Taking T large enough yields the result of the Lemma. ■

References:

[G.1] P. Grisvard, "A conectirisation de quelques espieces d'interpolation", Arch. Ret. Mech. & Anal., 25 (1967), pp. 40-63.

[L.1] J. Lagnese, "Uniform boundary stabilization of homogeneous isotropic plates", Lecture Notes in Control and Information Sciences. Distributed Parometer Systems, vol. 102, pp. 204-216, ed. (F. Kappel, K. Kunish, W. Shappacher), Springer Verlag 1987.

[L.2] I. Lasiecka, "Asymptotic behavior of the solutions to the plate equations with nonlinear dissipation occurring through the shear forces and bending moments", to appear Journal Differential Equations.

[L.3] J. Lagnese, "Uniform stabilization of the Kirkhoff system by nonlinear feedback", Manuscript, 1988.

[L-L] J. E. Lagnese, J. L. Lions, "Modelling analysis and control of thin plates", Masson, 1988.

STABILIZATION OF A VIBRATING STRING SYSTEM LINKED BY POINT MASSES

E. Bruce Lee and Yuncheng You
Department of Electrical Engineering
and
Center for Control Science and Dynamical Systems
University of Minnesota, Minneapolis, MN 55455 USA

Abstract

Controlled motion of a string system linked with several point masses is described. It is established that strong stabilization cannot be achieved by feedback of pure velocities of the vibrating masses in certain cases, but can be achieved by feedback combining the mass velocities with the interior distributed or pointwise velocity when designed appropriately.

1. Introduction

In [1,2], hybrid systems of string vibration and beam vibration with dynamical boundary controllers are investigated. An abstract formulation based on the specific properties of hybrid differential operators yields the result of strong stabilization.

The same approach is taken in this paper to study another vibrating system of strings linked by several point masses. The results are quite different from the case associated with dynamical boundary control [1,2]. A similar but essentially different configuration (without point masses) was described in [3].

Consider the vertical vibration of several elastic strings linked together by point masses as shown by the Figure.

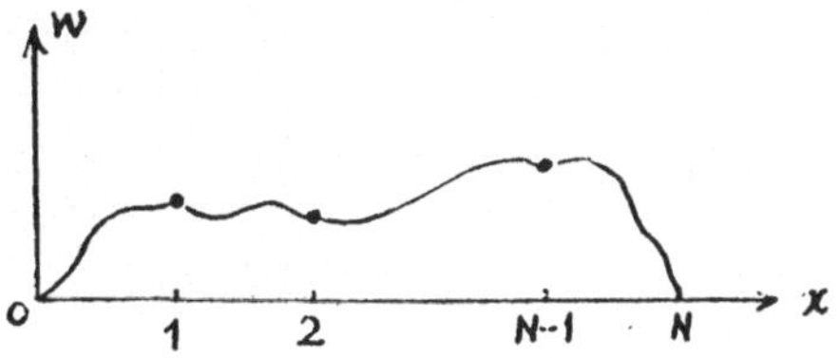

The mathematical model can be described by the equations of motion of a hybrid linear controlled system: (with $N>2$)

$$W_{tt}(t,x) = W_{xx}(t,x) \quad , \quad t > 0, \quad k-1 < x < k, \ k=1,\cdots,N$$

$$W(t,0) = W(t,N) \equiv 0, \qquad t > 0$$

$$W_{tt}(t,k) = -\beta_k[W_x^-(t,k)-W_x^+(t,k)] + f_k(t), \ t > 0, \ k=1,\cdots,N-1, \tag{1}$$

$$W(0,x) = W_0(x), \quad W_t(0,x) = W_1(x), \quad 0 < x < N;$$

where it is assumed that the tension coefficient τ is constant, $\beta_k = \tau/m_k$ and m_k is the mass at point $x = k$, and $\tau/\sigma = 1$ by time rescaling with σ the linear density of the strings. Besides, W_x^- and W_x^+ represent the left derivative and right derivative at specific points respectively. The control vector function is $f(t) = \text{col}\,(f_1(t),\cdots,f_{N-1}(t)), \ t > 0$.

2. Abstract Formulation

Set a real Hilbert space

$$H = L^2(0,N)\times R^{N-1} \ , \quad N > 2 \ , \tag{2}$$

with the inner product defined by

$$\langle\hat{W},\hat{V}\rangle_H = \sum_{k=1}^{N} \int_{k-1}^{k} W(x)V(x)dx + \sum_{k=1}^{N-1} \frac{1}{\beta_k} W_k V_k \quad , \tag{3}$$

where

$$\hat{W} = \begin{pmatrix} W(x) \\ W_1 \\ \vdots \\ W_{N-1} \end{pmatrix} \quad \text{and} \quad \hat{V} = \begin{pmatrix} V(x) \\ V_1 \\ \vdots \\ V_{N-1} \end{pmatrix} ,$$

$W(\cdot)$ and $V(\cdot)\in L^2(0,N)$.

Let

$$Z(t) = \begin{bmatrix} W(t,x), & x\varepsilon(0,N) \\ W(t,1) \\ \vdots \\ W(t,N-1) \end{bmatrix}, \quad t > 0. \tag{4}$$

Define an operator A by

$$A = \begin{bmatrix} \dfrac{d^2}{dx^2} & 0 & \cdots & 0 \\ \beta_1\left(\dfrac{d^+}{dx} - \dfrac{d^-}{dx}\right)\Big|_{x=1} & 0 & \cdots & 0 \\ \vdots & \vdots & & \vdots \\ \beta_{N-1}\left(\dfrac{d^+}{dx} - \dfrac{d^-}{dx}\right)\Big|_{x=N-1} & 0 & \cdots & 0 \end{bmatrix} \tag{5}$$

with domain

$$\mathcal{D}(A) = \left\{ \begin{pmatrix} W \\ W_1 \\ \vdots \\ W_{N-1} \end{pmatrix} \varepsilon H^2(0,N) x R^{N-1} : W(0) = W(N) = 0, \ W(k) = W_k, \ k = 1,\cdots,N-1 \right\}. \tag{6}$$

Define also an operator B by

$$B = \begin{pmatrix} 0 \\ I_{N-1} \end{pmatrix} \varepsilon \ \mathcal{L}(R^{N-1};H) \tag{7}$$

where $I_{N-1} = \text{diag}(1,\cdots,1)$.

Let $Z_o = \begin{bmatrix} W_o(x), & x\varepsilon(0,N) \\ W_o(1) \\ \vdots \\ W_o(N-1) \end{bmatrix}$ and $Z_1 = \begin{bmatrix} W_1(x), & x\varepsilon(0,N) \\ W_1(1) \\ \vdots \\ W_1(N-1) \end{bmatrix}$.

Thus the original system (1) can be recast as the following second order evolutionary equation with initial conditions:

$$\frac{d^2Z(t)}{dt^2} = AZ(t) + Bf(t), \quad t > 0 \tag{8}$$

$$Z(0) = Z_0, \qquad \dot{Z}(0) = Z_1 \quad .$$

The essential arguments of concern below depend on the properties of the differential operator A: $\mathcal{D}(A) \to H$, which will be detailed in the next section.

3. Hybrid Differential Operator

Lemma 1 A: $\mathcal{D}(A) \to H$ is a densely defined and closed operator. Moreover, A is self-adjoint and coercively dissipative.

Proof. The denseness of $\mathcal{D}(A)$ can be shown as in Lemma 1 in [1]. In order to show that A is closed, we need only to show A is self-adjoint; which will be done by the following steps:

1) A: $\mathcal{D}(A) \to H$ is symmetric: In fact, for any $\hat{W} = \begin{pmatrix} W \\ W_1 \\ \vdots \\ W_{N-1} \end{pmatrix}$ and $\hat{V} = \begin{pmatrix} V \\ V_1 \\ \vdots \\ V_{N-1} \end{pmatrix}$ in $\mathcal{D}(A)$, one has, by integration by parts, that

$$\langle A\hat{W},\hat{V}\rangle_H = \sum_{k=1}^{N} \int_{k-1}^{k} W''(x)V(x)dx + \sum_{k=1}^{N-1} (W'_+(k) - W'_-(k))V(k) \tag{9}$$

$$= \sum_{k=1}^{N} W'(x)V(x) \Big|_{k-1}^{k} - \sum_{k=1}^{N} \int_{k-1}^{k} (W'(x)V'(x)dx + \sum_{k=1}^{N-1} (W'_+(k) - W'_-(k))V(k)$$

$$= \sum_{k=1}^{N} (W'_-(k)V(k) - W'_+(k-1)V(k-1)] + \sum_{k=1}^{N-1} (W'_+(k) - W'_-(k))V(k)$$

$$- \sum_{k=1}^{N} \int_{k-1}^{k} W'(x)V'(x)dx$$

$$= - \sum_{k=1}^{N} \int_{k-1}^{k} W'(x)V'(x)dx = \langle\hat{W},A\hat{V}\rangle_H$$

Therefore A is symmetric and $\mathcal{D}(A) \subset \mathcal{D}(A^*)$.

2) A: $\mathcal{D}(A) \to H$ is coercively dissipative: since (9) implies that

$$\langle A\hat{W},\hat{V}\rangle_H = -\sum_{k=1}^{N}\int_{k-1}^{k} |W'(x)|^2dx = -\int_{0}^{N} |W'(x)|^2dx \;, \tag{10}$$

and

$$|W(1)|^2 < \Big(\int_0^1 |W'(x)|dx\Big)^2 < \int_0^1 |W'(x)|^2dx \;,$$

$$|W(2)|^2 < \Big(|W(1)| + \int_0^2 |W'(x)|dx\Big)^2 < \Big(\int_0^2 |W'(x)|dx\Big)^2$$

$$< 2\int_0^2 |W'(x)|^2dx, \tag{11}$$

$$\cdots$$

$$|W(N-1)|^2 < (N-1)\int_0^{N-1} |W'(x)|^2dx,$$

we have

$$\sum_{k=1}^{N-1} |W(k)|^2 < (1 + 2 +\ldots+ (N-1))\sum_{k=1}^{N}\int_{k-1}^{k} |W'(x)|^2dx \tag{12}$$

$$= \frac{N(N-1)}{2}\sum_{k=1}^{N}\int_{k-1}^{k} |W'(x)|^2dx \;.$$

Thus it follows that

$$\|\hat{W}\|_H^2 < \Big[1 + \frac{N(N-1)}{2}\Big]\sum_{k=1}^{N}\int_{k-1}^{k} |W'(x)|^2dx \;, \tag{13}$$

so that

$$\langle A\hat{W},\hat{W}\rangle_H < -\Big[1 + \frac{N(N-1)}{2}\Big]^{-1}\|\hat{W}\|_H^2 \;, \qquad \forall\hat{W}\in \mathcal{D}(A). \tag{14}$$

This indicates that A is coercively dissipative.

3) Range A = H. In fact, for any given $\hat{V} = \begin{pmatrix} V(x) \\ V_1 \\ \vdots \\ V_{N-1} \end{pmatrix} \varepsilon H$, we can find an element $\hat{W} = \begin{pmatrix} W(x) \\ W_1 \\ \vdots \\ W_{N-1} \end{pmatrix}$ in $\mathcal{D}(A)$, such that $A\hat{W} = \hat{V}$. This W(x) can be given by

$$W(x) = \int_{k-1}^{x} \int_{k-1}^{\eta} V(\xi)d\xi d\eta + a_k x + b_k, \quad x\varepsilon[k-1,k],\ k=1,\cdots,N; \tag{15}$$

where

$$\begin{aligned}
&b_1 = 0 \\
&-a_1 + a_2 = \int_0^1 V(\xi)d\xi + \frac{V_1}{\beta_1} \\
&\quad\vdots \\
&-a_{N-1} + a_N = \int_{N-2}^{N-1} V(\xi)d\xi + \frac{V_{N-1}}{\beta_{N-1}} \\
&b_2 = a_1 + \int_0^1 \int_0^{\eta} V(\xi)d\xi d\eta - a_2 \\
&\quad\vdots \\
&b_N = b_{N-2} + a_{N-1}(N-1) + \int_{N-2}^{N-1} \int_{N-2}^{\eta} V(\xi)d\xi d\eta - a_N(N-1) \\
&a_N N + b_N = -\int_{N-1}^{N} \int_{N-1}^{\eta} V(\xi)d\xi d\eta.
\end{aligned} \tag{16}$$

The simultaneous equations (16) can be solved as follows:

$$b_1 = 0$$

$$b_2 = \int_0^1 \int_0^{\eta} V(\xi)d\xi d\eta - \left(\int_0^1 V(\xi)d\xi + \frac{V_1}{\beta_1}\right),$$

$$b_3 = b_2 + \int_1^2\int_1^\eta V(\xi)d\xi d\eta - 2\left(\int_1^2 V(\xi)d\xi + \frac{V_2}{\beta_2}\right),$$

$$\vdots \tag{17}$$

$$b_N = b_{N-1} + \int_{N-2}^{N-1}\int_{N-2}^{\eta} V(\xi)d\xi d\eta - (N-1)\left(\int_{N-2}^{N-1} V(\xi)d\xi + \frac{V_{N-1}}{\beta_{N-1}}\right),$$

$$a_N = -\frac{1}{N}\left(b_N + \int_{N-1}^{N}\int_{N-1}^{\eta} V(\xi)d\xi d\eta\right),$$

$$a_{N-1} = a_N - \left(\int_{N-2}^{N-1} V(\xi)d\xi + \frac{V_{N-1}}{\beta_{N-1}}\right)$$

$$a_1 = a_2 - \left(\int_0^1 V(\xi)d\xi + \frac{V_1}{\beta_1}\right).$$

Besides, let

$$W_1 = W(1),\cdots,W_{N-1} = W(N-1). \tag{18}$$

Then, $\hat{W} = \begin{pmatrix} W(x) \\ W_1 \\ \vdots \\ W_{N-1} \end{pmatrix}$ with $W(\cdot),W_1,\cdots,W_{N-1}$ defined by (15), (17) and (18) satisfies $A\hat{W} = \hat{V}$. So Range $A = H$.

From the theory of operators in Hilbert spaces, the two facts that A is symmetric and Range $A = H$ imply that A is self-adjoint, so that A is closed. Q.E.D.

Lemma 2 $A^{-1} \in \mathcal{L}(H)$ is compact.

Proof. Since A is self-adjoint and coercively dissipative, A is invertible and $A^{-1} \in \mathcal{L}(H)$.

By the Sobolev embedding theorem, we need only to show that

$$\|\hat{W}\|_{H^1(0,N)\times R^{N-1}} \leq \text{const}\|A\hat{W}\|_H, \quad \text{for any } \hat{W} \in \mathcal{D}(A). \tag{19}$$

In fact, for any $\hat{W} = \text{col}(W(x),W_1,\cdots W_{N-1}) \in \mathcal{D}(A)$, we have

$$\|\hat{W}\|_{H^1(0,N)\times R^{N-1}} = \sum_{k=1}^{N} \left(\int_{k-1}^{k} |W'(x)|^2 dx + \int_{k-1}^{k} |W(x)|^2 dx\right) + \sum_{k=1}^{N-1} |W(k)|^2$$

$$\leq \frac{N(N-1)}{2} \sum_{k=1}^{N} \int_{k-1}^{k} |W'(x)|^2 dx + \sum_{k=1}^{N-1} |W(k)|^2$$

$$\leq N(N-1) \sum_{k=1}^{N} \int_{k-1}^{k} |W'(x)|^2 dx = -N(N-1)\langle A\hat{W},\hat{W}\rangle_H \tag{20}$$

$$\leq N(N-1)\|A\hat{W}\|_H \|\hat{W}\|_H \ ,$$

where the deduction is similar to (11) and (12). However it is clear that

$$\|\hat{W}\|_H^2 \leq \left(1 + \frac{N(N-1)}{2}\right)\langle -A\hat{W},\hat{W}\rangle_H \leq \left(1 + \frac{N(N-1)}{2}\right)\|A\hat{W}\|_H \|\hat{W}\|_H \ . \tag{21}$$

Thus we have, by (20) and (21), that

$$\|\hat{W}\|^2_{H^1(0,N)\times R^{N-1}} \leq N(N-1)\left(1 + \frac{N(N-1)}{2}\right)\|A\hat{W}\|_H^2 \ . \tag{22}$$

Therefore (19) holds. Q.E.D.

Since $A: \mathcal{D}(A)\to H$ is self-adjoint, coercively dissipative, and with compact resolvent, the standard spectral theory asserts that

$$\sigma(A) = \sigma_p(A) = \{-\lambda_n\}_{n=1}^{\infty} \subset R^- \quad \text{with} \quad \lambda_n \to +\infty \quad \text{as} \quad n\to \infty,$$

and $\{\lambda_n\}_{n=1}^{\infty}$ admits no finite accumulation points and each λ_n has a finite multiplicity. Besides, we know that the corresponding complete normalized eigenvectors $\{\phi_n\}_{n=1}^{\infty}$ form an orthonormal basis for H.

4. Main Results

As usual (cf. [1], [2]), we define a product space

$$X = \mathcal{D}(\sqrt{-A})xH \tag{23}$$

where $\mathcal{D}(\sqrt{-A})$ with the inner product $<v_1,v_2>(\mathcal{D}(\sqrt{-A})) = <\sqrt{-A}v_1, \sqrt{-A}v_2>_H$ is a Hilbert space. Then define operators $\mathcal{A}$ and $\mathcal{B}$ by

$$\mathcal{A} = \begin{pmatrix} 0 & I \\ A & 0 \end{pmatrix}, \quad \mathcal{D}(\mathcal{A}) = \mathcal{D}(A)x\mathcal{D}(\sqrt{-A}), \quad \mathcal{A} : \mathcal{D}(\mathcal{A}) \to X; \tag{24}$$

and

$$\mathcal{B} = \begin{pmatrix} 0 \\ B \end{pmatrix} \in \mathcal{L}(R^{N-1};X) \quad . \tag{25}$$

Let

$$u(t) = \begin{pmatrix} Z(t) \\ \dot{Z}(t) \end{pmatrix} \tag{26}$$

where $\dot{Z}(t)$ is the H-strong derivative of $Z(t)$ with respect to the variable t. Denote by

$$u_0 = \begin{pmatrix} Z_0 \\ Z_1 \end{pmatrix} \tag{27}$$

Then the second-order system (8) can be described by following first-order evolutionary equation with initial data:

$$\frac{du(t)}{dt} = \mathcal{A}u(t) + \mathcal{B}f(t),$$

$$u(0) = u_0 \quad . \tag{28}$$

Based on Lemma 1 and Lemma 2, it follows that:

1) $\mathcal{A}$ generates a strongly continuous unitary group T(t) given by

$$T(t) = \begin{bmatrix} \cos(\sqrt{-A}t) & \sqrt{-A}^{-1}\sin(\sqrt{-A}t) \\ -\sqrt{-A}\sin(\sqrt{-A}t) & \cos(\sqrt{-A}t) \end{bmatrix}, \ t \in R; \tag{29}$$

2) $\mathcal{A}$ is skew-adjoint, i.e. $\mathcal{A} = -\mathcal{A}^*$; and $\mathcal{A}^{-1} \in \mathcal{L}(X)$ is compact; and

3) $\sigma(\mathcal{A}) = \sigma_p(\mathcal{A}) = \{\pm j\sqrt{\lambda_n} \mid -\lambda_n \varepsilon \sigma_p(A)\}$ where $j = \sqrt{-1}$. (30)

The main results are given as follows.

Theorem 1. The abstract evolutionary system (28) cannot be weakly or strongly stabilized by any bounded linear feedback control f(t).

Proof. Since T(t) generated by $\mathcal{A}$ is a unitary group, by the Benchimol decomposition (cf. [4] and [2]), we can assert that the evolutionary system (28) is weakly stabilizable (equivalently strongly stabilizable by the property of compact resolvent) if and only if (28) is approximately controllable, i.e.

$$C\{\mathcal{A}, \mathcal{B}\} = X, \tag{31}$$

where $C\{\mathcal{A}, \mathcal{B}\}$ is the controllable subspace associated with the system (28).

It is easy to see that (31) amounts to

$$C\{\mathcal{A}, \mathcal{B}\}^{\perp} = \bigcap_{t \geq 0} \ker[\mathcal{B}^{*}T^{*}(t)] = \{0\}. \tag{32}$$

For an element $\begin{pmatrix} \hat{v} \\ \hat{W} \end{pmatrix} \varepsilon X$, the direct calculation shows that

$$\mathcal{B}^{*}T^{*}(t)\begin{pmatrix} \hat{v} \\ \hat{W} \end{pmatrix} = [0, I_{N-1}](\sqrt{-A}\sin(\sqrt{-A}t)\hat{v} + \cos(\sqrt{-A}t)\hat{W}) \tag{33}$$

$$= \sum_{n=1}^{\infty} \{\mu_n \sin(\mu_n t)\langle\hat{\phi}_n, \hat{v}\rangle_H + \cos(\mu_n t)\langle\hat{\phi}_n, \hat{W}\rangle_H\} \begin{bmatrix} \phi_n(1) \\ \vdots \\ \phi_n(N-1) \end{bmatrix} .$$

where $\mu_n = \sqrt{\lambda_n}$ and $\hat{\phi}_n = \mathrm{col}(\phi_n(x), \phi_n(1), \cdots, \phi_n(N-1))$, $n=1,2,\cdots$, are the normalized orthogonal basis for H, with

$$A\hat{\phi}_n = -\lambda_n\hat{\phi}_n \quad , \quad n=1,2,\cdots. \tag{34}$$

Denote by $v_n = \langle\hat{\phi}_n, \hat{v}\rangle$ and $W_n = \langle\hat{\phi}_n, \hat{W}\rangle$, $n \geq 1$.

Then (32) holds if and only if

$$\sum_{n=1}^{\infty} \{\mu_n v_n \sin(\mu_n t) + W_n \cos(\mu_n t)\} \begin{pmatrix} \phi_n(1) \\ \vdots \\ \phi_n(N-1) \end{pmatrix} = 0, \quad \text{for all } t > 0, \tag{35}$$

implies that

$$\{v_n\}_{n=1}^{\infty} = \{W_n\}_{n=1}^{\infty} = 0 \text{ in } \ell^2. \tag{36}$$

According to [2; Lemma 8], which provided a useful property of the B^2-class almost periodic functions (cf. [5]), the relation (35) implies that

$$\begin{aligned} \mu_n v_n \phi_n(k) &= 0, \\ & \qquad\qquad k=1,\cdots,N-1; \quad n=1,2,\cdots. \\ W_n \phi_n(k) &= 0, \end{aligned} \tag{37}$$

However, we find that

$$-\lambda_{n_m} = -m^2\pi^2 \quad (m = \text{positive integers}) \in \sigma_p(A) \tag{38}$$

and the corresponding eigenvector associated with λ_{n_m} is

$$\hat{\phi}_{n_m} = \begin{pmatrix} \phi_{n_m}(x) = \sin(m\pi x), \quad x\in[0,N] \\ 0 \\ \vdots \\ 0 \end{pmatrix}, \qquad m = 1,2,\cdots. \tag{39}$$

Hence letting $n = n_m$ (for $m = 1,2,\cdots$) in (37), we have a pair of $\{\mu_n v_n\}$ and $\{W_n\}$ in $\ell^2 \times \ell^2$, such that

$$\{v_n\}_{n=1}^{\infty} \neq 0 \quad \text{and} \quad \{W_n\}_{n=1}^{\infty} \neq 0 , \tag{40}$$

since at at least $\phi_{n_m}(k) = 0$, $k=1,\cdots,N$; $m=1,2,\cdots$ makes it possible to have

$$\{v_{n_m}\}_{m=1}^{\infty} \neq 0 \quad \text{and} \quad \{W_{n_m}\}_{m=1}^{\infty} \neq 0 . \tag{41}$$

This means that

$$C\{\mathcal{A}, \mathcal{B}\} \neq X, \tag{42}$$

so that the system (28) is not weakly (and strongly) stabilizable by any bounded linear feedback control $f(t)$. Q.E.D.

In contrast to this negative result with respect to the configuration with finite controllers only designed at the positions where the point-masses are located, we try to present some affirmative answer to the question how to stabilize this evolutionary system by other control configurations.

If we exert additional control in a distributed manner as well, then the original system is replaced by the following,

$$\begin{aligned} &W_{tt}(t,x) = W_{xx}(t,x) + g(x)f_0(t), \quad t > 0, \quad x \varepsilon (k-1,k),\ k=1,\cdots,N, \\ &W(t,0) = W(t,N) \equiv 0, \qquad t > 0 \\ &W_{tt}(t,k) = -\beta_k[W_x^-(t,k) - W_x^+(t,k)] + f_k(t),\ t > 0,\ k=1,\cdots,N-1, \\ &W(0,x) = W_0(x), \quad W_t(0,x) = W_1(x), \quad 0 < x < N; \end{aligned} \tag{43}$$

where $f_0(t)$ is a scalar control function, so that the control vector defined by

$$f(t) = \begin{bmatrix} f_0(t) \\ f_1(t) \\ \vdots \\ f_{N-1}(t) \end{bmatrix}, \qquad t > 0, \tag{44}$$

is still finite dimensional. There are some possible choices for the function $g(x)$, namely,

Case 1 $g(x) \varepsilon L^2(0,N)$; and

Case 2 $g(x) = \delta(x-b)$, where $b \varepsilon (i-1,i)$ and $1 < i < N$ is fixed.

Now consider the case 1 by defining, instead of B and $\mathcal{B}$, two operators

$$B_1 = \begin{pmatrix} g(\cdot) & 0 \\ 0 & I_{N-1} \end{pmatrix} \varepsilon\, \mathcal{L}(R^N;H) \tag{45}$$

and

$$\mathcal{B}_1 = \begin{pmatrix} 0 \\ B_1 \end{pmatrix} \varepsilon\, \mathcal{L}(R^N;\ X) \quad . \tag{46}$$

Then we can consider the abstract version of the system (43) as follows

$$\frac{du(t)}{dt} = \mathcal{A}\, u(t) + \mathcal{B}_1\ f(t) \quad \text{in } X \ , \quad t \geq 0, \tag{47}$$

where $u(t)$ is still defined by (26), $\mathcal{A}$ is still defined by (24), and $f(t)$ is defined by (44).

Theorem 2 For any given $g(x) \varepsilon L^2(0,N)$ such that

$$\langle g(x), \sin(\frac{m\pi x}{N})\rangle_{L^2(0,N)} \neq 0, \quad \forall \text{ positive integers } m,$$

the evolutionary system (47) is strongly stabilized by the feedback

$$f(t) = -\begin{bmatrix} \langle g, W_t(t,\cdot)\rangle_{L^2(0,N)} \\ W_t(t,1) \\ \vdots \\ W_t(t,N-1) \end{bmatrix}, \quad t > 0. \tag{48}$$

Proof. By the argument in proving Theorem 1, the system (47) is strongly stabilizable if and only if

$$C\{\mathcal{A}, \mathcal{B}_1\} = X \quad . \tag{49}$$

Besides, (49) holds if and only if (cf. (33))

$$\mathcal{B}_1^* T^*(t)\begin{pmatrix} \hat{v} \\ \hat{W} \end{pmatrix} = \sum_{n=1}^{\infty} [\mu_n v_n \sin(\mu_n t) + W_n \cos(\mu_n t)] \begin{bmatrix} \langle g, \phi_n\rangle_{L^2} \\ \phi_n(1) \\ \vdots \\ \phi_n(N-1) \end{bmatrix} = 0, \ t > 0, \tag{50}$$

implies that $\{v_n\}_{n=1}^{\infty} = \{W_n\}_{n=1}^{\infty} = 0$ in ℓ^2. Here also $v_n = \langle\hat{\phi}_n, \hat{v}\rangle$ and $W_n = \langle\hat{\phi}_n, W\rangle$, with μ_n, ϕ_n the same as above.

By the same reasoning as in Theorem 1, (50) implies that

$$\begin{aligned} \mu_n v_n \langle g, \phi_n\rangle_{L^2} &= 0, \\ \mu_n v_n \phi_n(1) &= 0, \\ &\vdots \\ \mu_n v_n \phi_n(N-1) &= 0; \end{aligned} \quad \text{and} \quad \begin{aligned} W_n \langle g, \phi_n\rangle_{L^2} &= 0, \\ W_n \phi_n(1) &= 0, \\ &\vdots \\ W_n \phi_n(N-1) &= 0; \end{aligned} \quad n \geq 1 \ . \tag{51}$$

It suffices to prove that any $\{\mu_n v_n\}_{n=1}^{\infty} \in \ell^2$ and $\{W_n\}_{n=1}^{\infty} \in \ell^2$ satisfying (51) must be identically zero vectors of ℓ^2. This can be done as follows.

First, if there is a positive integer k such that $1 \leq k \leq N$ and $\phi_n(k) \neq 0$, then it follows that $\mu_n v_n = W_n = 0$ so that $v_n = W_n = 0$ since μ_n is always positive.

Next, if there is an eigenvector $\phi_n = \begin{bmatrix} \phi_n(x) \\ \phi_n(1) \\ \vdots \\ \phi_n(N-1) \end{bmatrix}$ of A associated with an eigenvalue $-\lambda_n \varepsilon \sigma_p(A)$ and $\mu_n = \sqrt{\lambda_n}$, such that $\phi_n(1) = \ldots = \phi_n(N-1) = 0$, then $\phi_n(x)$ must satisfy

$$\begin{aligned} &\phi_n''(x) = -\lambda_n \phi_n(x), \qquad x\varepsilon(0,1)x\cdots x(N-1,N), \\ &\phi_n(0) = \phi_n(N) = 0, \\ &\phi_n'(k-o) = \phi_n'(k+o), \quad k=1,\cdots,N-1. \end{aligned} \tag{52}$$

The only possible candidates which satisfy (52) will be

$$\phi_{n_m} = C_{n_m} \sin(\frac{m\pi}{N} x), \qquad m = \text{positive integers}, \tag{53}$$

where C_{n_m} is a normalized constant, and the corresponding eigenvalues will be

$$-\lambda_{n_m} = -(\frac{m\pi}{N})^2, \qquad m = \text{positive integers}. \tag{54}$$

By assumption, we have

$$\langle g, \sin(\frac{m\pi}{N} \cdot)\rangle_{L^2} \neq 0, \quad \text{for all } m = \text{positive integers}, \tag{55}$$

thus $\langle g, \phi_{n_m}\rangle_{L^2} \neq 0$, so that $\mu_{n_m} v_{n_m} = W_{n_m} = 0$, hence $v_{n_m} = W_{n_m} = 0$ for all m positive integers.

In other words, we have shown that for each eigenvector $\phi_n = \begin{bmatrix} \phi_n(x) \\ \phi_n(1) \\ \vdots \\ \phi_n(N-1) \end{bmatrix}$,

$$\mu_n \begin{bmatrix} \langle g, \phi_n\rangle_{L^2} \\ \phi_n(1) \\ \vdots \\ \phi_n(N-1) \end{bmatrix} \neq 0, \quad n=1,2,\cdots. \tag{56}$$

Therefore (51) and (56) lead to

$$\{v_n\}_{n=1}^{\infty} = \{W_n\}_{n=1}^{\infty} = 0. \tag{57}$$

Hence the system is strongly stabilizable. Moreover, by the Benchimal theory (cf. [4]), as long as the system (47) is strongly stabilizable, it can be stabilized by a special feedback control

$$f(t) = -\mathcal{B}_1^* u(t), \quad t > 0. \tag{58}$$

Finally the concrete form of the feedback (58) is

$$f(t) = -\mathcal{B}_1^* u(t) = -B_1^* \dot{Z}(t) = -\begin{bmatrix} \langle g, W_t(t,\cdot)\rangle_{L^2(0,N)} \\ W_t(t,1) \\ \vdots \\ W_t(t,N-1) \end{bmatrix}, \quad t > 0. \tag{59}$$

Thus the proof is completed. Q.E.D.

Remark 1 There certainly exists a function $g(x)$ in $L^2(0,N)$ such that the condition in Theorem 2,

$$\langle g(x), \sin(\frac{m\pi x}{N})\rangle_{L^2(0,N)} \neq 0, \qquad \forall \text{ positive integers } m, \tag{60}$$

is satisfied. Such a $g(x)$ can be simply taken as

$$g(x) = \sum_{m=1}^{\infty} \frac{1}{m} \sin(\frac{m\pi x}{N}) \in L^2(0,N), \tag{61}$$

since $\{\sin(\frac{m\pi x}{N})\}_{m=1}^{\infty}$ forms an orthogonal basis for $L^2(0,N)$ and the Fourier expansion in (61) converges in $L^2(0,N)$. It is obvious that for this $g(x)$,

$$\langle g(x), \sin(\frac{m\pi x}{N})\rangle_{L^2(0,N)} = \frac{N}{2m} \neq 0, \quad \forall \text{ positive integers } m. \tag{62}$$

Thus (60) is satisfied.

Remark 2 Although $\{\sin(\frac{m\pi x}{N})\}_{m=1}^{\infty}$ forms a basis for $L^2(0,N)$, the corresponding vector $\left\{\begin{pmatrix} \sin(\frac{m\pi x}{N}) \\ 0 \\ \vdots \\ 0 \end{pmatrix}\right\}_{m=1}^{\infty}$ does not form a basis of the the space H.

5. Pointwise Control Case

Now we investigate the pointwise configurations: Case 2. In practical applications, these configurations have some advantage in design. Since the involved Dirac functions belong to $H^{-1}(0,N)$ but not $L^2(0,N)$, the theoretical framework should be generalized to deal with this case. We refer to [6] and [7] as the conceptional and methodological source in this connection.

Define V to be the space

$$V = \mathcal{D}(\sqrt{-A}) \subset H_o^1(0,N) x R^{N-1} \tag{63}$$

with the norm

$$\|\hat{W}\|_V = \left(\int_0^N |W'(x)|^2 dx\right)^{1/2}, \qquad \text{for } \hat{W} = \begin{bmatrix} W(x) \\ W_1 \\ \vdots \\ W_{N-1} \end{bmatrix} \varepsilon V. \tag{64}$$

By (10), we see that

$$\|\hat{W}\|_V = \|\sqrt{-A}\,\hat{W}\|_H^2 , \quad \forall\ \hat{W} \varepsilon V. \tag{65}$$

Denote by V^* the dual space of V. Then we can assert that

$$H^{-1}(0,N) x R^{N-1} \subset V^*, \tag{66}$$

so that $V \subset H = H^* \subset V^*$.

Define an operator B_2 by

$$B_2 = \begin{pmatrix} \delta(\cdot - b_i) & 0 \\ 0 & I_{N-1} \end{pmatrix} \varepsilon\ \mathcal{L}(R_N; V^*), \tag{67}$$

and accordingly define

$$\mathcal{B}_2 = \begin{pmatrix} 0 \\ B_2 \end{pmatrix} \varepsilon\ \mathcal{L}(R_N; H x V^*). \tag{68}$$

Denote a new space by

$$Y = H \times V^*. \tag{69}$$

We want to establish an evolutionary equation in this extension space Y, $Y \supset X$, to remodel the original control system (43) with $g(x) = \delta(x-b_i)$ given.

<u>Lemma 3</u> Following statements hold:

1) There exists a unique operator $\tilde{A} \in \mathcal{L}(V,V^*)$, such that
$\langle \hat{u},\hat{v}\rangle_V = -(\tilde{A}\hat{u})(\hat{v})$, for $\hat{u}\in V$ and any $\hat{v}\in V$. (70)
2) $A = \tilde{A}|\mathcal{D}(A)$ and $\mathcal{D}(A) = \{\hat{u}\in V | \tilde{A}\hat{u}\in H\}$.
3) $\tilde{A}$ is invertible on V^*, and $\tilde{A}^{-1}\in \mathcal{L}(V^*)$ is compact.
4) $\tilde{A}$: $\mathcal{D}(\tilde{A})(= V \subset V^*) \to V^*$ is self-adjoint.
5) $\sigma(\tilde{A}) = \sigma_p(\tilde{A}) = \sigma(A) = \sigma_p(A)$, and the eigenspace $N_\lambda(\tilde{A}) = N_\lambda(A)$ for each $\lambda\in\sigma(\tilde{A})$.

6) Let $\{-\lambda_n\}_{n=1}^\infty$ and $\{\phi_n\}_{n=1}^\infty$ be as above the complete eigenvalues and the corresponding normalized eigenvectors of the operator A. Then,

$$\tilde{A}\psi = \sum_{n=1}^{\infty} -\lambda_n \psi(\hat{\phi}_n)\hat{\phi}_n, \quad \text{for any } \psi\in \mathcal{D}(\tilde{A}) = V, \tag{71}$$

where $\psi(\hat{\phi}_n)$ is the action of the functional $\psi\in V^*$ on the element $\hat{\phi}_n\in V$, and the series in (71) converges in the V^*-norm.

The proof of this Lemma is similar to the argument contained in Lemma 2 through Lemma 6 of [6]. Here this detail is omitted, while the existence and uniquess of $\tilde{A}$ in (70) is due to the Lax-Milgram theorem.

Thus the original system (43) in the Case 2 can be formulated as follows,

$$\frac{du}{dt} = \tilde{\mathcal{A}}u(t) + \mathcal{B}_2 f(t), \qquad \text{in } Y, \quad t > 0, \tag{72}$$

where $\tilde{\mathcal{A}}$ is defined by

$$\tilde{\mathcal{A}} = \begin{pmatrix} 0 & I_H \\ \tilde{A} & 0 \end{pmatrix} : \mathcal{D}(\tilde{\mathcal{A}}) \to Y, \tag{73}$$

$$\mathcal{D}(\tilde{\mathcal{A}}) = V \times H \subset Y ;$$

the operator $\mathcal{B}_2$ is defined by (68), and f(t) is defined by (44).

Based on Lemma 3, we know that $\tilde{\mathcal{A}}$ also generates a strongly continuous unitary group $\tilde{T}(t)$ on the space Y, given by

$$\tilde{T}(t) = \begin{bmatrix} \cos(\sqrt{-\tilde{A}}t) & \sqrt{-\tilde{A}}^{-1}\sin(\sqrt{-\tilde{A}}t) \\ -\sqrt{-\tilde{A}}\sin(\sqrt{-\tilde{A}}t) & \cos(\sqrt{-\tilde{A}}t) \end{bmatrix}, \quad t \in R. \tag{74}$$

In this case, we regard the mild solution u(t) of (72) for any $f \in L_{loc}(R^+;R^N)$ as the state function.

Now we can describe the stabilization result as follows.

Theorem 3 Let $i \in \{1,\cdots,N\}$ be arbitrarily fixed. For any given point $b \in (i-1,i)$ such that b is an irrational number, the evolutionary system (72), as an abstract model in the Case 2, is strongly stabilized in Y by the feedback

$$F(t) = -\begin{bmatrix} \langle \delta(\cdot - b), W_t(t,\cdot)\rangle_{H^{-1}(0,N)} \\ W_t(t,1) \\ \vdots \\ W_t(t,N-1) \end{bmatrix}, \quad t > 0. \tag{75}$$

Proof. Similarly as in the proof of Theorem 2, it suffices to show that

$$C\{\tilde{\mathcal{A}}, \mathcal{B}_2\} = Y. \tag{76}$$

Equivalently, it suffices to show that

$$\mathcal{B}^*_2\tilde{T}^*(t)\begin{pmatrix}\hat{v}\\ \hat{W}\end{pmatrix} = \sum_{n=1}^{\infty} [\mu_n v_n \sin(\mu_n t) + W_n\cos(\mu_n t)]\begin{bmatrix} \langle g, \tilde{\phi}_n\rangle_{H^{-1}} \\ \tilde{\phi}_n(1) \\ \vdots \\ \tilde{\phi}_n(N-1) \end{bmatrix} = 0, \quad t > 0 \tag{77}$$

implies that

$$\{v_n\}_{n=1}^{\infty} = \{W_n\}_{n=1}^{\infty} = 0 \quad \text{in } \ell^2 , \tag{78}$$

where $\begin{pmatrix}\hat{v}\\ \hat{W}\end{pmatrix} \in Y = H \times V^*$, $v_n = \langle \hat{v}, \phi_n\rangle_{V^*}$ and $W_n = \langle \hat{W}, \phi_n\rangle_{V^*}$, with $\mu_n = \sqrt{\lambda_n}$ as above, but

$$\tilde{\phi}_n = \begin{bmatrix} \tilde{\phi}_n(x) \\ \tilde{\phi}_n(1) \\ \vdots \\ \tilde{\phi}_n(N-1) \end{bmatrix}$$ is the corresponding eigenvector normalized in V^*, so $\tilde{\phi}_n$ and ϕ_n

only differentiate from one another by a constant c_n which depends on the normalization in different spaces H and V^*.

By Lemma 8 of [2], the specific property of the B^2-class almost periodic functions, from (77) it follows that for all $n=1,2,\cdots$,

$$\begin{aligned} &\mu_n v_n \langle \delta(\cdot-b), \tilde{\phi}_n \rangle_{H^{-1}} = 0 \qquad \text{and} \qquad && W_n \langle \delta(\cdot-b), \tilde{\phi}_n \rangle_{H^{-1}} = 0, \\ &\mu_n v_n \tilde{\phi}_n(k) = 0, \; k=1,\cdots,N-1; && W_n \tilde{\phi}_n(k) = 0, \; k=1,\cdots,N-1 . \end{aligned} \tag{79}$$

Similarly as in the proof of Theorem 2, we only need to show the following fact: for each V^*-normalized eigenvector $\tilde{\phi}_{n_m} = \begin{bmatrix} \tilde{\phi}_{n_m}(x) \\ \tilde{\phi}_{n_m}(1) \\ \vdots \\ \tilde{\phi}_{n_m}(N-1) \end{bmatrix}$ with the first component function being

$$\tilde{\phi}_{n_m}(x) = \tilde{C}_{n_m} \sin(\frac{m\pi x}{N}) \quad , \qquad m = 1,2,\cdots, \tag{80}$$

where $\tilde{C}_{n_m}$ is the V^*-normalizing constant, it holds that

$$\langle \delta(\cdot-b), \; \tilde{\phi}_{n_m}(\cdot) \rangle_{H^{-1}(0,N)} \neq 0, \qquad m = 1,2,\cdots. \tag{81}$$

It is easy to see that $\{\sin(\frac{m\pi x}{N})\}_{m=1}^{\infty}$ also forms an orthogonal (but not normalized) basis for the space $H_0^1(0,N)$. Then by the definition of the dual inner-product, we have

$$\langle \delta(\cdot-b), \; \tilde{\phi}_{n_m}(\cdot) \rangle_{H^{-1}(0,N)}$$

$$= \sum_{j-1}^{\infty} p_j \, \delta(\cdot-b)[\sin(\frac{j\pi x}{N})] \tilde{\phi}_{n_m}(\cdot)[\sin(\frac{j\pi x}{N})], \tag{82}$$

where $p_j = \| \sin(\frac{j\pi x}{N}) \|^2_{H_0^1(0,N)}$, and $\delta(\cdot-b)[\sin(\frac{j\pi x}{N})]$ and $\tilde{\phi}_{n_m}(\cdot)[\sin(\frac{j\pi x}{N})]$ are the actions of the functionals $\delta(\cdot-b)\varepsilon H^{-1}(0,N)$ and $\tilde{\phi}_{n_m}(\cdot)\varepsilon H^{-1}(0,N)$ on the element $\sin(\frac{j\pi x}{N})\varepsilon H_0^1(0,N)$ respectively.

However, due to (80) we have $\tilde{\phi}_{n_m}$ itself belonging to $L^2(0,N)$. Thus the canonical triplet structure

$$H_0^1(0,N) \subset L^2(0,N) \subset H^{-1}(0,N) \tag{83}$$

possesses the property that

$$\psi(\phi) = \langle\psi,\phi\rangle_{L^2(0,N)} \quad \text{for } \psi\in L^2(0,N) \text{ and } \phi\varepsilon H_0^1(0,N),$$

where $\psi(\phi)$ is the action of ψ on ϕ, or the dual product between $H^{-1}(0,N)$ and $H_0^1(0,N)$. According to property, it follows that

$$\tilde{\phi}_{n_m}(\cdot)\,[\sin(j\pi x/N)] = \langle \tilde{C}_{n_m}\sin(m\pi x/N),\ \sin(j\pi x/N)\rangle_{L^2(0,N)} \tag{84}$$

$$= \begin{cases} \frac{N}{2}\tilde{C}_{n_m}, & \text{if } j = m, \\ 0, & \text{if } j \neq m. \end{cases} \tag{85}$$

Substitution of (85) into (82) leads to

$$\langle\delta(\cdot-b),\tilde{\phi}_{n_m}(\cdot)\rangle_{H^{-1}(0,N)} = \frac{N}{2}\tilde{C}_{n_m}\, p_m\delta(\cdot-b)[\sin(\frac{m\pi x}{N})]. \tag{86}$$

By the assumption that $b\varepsilon(i-1,i)$ is irrational, it follows that

$$\delta(\cdot-b)[\sin(\frac{m\pi x}{N})] = \sin(\frac{m\pi b}{N}) \neq 0, \quad \text{for all } m=1,2,\cdots, \tag{87}$$

because $\frac{mb}{N}$ will never be integer for all $m=1,2,\cdots$.

Therefore (86) and (87) imply that (81) holds. Then in turn (81) and (79) imply that (78) holds under the condition (77), so that (76) holds. Thus the system (72) is strongly stabiliable in Y. Moreover, by the Benchimal theory as above, the system (72) can be strongly stabilized by the feedback

$$f(t) = -\mathcal{B}_2^*U(t) = -B_2^*Z(t) = -\begin{bmatrix} \langle\delta(\cdot-b),\, W_t(t,\cdot)\rangle_{H^{-1}(0,N)} \\ W_t(t,1) \\ \vdots \\ W_t(t,N-1) \end{bmatrix}, \quad t > 0. \tag{88}$$

This completes the proof of this theorem. G.E.D.

Remark 3 If the string lengths are more general, i.e. $(0,a_1),\cdots,(a_{N-1},a_N)$, with $0 < a_1 <\cdots< a_N$, then we have the following results.

Theorem 4 If the following condition is satisfied,

$$\text{at least one of } \{\frac{a_1}{a_N},\cdots,\frac{a_{N-1}}{a_N}\} \text{ is irrational,} \tag{89}$$

then the corresponding system (28) can be strongly stabilized by the feedback

$$f(t) = -\begin{bmatrix} W_t(t,1) \\ \vdots \\ W_t(t,N-1) \end{bmatrix}, \quad t > 0. \tag{90}$$

If (89) does not hold, then (28) cannot be stronly stabilized by any bounded linear feedback control. In this latter case, the conclusions of Theorem 2 and 3 remain valid except that the conditions are replaced by

$$\langle g(x),\, \sin(\frac{m\pi x}{a_N})\rangle_{L^2(0,a_N)} \neq 0, \qquad \forall m=1,2,\cdots, \tag{91}$$

and

$$b\,\varepsilon(a_{i-1},a_i) \text{ such that } \frac{b}{a_N} \text{ is irrational.} \tag{92}$$

6. Conclusions

For the string system linked by point-masses, we have used the abstract framework of evolutionary equations to obtain the following results: (1) Different from the dynamical boundary control as in [1] and [2], the feedback control exerted only at the point-masses may not strongly nor weakly stabilize the system. (2) The system can be strongly stabilized by a combination feedback control implemented both at point-masses and in the interior with the distribution g(x) not orthogonal to any eigen-mode. (3) The system can also be strongly stabilized in an extension space by a combination feedback control implemented both at point-masses and pointwise located in the interior at irrational points.

There are many open research questions associated with linked flexural structures, e.g. modeling and control of several elastic plates or membranes linked by beams with some kind of control exerted at the joint points. Also, of interest is a study of vibrations of complex frame structures, etc.

References

[1] E. B. Lee and Y. You, On stabilization of a hybrid (string/point-mass) system, Proceedings of the 5th International Conference on Systems Engineering, Dayton, OH, USA, September 1987, pp. 109-112.

[2] W. Littman, L. Markus, and Y. You, A note on stabilization and controllability of a hybrid elastic system with boundary control, Math Report #807-103, University of Minnesota.

[3] G. Chen, M. Coleman and H. H. West, Pointwise stabilization in the middle of the span for second order systems: nonuniform and uniform exponential decay of solutions, SIAM J. Appl. Math, 47(4), (1987), pp. 751-780.

[4] C. D. Benchimal, A note on weak stabilizability of contraction semigroups, SIAM J. Control and Optimization, Vol. 16, No. 3, (1978), pp. 373-379.

[5] A. S. Besicovitch, Almost Periodic Functions, Cambridge, 1932.

[6] Y. C. You, Controllability and stabilizability of vibrating simply supported plate with pointwise control, Math Report #87-105, University of Minnesota, 1987, to appear in Advances in Applied Mathematics.

[7] Y. C. You, Indefinite quadratic cost optimal control for distributed parameter systems with unbounded control operator, Proceedings of the 3rd IFAC Symposium on Control of Distributed Parameter Systems, Toulouse, France, July 1982, pp. 619-625.

Acknowledgement

The research reported on here was supported by the National Science Foundation under Grants DMS8607687 and DMS8722402.

SOME IDENTIFICATION PROBLEMS
IN PETROLEUM GEOPHISICS

Li Ta-tsien and Tan Yong-ji

Fudan University

Shanghai, China

§1. Introduction

In petroleum exploitation one often uses various methods of well-logging, among which resistivity well-logging is one of the most common and important techniques.

After a well has been drilled, one puts a log tool into the well. The log tool is often an insulating rod, a part of the lateral surface of which is covered by metal membranes as electrodes. In operation, the electrode discharges a steady electric current to the ground. The goal of the resistivity well-logging is to determine the resistivity of the objective layer and then to estimate the capacity of the oil field by measuring the resulting potential on the electrode.

Without loss of generality and for simplicity, in what follows we shall only consider the corresponding two dimensional problem and one electrode case.

According to the convention in electric well-logging, we may suppose that the layer is symmetric about the well axis and the central plane, and the resistivity of the earth is piecewise constant:

$$(1.1) \qquad R=\begin{cases} R_m & \text{in } \Omega_m , \\ R_s & \text{in } \Omega_s , \\ R_{x_o} & \text{in } \Omega_{x_o} , \\ R_t & \text{in } \Omega_t \end{cases}$$

(see Figure 1). In Figure 1, the shaded part is the area occupied by the log tool; Ω_m is the well-bore domain filled with mud; Ω_s is the surrounding rock; Ω_t and Ω_{x_o} are two parts of the objective layer, which is the main object of electric

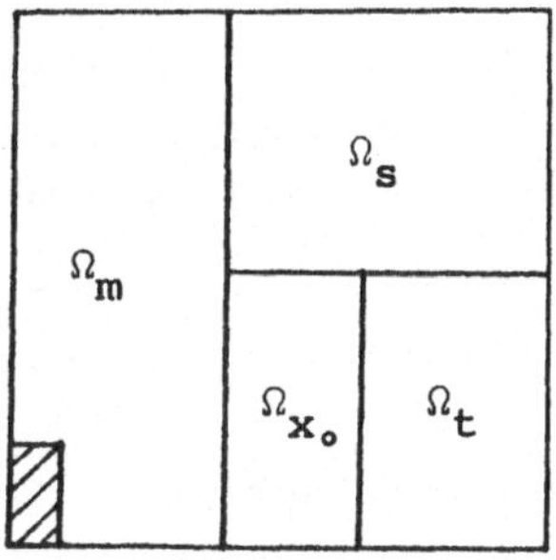

Figure 1.

well-logging when the log tool is located in this position. Since the objective layer is usually composed of porous sandstone, the liquid from the mud penetrates into the porous region and changes the resistivity in the domain Ω_{x_o} which is then called invaded area. Since the invasion of this liquid decreases the resistivity, we always have

(1.2) $$R_{x_o}<R_t \; .$$

If the geometrical structure of the layer, the resistivity in each subdomain and the total steady electric current I emitted from the electrode are known, then the potential $u=u(x,y)$ is the solution to the following non-local boundary value problem (Cf. [1]. [2]):

In the domaim $\Omega=\Omega_m \cup \Omega_s \cup \Omega_{x_o} \cup \Omega_t$, u satisfies the quasiharmonic equation

(1.3) $$\frac{\partial}{\partial x}\left(\frac{1}{R}\frac{\partial u}{\partial x}\right)+\frac{\partial}{\partial y}\left(\frac{1}{R}\frac{\partial u}{\partial y}\right)=0.$$

On the surface of the earth (the upper boundary in Figure 1) and on the right boundary which is usually far from the well axis and is called the infinitely distant boundary, the potential can be assumed to be zero. Let Γ_1 denote these two parts of the boundary, we have

(1.4) $$u=0 \quad \text{on } \Gamma_1.$$

On the well axis (the left boundary in Figure 1), the line of symmetry (the bottom boundary in Figure 1) and the rubber surface of the log tool, the normal derivative of the potential must vanish. Denoting these three parts of the boundary by Γ_2, we have

(1.5) $$\frac{\partial u}{\partial n} = 0 \quad \text{on } \Gamma_2 ,$$

where $\vec{n}$ is the unit outward normal vector to Γ_2 .

On each segment of the interface Γ_3 between layers of different resistivity, we have the conditions

(1.6) $$\begin{cases} u_+=u_- , \\ \left(\frac{1}{R}\frac{\partial u}{\partial n}\right)_- = \left(\frac{1}{R}\frac{\partial u}{\partial n}\right)_+ , \end{cases}$$

where the subscripts + and - stand for the values on both sides of Γ_3 respectively and the unit normal vector $\vec{n}$ takes the same dirction on both sides of Γ_3. (1.6) simply means that both the potential and the electric current are continuous on Γ_3.

Since the metal electrode is a good conductor, the potential must be constant on the surface Γ_0 of the electrode, but this constant is to be determined. On the other hand, the total electric current emitted from the electrode is a given constant $I>0$. Thus, on the surface Γ_0 of the electrode, we have the following non-local boundary conditions:

$$(1.7)\qquad \begin{cases} u=U\text{(unknown constant) on } \Gamma_o\ , \\ \int_{\Gamma_o} \frac{1}{R}\frac{\partial u}{\partial n}\, ds = I \text{ (given constant)}, \end{cases}$$

where $\vec{n}$ is the unit outward normal vector to Γ_o.

Problem (1.1)-(1.7) is a direct problem. By means of a variational principle, it is easy to see that this problem admits a unique solution $u \in H^1(\Omega)$ and the finite element method can be used to obtain the corresponding numerical solution.

Now if the value $U=\bar{U}$ of the potential on the electrode Γ_o can be measured and we want to determine from this information a geometrical parameter (e.g. the depth $h=\bar{h}$ of the invaded area), the resistivity on a subdomain (e.g. the resistivity R_t of the objective layer Ω_t) or the total electric current I emitted from the electrode, then we correspondingly get an identification problem. The first aim of this paper is to solve this kind of identification problem and to prove their well-posedness.

The second aim of this paper is to consider the corresponding problem of homogenization of boundary conditions. The problem arises as follows. In order to detect the resistivity of the layer near by the well-bore (i.e. the resistivity of the invaded area), various well-logging techniques have been used in the field. Among them we can list neighbour-lateral logging, micro-lateral logging and micro-spheric-focusing logging etc. For these techniques, it is very important to make a perfect contact of the electrode with the well wall, since any mud between electrode and well wall will cause a serious loss of accuracy of the instrument. Recently, in order to fit the well wall more closely, a so-called piecewise electrode has been used. A piecewise electrode is made by separating the electrode into many connected pieces (cells), embedding them into rubber then connecting them with wire behind the rubber. We show the original electrode and the corresponding piecewise electrode respectively in Figure 2.

Thus, the surface Γ_o of the original electrode is divided into two regular subsets Γ_o^ε and $\tilde{\Gamma}_o^\varepsilon$, where $\tilde{\Gamma}_o^\varepsilon$ is composed of insulating material, while Γ_o^ε, the surface of the piecewise electrode, is composed of a number of connected surfaces $\Gamma_{0,i}^\varepsilon$ $(i=1,\dots,m(\varepsilon))$:

$$(1.8)\qquad \Gamma_o^\varepsilon = \bigcup_{i=1}^{m(\varepsilon)} \Gamma_{0,i}^\varepsilon\ ,$$

in which every $\Gamma_{0,i}^\varepsilon$ denotes the surface of a connected piece of electrode. Since there is a short-circuit between these connected pieces of electrode, the potential u must be constant on the whole electrode Γ_o^ε. On the other hand, the total electric current emitted from Γ_o^ε is still the same given constant I. Therefore, the boundary condition on $\Gamma_o=\tilde{\Gamma}_o^\varepsilon \cup \Gamma_o^\varepsilon$ should be changed from (1.7) to

$$(1.9)\quad \begin{cases} \dfrac{\partial u}{\partial n} = 0, \text{ on } \tilde{\Gamma}_o^\varepsilon, \\ u=U \text{ (unknown constant) on } \Gamma_o^\varepsilon, \\ \int_{\Gamma_o^\varepsilon} \dfrac{1}{R}\dfrac{\partial u}{\partial n} ds = I \text{ (given constant).} \end{cases}$$

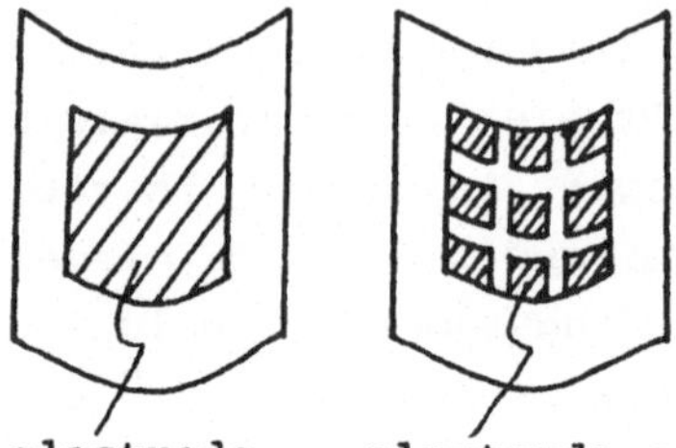

Figure 2.

The problem (1.1)-(1.6), (1.9) is still well-posed. Noting (1.8), however, it is almost impossible to solve this problem driectly by, say, the finite element method, since the boundary condition on Γ_o rapidly changes type, and then we need a great number of refined elements in the neighbourhood of Γ_o. In order to reduce this complexity of computation it is natural to ask if this complicated boundary condition (1.9) on Γ_o can be replaced approximately by a much simpler and unified boundary condition and, if it exists, what this reduced boundary condition may be. We refer to this problem as the problem of homogenization of boundary conditions.

Let

$$(1.10)\qquad \chi_\varepsilon = \begin{cases} 1 & \text{on } \Gamma_o^\varepsilon, \\ 0 & \text{on } \tilde{\Gamma}_o^\varepsilon \end{cases}$$

be the characteristic function of Γ_o^ε on Γ_o. We make the following hypothesis

(H): For any weak * convergent subsequence $\{\chi_{\varepsilon'}\}$ of $\{\chi_\varepsilon\}$ in $L^\infty(\Gamma_o)$, its limit function is always different from zero almost everywhere on Γ_o; namely,

if

$$(1.11)\qquad \chi_{\varepsilon'} \to \chi \text{ weak * in } L^\infty(\Gamma_o),$$

then

$$(1.12)\qquad \chi \neq 0 \quad \text{a.e. on } \Gamma_o .$$

Hypothesis (H) actually gives a restriction on the geometrical structure of Γ_o^ε as $\varepsilon\to 0$. In fact, it follows from (H) that there exists a constant a>0 such that for any $\varepsilon>0$,

$$(1.13)\qquad \operatorname*{meas}_{\Gamma_o} \Gamma_o^\varepsilon \geq a > 0.$$

In particular, in the case of a periodic structure as shown in Figure 3, the whole sequence

$$(1.14)\qquad \chi_\varepsilon \to \theta\neq 0 \text{ weak * in } L^\infty(\Gamma_o) \text{ as } \varepsilon\to 0,$$

so that (H) is satisfied.

Under hypothesis (H) it has been proved by A. Damlamian and Li Ta-tsien (Cf. [3]) that as $\varepsilon \to 0$, the solution $u^{\varepsilon}(x.y)$ of the direct problem (1.1)-(1.6), (1.9) converges strongly in $H^1(\Omega)$ to the solution u of the direct problem (1.1)-(1.7). Thus the homogenized boundary condition on Γ_0 is nothing but the boundary condition (1.7). That is to say, in resistivity well-logging, instead of the piecewise electrode Γ_0^{ε}, we can approximately regard the whole surface Γ_0 as an electrode and then there will be no difficulty in using the finite element method to get a numerical solution.

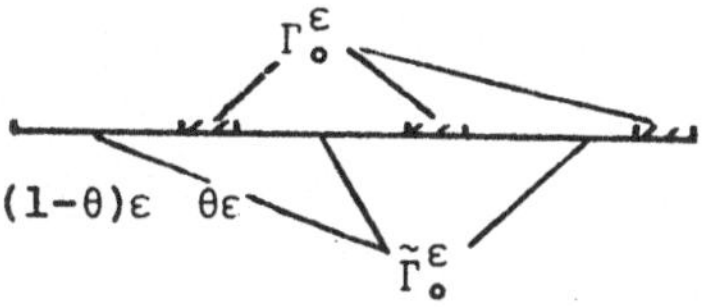

Figure 3.

Now if the potential on the original electrode Γ_0 and on the piecewise electrode Γ_0^{ε} can be measured and take the same value $U=\bar{U}$, the second aim of this paper is then to prove that as $\varepsilon \to 0$, the solution to any of the identification problems mentioned above for the piecewise electeode Γ_0^{ε} must converge to the solution to the same kind of identification problem for the original electrode Γ_0. That is to say, for the identification problems under consideration, the homogenization of boundary conditions is still valid.

Since it is always important to determine the depth of the invaded area in well-logging, we only consider the identification of the depth of the invaded area in what follows. Other identification problems mentioned above can be similarly discussed (also see Tan Yong-Ji [4],[5]). Moreover, for simplicity we may assume that the domain Ω is only composed of two subdomains Ω_{x_0} and Ω_t as shown in Figure 4.

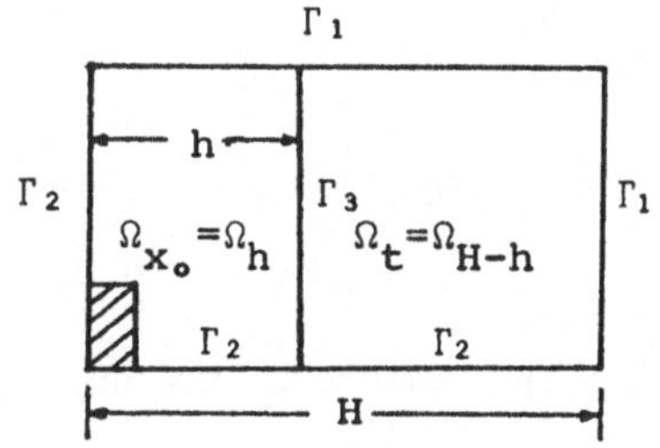

Figure 4.

In order to express the dependence of the resistivity on the depth h of the invaded area explicitly, set

$$(1.15) \qquad \begin{cases} \Omega_h = \Omega_{x_0}, \\ \Omega_{H-h} = \Omega_t , \end{cases}$$

then

$$(1.16)\qquad R = \begin{cases} R_{x_0} & \text{in } \Omega_h, \\ R_t & \text{in } \Omega_{H-h}, \end{cases} \quad (R_{x_0} < R_t).$$

§2. Identification problem (P_h)

We now consider the following identification problem (P_h): If the potential U on the electrode Γ_0 is measured: $U=\bar{U}$, we want to determine the depth $h=\bar{h}\varepsilon[0,H]$ of the invaded area such that the solution $u=u_{\bar{h}}(x,y)$ to the corresponding direct problem (1.3)-(1.7),(1.16) subject to $h=\bar{h}$ takes the measured value $\bar{U}$ on the electrode Γ_0:

$$(2.1)\qquad u_{\bar{h}}|_{\Gamma_0} = \bar{U}.$$

To solve this problem, we first consider the direct problem (1.3) -(1.7), (1.16) for any fixed $h\varepsilon[0,H]$.

Let

$$(2.2)\qquad V=\{v|v\varepsilon H^1(\Omega),\ v|_{\Gamma_1}=0,\ v|_{\Gamma_0}=\text{const}\}\ .$$

Obviously,V is a closed subspace of $H^1(\Omega)$. We minimize the following functional on V:

$$(2.3)\qquad J_h(v)=\frac{1}{2}\left(\int_{\Omega_h}\frac{1}{R_{x_0}}|\nabla v|^2dx+\int_{\Omega_{H-h}}\frac{1}{R_t}|\nabla v|^2dx\right)-Iv|_{\Gamma_0}\ ,$$

where the subscript h in J_h (v) shows that this functional depends on h.

For any fixed $h\varepsilon[0,H]$, there exists a unique $u_h\varepsilon V$ such that

$$(2.4)\qquad J_h(u_h)=\underset{v\varepsilon V}{\text{Min}}\ J_h(v).$$

u_h is the solution to the direct problem (1.3)-(1.7),(1.16). Moreover,

$$(2.5)\qquad \int_{\Omega_h}\frac{1}{R_{x_0}}\nabla u_h\nabla v dx+\int_{\Omega_{H-h}}\frac{1}{R_t}\nabla u_h\nabla v dx=Iv|_{\Gamma_0}\ ,\ \forall\ v\varepsilon V$$

and

$$(2.6)\qquad J_h(u_h)=-\frac{1}{2}Iu_h|_{\Gamma_0}=-\frac{1}{2}\left(\int_{\Omega_h}\frac{1}{R_{x_0}}|\nabla u_h|^2dx+\int_{\Omega_{H-h}}\frac{1}{R_t}|\nabla u_h|^2dx\right).$$

Setting

$$(2.7)\qquad U = u_h|_{\Gamma_0}\ ,$$

U is a function of $h\varepsilon[0,H]$:

$$(2.8)\qquad U = U(h),$$

provided that all other factors are fixed.

Since $I>0$, it is easy to see that

(2.9) $\quad U(h)>0,\ \forall\ h\varepsilon[0,H]$

and u_h attains its maximum $U(h)$ on $\Gamma_\circ$. Moreover, we have

LEMMA 2.1: $U=U(h)$ is a continuous, strictly decreasing function of $h\varepsilon[0,H]$.

PROOF: Let $h,\ \tilde{h}\varepsilon[0,\ H]$ and $\tilde{h}>h$. Suppose that $u_h\varepsilon V$ and $u_{\tilde{h}}\varepsilon V$ are the solutions to the corresponding direct problems (1.3)-(1.7), (1.16):

$$J_h(u_h)=\underset{v\varepsilon V}{\mathrm{Min}}\ J_h(v),$$
$$(2.10)\qquad J_{\tilde{h}}(u_{\tilde{h}})=\underset{v\varepsilon V}{\mathrm{Min}}\ J_{\tilde{h}}(v).$$

By (1.2) it is easy to see that

$$(2.11)\qquad U(\tilde{h})-U(h)=\frac{2}{I}(J_h(u_h)-J_{\tilde{h}}(u_{\tilde{h}}))\le\frac{2}{I}(J_h(u_{\tilde{h}})-J_{\tilde{h}}(u_{\tilde{h}}))$$
$$=\frac{2}{I}\left(\frac{1}{R_t}-\frac{1}{R_{x_\circ}}\right)\int_{\Omega_{\tilde{h}}\cap\Omega_{H-h}}|\nabla u_{\tilde{h}}|^2dx\le 0.$$

But, according to Holmogren's Theorem (cf. [6]), it is impossible to have

$$\int_{\Omega_{\tilde{h}}\cap\Omega_{H-h}}|\nabla u_{\tilde{h}}|^2dx=0,$$

we therefore get

(2.12) $\quad U(\tilde{h})<U(h),\ \forall\ h,\tilde{h}\varepsilon[0,H],\ \tilde{h}>h,$

namely, $U(h)$ is a strictly decreasing function of $h\varepsilon[0,H]$. It is then easy to see from (2.6) that

(2.13) $\quad \int_\Omega|\nabla u_h|^2dx\le C,\ \forall\ h\varepsilon[0,H],$

where C is a constant independent of h.

On the other hand, we have

$$(2.14)\qquad U(\tilde{h})-U(h)=\frac{1}{I}\left(\frac{1}{R_t}-\frac{1}{R_{x_\circ}}\right)\int_{\Omega_{\tilde{h}}\cap\Omega_{H-h}}\nabla u_h\nabla u_{\tilde{h}}dx\quad(\tilde{h}>h),$$

thus by the absolute continuity of Lebesgue integration, for any fixed $h\varepsilon[0,H)$ we get

(2.15) $\quad U(\tilde{h})\to U(h)$ as $\tilde{h}\to h^+$.

Similarly, for any fixed $h\varepsilon(0,H]$, we have

(2.16) $\quad U(\tilde{h})\to U(h)$ as $\tilde{h}\to h^-$.

Hence $U(h)$ is a continuous function of $h\varepsilon[0,H]$. Q.E.D.

By Lemma 2.1, $U=U(h)$ has a continuous, strictly decreasing inverse function $h=h(U)$ with $U\varepsilon[U(H),\ U(0)]$. Therefore we get

THEOREM 2.1: For any given $\bar{U}\varepsilon[U(H),U(0)]$, the identification problem (P_h) admits a unique solution $\bar{h}=h(\bar{U})\varepsilon[0,H]$. Moreover, $\bar{h}$ depends continuously on $\bar{U}$.

3. Homogenization of boundary conditions for the identification problem (P_h^ε)

In the case of a piecewise electrode:

(3.1) $$\Gamma_o=\Gamma_o^\varepsilon \cup \tilde{\Gamma}_o^\varepsilon$$

and

(3.2) $$\Gamma_o^\varepsilon=\bigcup_{i=1}^{m(\varepsilon)} \Gamma_{o,i}^\varepsilon ,$$

if the potential measured on the electrode Γ_o^ε still has the same value $U=\bar{U}$, we can similarly solve the corresponding identification problem (P_h^ε): Determine the depth $h=\bar{h}^\varepsilon \in[0,H]$ of the invaded area such that the solution $u^\varepsilon=u^\varepsilon_{\bar{h}^\varepsilon}(x,y)$ to the corresponding direct problem (1.3)-(1.6), (1.9), (1.16) subject to $h=\bar{h}^\varepsilon$ takes the measured value $\bar{U}$ on the piecewise electrode Γ_o^ε, i.e.

(3.3) $$u^\varepsilon_{\bar{h}^\varepsilon}=\bar{U} \quad \text{on} \quad \Gamma_o^\varepsilon .$$

Following our previous treatment we first consider the corresponding direct problem (1.3)-(1.6), (1.9), (1.16) for any fixed $\varepsilon>0$ and $h\in[0,H]$.

Let

(3.4) $$V_\varepsilon=\{v \mid v\in H^1(\Omega),\ v|_{\Gamma_1}=0,\ v|_{\Gamma_0^\varepsilon}=\text{const}\}.$$

For any fixed $\varepsilon>0$, V_ε is still a closed subspace of $H^1(\Omega)$ and we have

(3.5) $$V\subset V_\varepsilon .$$

For any fixed $\varepsilon>0$, we minimize the following functional on V_ε:

(3.6) $$J_h^\varepsilon(v)=\frac{1}{2}\Big(\int_{\Omega_h}\frac{1}{R_{x_o}}|\nabla v|^2dx+\int_{\Omega_{H-h}}\frac{1}{R_t}|\nabla v|^2dx\Big)-Iv|_{\Gamma_o^\varepsilon} ,$$

where the subscript h in $J_h^\varepsilon(v)$ shows that this functional depends not only on ε but also on h.

For any fixed $h\in[0,H]$, there exists a unique $u_h^\varepsilon\in V_\varepsilon$ such that

(3.7) $$J_h^\varepsilon(u_h^\varepsilon) = \operatorname*{Min}_{v\in V_\varepsilon} J_h^\varepsilon(v).$$

u_h^ε is the solution to the direct problem (1.3)-(1.6), (1.9), (1.16) and so

(3.8) $$\int_{\Omega_h}\frac{1}{R_{x_o}}\nabla u_h^\varepsilon\nabla v dx+\int_{\Omega_{H-h}}\frac{1}{R_t}\nabla u_h^\varepsilon\nabla v dx=Iv|_{\Gamma_o^\varepsilon} ,\ \forall\ v\in V_\varepsilon$$

and

(3.9) $$J_h^\varepsilon(u_h^\varepsilon)=-\frac{1}{2}Iu_h^\varepsilon|_{\Gamma_o^\varepsilon}=-\frac{1}{2}\Big(\int_{\Omega_h}\frac{1}{R_{x_o}}|\nabla u_h^\varepsilon|^2dx+\int_{\Omega_{H-h}}\frac{1}{R_t}|\nabla u_h^\varepsilon|^2dx\Big).$$

Setting

(3.10) $$U = u_h^\varepsilon|_{\Gamma_o^\varepsilon} \quad ,$$

U is a function of $h\varepsilon[0,H]$:

(3.11) $$U = U^\varepsilon(h),$$

provided that all other factors are fixed.

In a similar way to §2, we know that for any fixed $\varepsilon>0$, $U^\varepsilon(h)$ is a continuous, strictly decreasing function of $h\varepsilon[0,H]$ and

(3.12) $$U^\varepsilon(h)>0, \ \forall\ h\varepsilon[0,H].$$

In fact,we have a stronger result than (3.12) as follows.

LEMMA 3.1: For any $h\varepsilon[0,H]$ and $\varepsilon>0$,

(3.13) $$U^\varepsilon(h)>U(h),$$

where U(h) is defined by (2.7)-(2.8). In other words, if the total electric current I emitted from the electrode is the same, then for any fixed $h\varepsilon[0,H]$, the potential on the piecewise electrode Γ_o^ε is always bigger than the potential on the original electrode Γ_o.

PROOF: Noting $V \subset V_\varepsilon$ and $u_h \varepsilon V$, it is easy to see that

(3.14) $$-\frac{1}{2}IU(h)=J_h(u_h)=J_h^\varepsilon(u_h)\geq \underset{v\varepsilon V_\varepsilon}{\text{Min}}\ J_h^\varepsilon(v)=J_h^\varepsilon(u_h^\varepsilon)=-\frac{1}{2}IU^\varepsilon(h).$$

However, equality in (3.14) is impossible. For if not, we have

(3.15) $$J_h^\varepsilon(u_h) = \underset{v\varepsilon V_\varepsilon}{\text{Min}}\ J_h^\varepsilon(v),$$

then

(3.16) $$u_h = u_h^\varepsilon$$

is also the solution to problem (1.3)-(1.6), (1.9),(1.16). Since u_h attains its maximum U(h) on $\Gamma_o=\Gamma_o^\varepsilon \cup \tilde{\Gamma}_o^\varepsilon$, noticing the boundary condition (1.9) (i.e. $\frac{\partial u}{\partial n}|_{\tilde{\Gamma}_o^\varepsilon}=0$), the maximum principle leads to a contradiction. Q.E.D.

According to results on homogenization of boundary conditions for the direct problem (see A. Damlamian and Li Ta-tsien [3]), we have

LEMMA 3.2: For any fixed $h\varepsilon[0,H]$, as $\varepsilon\to 0$, we have

(3.17) $$u_h^\varepsilon \to u_h \quad \text{strongly in } H^1(\Omega)$$

and

(3.18) $$U^\varepsilon(h) \to U(h).$$

By Lemmas 3.1 and 3.2, in a similar way to §2, it is easy to see that for $\varepsilon>0$ small enough, function (3.11) has a continuous, strictly decreasing inverse function

(3.19) $$h=h^\varepsilon(U) \quad \text{for} \quad U\varepsilon(U(H),\ U(0)].$$

Thus, the solution to the identification problem (P_h^ε) must be $\bar{h}^\varepsilon=h^\varepsilon(\bar{U})$

for any fixed $\bar{U}\varepsilon(U(H),\ U(0)]$.

By Lemmas 3.1 and 2.1, we get easily

LEMMA 3.3: For any $\varepsilon>0$ small enongh,

$$\bar{h}^{\varepsilon}>\bar{h}\ , \tag{3.20}$$

where

$$\begin{cases} \bar{h}^{\varepsilon}=h^{\varepsilon}(\bar{U}) \\ \bar{h}\ =h(\bar{U})\ , \end{cases} \tag{3.21}$$

in which $h=h(U)$ is the inverse function of $U=U(h)$ and $\bar{U}\varepsilon(U(H),\ U(0)]$. In other words, for the same total electric current I emitted from the electrode and the same measured potential $\bar{U}$ on the electrode, the depth $\bar{h}^{\varepsilon}$ of the invaded area as the solution to the identification problem (P_h^{ε}) for the piecewise electrode Γ_o^{ε} is always bigger than the depth $\bar{h}$ of the invaded area as the solution to the identification problem (P_h) for the original electrode Γ_o.

Now we prove

LEMMA 3.4: Under hypothesis (H), For any fixed $\bar{U}\varepsilon(U(H),\ U(0)]$, as $\varepsilon\to0$,

$$\bar{h}^{\varepsilon}\to\bar{h} \tag{3.22}$$

and

$$u^{\varepsilon}_{\bar{h}^{\varepsilon}}\to u_{\bar{h}} \quad \text{strongly in } H^1(\Omega), \tag{3.23}$$

where $\bar{h}^{\varepsilon}$ and $\bar{h}$ are still defined by (3.21), $u^{\varepsilon}_{\bar{h}^{\varepsilon}}$ and $u_{\bar{h}}$ are respectively the corresponding potential functions.

PROOF: From (3.9) and noting $u^{\varepsilon}_{\bar{h}^{\varepsilon}}|_{\Gamma_o^{\varepsilon}}=\bar{U}$, it is easy to see that

$$||\, u^{\varepsilon}_{\bar{h}^{\varepsilon}}\, ||_{H^1(\Omega)}\leq C, \tag{3.24}$$

where C is a constant independent of ε. Then by weak compactness and noting that $0\leq\bar{h}^{\varepsilon}\leq H$, there exists a subsequence $\{\varepsilon_n\}$ such that as $\varepsilon_n\to0$,

$$\bar{h}^{\varepsilon_n}\to\tilde{h}, \tag{3.25}$$

$$u^{\varepsilon_n}_{\bar{h}^{\varepsilon_n}}\to\tilde{u} \quad \text{weakly in } H^1(\Omega), \tag{3.26}$$

$$u^{\varepsilon_n}_{\bar{h}^{\varepsilon_n}}|_{\Gamma_o}\to\tilde{u}|_{\Gamma_o} \quad \text{strongly in } L^2(\Gamma_o). \tag{3.27}$$

It is then easy to prove from the property satisfied by $u^{\varepsilon_n}_{\bar{h}^{\varepsilon}}$ that

$$\int_{\Omega_{\tilde{h}}}\frac{1}{R_{x_o}}\nabla\tilde{u}\nabla v dx+\int_{\Omega_{H-\tilde{h}}}\frac{1}{R_t}\nabla\tilde{u}\nabla v dx=Iv|_{\Gamma_o}\ ,\ \ \forall\ v\varepsilon V. \tag{3.28}$$

On the other hand we have (at least for a subsequence of $\{\varepsilon_n\}$, still denoted by $\{\varepsilon_n\}$)

(3.29) $\chi_{\varepsilon_n} u_{\bar{h}^{\varepsilon_n}}^{\varepsilon_n}|_{\Gamma_0} \to \chi\tilde{u}|_{\Gamma_0}$ weakly in $L^2(\Gamma_0)$

and

(3.30) $\chi_{\varepsilon_n} u_{\bar{h}^{\varepsilon_n}}^{\varepsilon}|_{\Gamma_0} = \chi_\varepsilon \bar{U} \to \chi\bar{U}$ weak * in $L^\infty(\Gamma_0)$,

so by hypothesis (H), we get

(3.31) $\tilde{u}|_{\Gamma_0} = \bar{U}$.

Thus $\tilde{u}$ is actually the solution to the direct problem (1.3)-(1.7), (1.16) subject to $h=\tilde{h}$, so according to the uniqueness of solutions to the identification problem (P_h) we have

(3.32) $\tilde{h} = \bar{h} = h(\bar{U})$

and

(3.33) $\tilde{u} = u_{\bar{h}}$.

Regarding the convergence of an equivalent norm in $H^1(\Omega)$, we can easily get strong convergence of $u_{\bar{h}^{\varepsilon_n}}^{\varepsilon_n}$ in $H^1(\Omega)$.

Thus, as $\varepsilon \to 0$, for the whole sequence, (3.22)-(3.23) hold.

The combination of Lemmas 3.3 and 3.4 gives

THEOREM 3.1: Under hypothesis (H), we have

(3.34) $\bar{h}^\varepsilon > \bar{h}$,

and, as $\varepsilon \to 0$,

(3.25) $\bar{h}^\varepsilon \to \bar{h}$

and

(3.26) $u_{\bar{h}^\varepsilon}^{\varepsilon} \to u_{\bar{h}}$ strongly in $H^1(\Omega)$,

in which for any fixed $\bar{U} \in (U(H), U(0)]$, $\bar{h}=h(\bar{U})$ and $\bar{h}^\varepsilon = h^\varepsilon(\bar{U})$ stand for the solutions to the identification problems (P_h) and (P_h^ε) respectively; $u_{\bar{h}}$ and $u_{\bar{h}^\varepsilon}^{\varepsilon}$ are the corresponding electric potential functions in Ω.

References

[1] Li Ta-tsien et al., Boundary value problems with equi-valued surface boundary conditions for self-adjoint elliptic defferential equations (1), Fudan Journal (Natural Science), 1(1976), 61-71.

[2] Li Ta-tsien et al., Applications of the finite element methods in the electric well-logging, Oil Industry Press, Beijing (1979).

[3] A. Damlamian and Li Ta-tsien, Boundary homogenization for elliptic

problems, J. Math. pures et appl., 66(1987), 351-361.

[4] Tan Yong-ji, An inverse problem for nonlocal elliptic BVP and resistivity identification, Partial Differential Equations (Proceedings, Tianjin 1986, edited by S.S. Chern), Lecture Notes in Mathematics 1306, Springer-Verlag, 1988, 149-159.

[5] Tan Yong-ji, Boundary homogenization of an inverse non-local elliptic boundary value problem, to appear in IMA Journal of Applied Mathematics.

[6] F. Treves, Basic linear partial differential equations, Academic Press, Inc., 1975.

DEVELOPMENT OF A TWO/THREE DIMENSIONAL SHAPE OPTIMIZATION PROGRAM WITH SOLID MODELLING, SEMI-AUTOMATIC MESH GENERATION AND ADAPTIVE MESH REFINEMENT

C.V. RAMAKRISHNAN*
Professor and Head
Department of Applied Mechanics
Indian Institute of Technology
NEW DELHI - 110016, INDIA

1. INTRODUCTION

Optimum design of structural shapes is an important branch of structural optimization relevant to the fields of mechanical, aeronautical, automotive and civil engineering disciplines. Here, the best structural shape from the point of view of a specific design function is to be obtained in a design environment consisting of other behavioural, geometric and technological constraints. Generally, mechanical components require minimization of stress concentration from considerations of fatigue strength and fracture toughness. Material volume minimization is usually unimportant is these cases. Aerospace components invoke minization of weight for a specified strength while civil engineering structures require total cost minimization.

In the past there have been several attempts at shape optimization [1-5]. The author and his coworkers obtained the solution of several practial engineering problems [1-3] and the following general conclusions were drawn.

Two dimensional shape optimization problems can be conveniently solved using finite element analysis and improved move limit method of sequential linear programming.

----

* Senior Visitor, University College of Swansea, Swansea, UK

- Structural shape control should be exercised by using polynomial equation definition for curves. Use of nodal coordinates of finite element mesh as variables is unsatisfactory.

- The variation of finite element gridding with the progress of optimization should be checked from the point of view of accuracy.

- Visualization of intermediate designs and mesh refinement should be investigated

- The possible use of approximate analyses and coarse meshes should be studied. Use of higher order elements may be explored.

While the results were excellent from an engineering point of view, the following difficulties were present from a computational point of view:

- The solution involved user supplied subroutines for the definition of geometry, mesh generation, side constraints and the evaluation of objective function of behavioural constraints.

- The optimum curve is unique only among a class of curves described by the specified procedure.

- The finite element analysis was frequently suspect and the design derivatives were even more so.

- The practical implementation for 3-D problems was very difficult.

The following developments in the last decade have created the necessary tools for the development of general purpose shape optimization programs.

a) Parametric representation of curves and surfaces.

b) Use of these curves for modelling and manufacture has become common and hence their use in design and synthesis is inevitable from a CADCAM point of view.

c) Powerful mesh generation techniques have been developed.

d) Availability of powerful computer hardware and computer graphics facilities.

e) Adaptive mesh refinement and better understanding of the methods for the computation of design derivatives.

The present paper is a summarised presentation of the development of a complete two dimensional optimization program and the modules for the three dimensional version. Similar parallel developments have been reported by Botkin et al. (1986), Choi (1987) and Shephard and Yerry (1986). Botkin uses a NASTRAN packages along with a program for generating sheet metal assemblies. Choi uses ANSYS package along with a user developed program for generating Bezier surfaces. Shephard and Yerry give general ideas on mesh generation and solid modellers and integration of the two in the context of shape optimization.

2. MATHEMATICAL CONCEPTS

Shape optimization is a nonlinear mathematical programming problem:

$$z = \min \quad W(\bar{x}) \tag{1}$$

where $\bar{x}$ is the vector of design variables controlling shape, 'W' is the function to be minimized

The problem has a set of constraints

$$g_j(x) \leq 0 \quad (j = 1,2,....,m) \tag{2}$$

These functions represent limitation on stress, deflection etc., for the elasticity problem under consideration. Only static loading is considered using a finite element description, the load deflection relation is

$$[K]\,\bar{a} = \bar{F} \tag{3}$$

Where $\bar{F}$ is the load vector, $\bar{a}$ is the vector of nodal displacements

DEVELOPMENT OF THE SHAPE OPTIMIZATION PROGRAM

THE SHAPE OPTIMIZATION PROGRAM DEVELOPED (FOR 2-D)

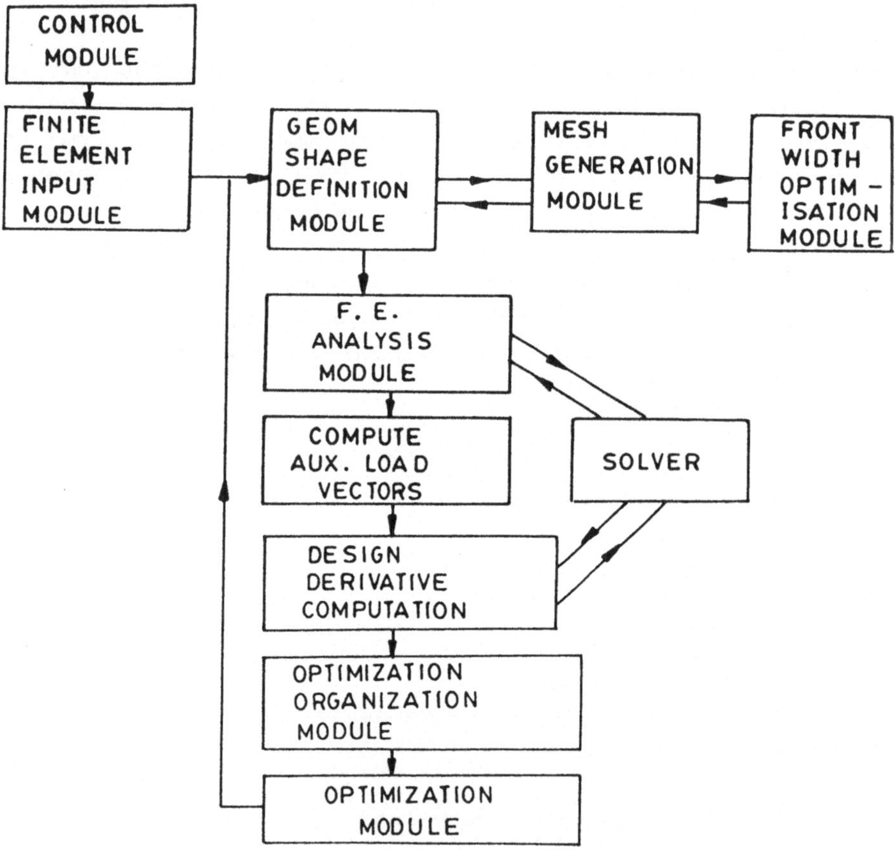

FIG.1 THE SALIENT FEATURES OF THE PROGRAM

and [k] is the stiffness matrix. Here [K] is a function of the design vector $\overline{x}$ and similarly $\overline{F}$ and $\overline{a}$ are also dependent on $\overline{x}$

To obtain displacement derivatives with respect to design variables (x's) we write

$$[K]\ \overline{a}_i' = F_i' - [k]_i'\ \overline{a} \tag{4}$$

where (p_i') denotes differentiation of the entity (p) with respect to the design variable x_i. For two and three dimensional problems the computation of $[K]_i'$ and F_i' and hence a_i' can be carried out accurately using a semi analytical procedure, which is currently known as the Implicit Differentiation Algorithm. This method, originally proposed by the author [1] has been used in the present program development. The accuracy of the procedure and comparison with finite difference computations are presented later.

3. DEVELOPMENT OF THE SHAPE OPTIMIZATION PROGRAM

The shape optimization program developed with 2-D F.E. analyser has the following organisational structure. The salient features of the program are described below, while clearly bringing out the implications in a three dimensional finite element analysis context which is under development. Fig. 1 shows the overall organisation of the complete software.

4. GEOMETRIC SHAPE DEFINITION

In order to describe any two dimensional design completely, it is identified as a simple assembleage of four noded curvilinear patches. Each patch can be obtained as a simple transformation of a rectangular domain using a mapping function. This facilitates both automated mesh generation and automatic recognition of design variables. The boundaries of the patch can be defined using lines, circular arcs, cubic splines or B-spline curves. Thus for the complete definition of a domain, the following are required,

- A list of all points and their spatial coordinates.
- A list of curves giving their type and control points for each.

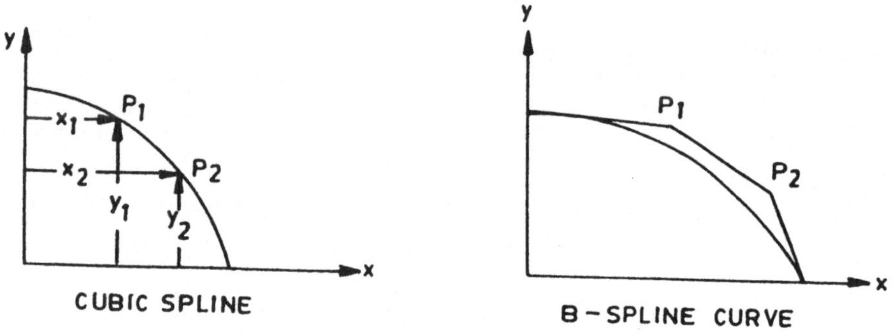

FIG. 2 CURVES

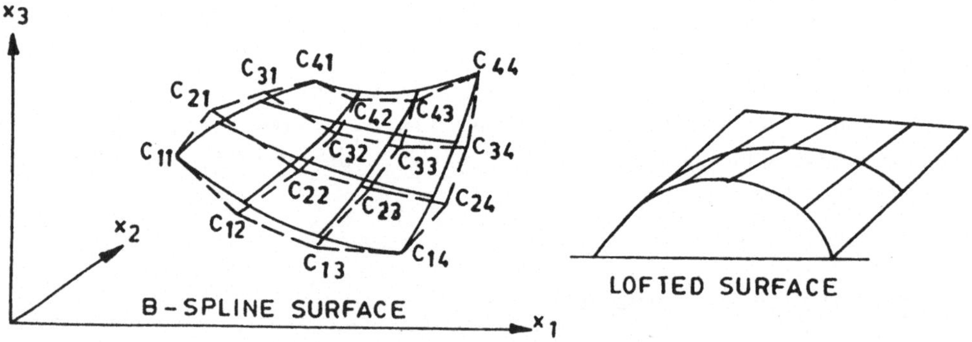

FIG. 3 SURFACES

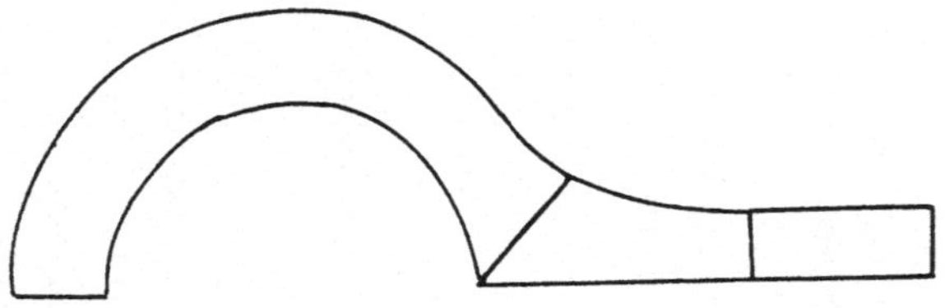

FIG.4 CONNECTING ROD PROBLEM

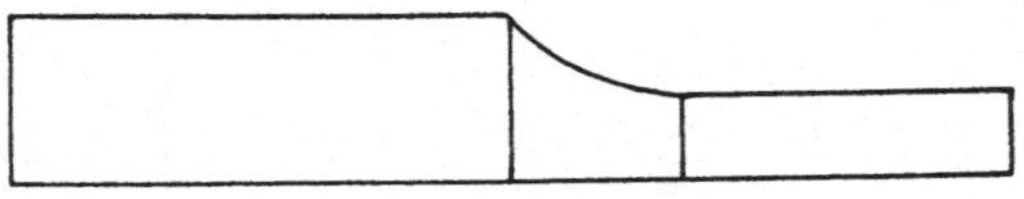

FIG.5 FILLET PROBLEM

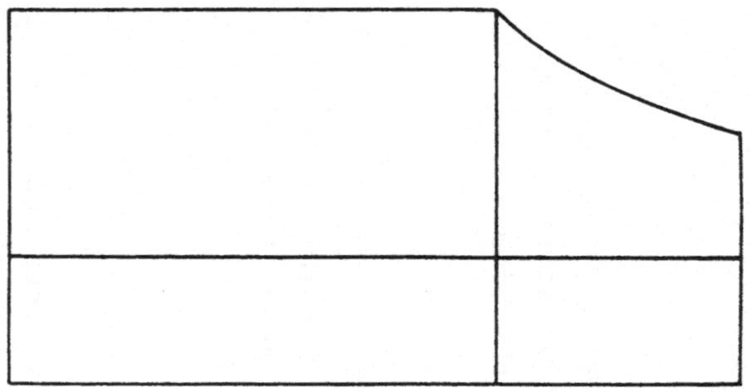

FIG.6 FOUR PATCH FILLET DOMAIN

A list of all patches alongwith details of boundary curves for each patch.

As shown below, for each curve control points are necessary in addition to the end points. If the control points do not constitute the design variables, the coordinates are to the supplied as data (Fig. 2).

The present software is general and can generate surfaces in space. Cubic spline and B-spline surfaces are easily generated in addition to lofted surfaces (Fig. 3). These are useful for the generation of plate and shell surfaces. This approach is naturally useful both for solid modelling and subsequent automated 3-D mesh generation.

Some of the domains analysed are shown in Fig.4 - Fig.6.

4.1 Three Dimensional Solid Modeller

A simple three dimensional solid modeller has been developed [9] for the purpose of representation of 3-D objects with the ultimate aim of automated mesh generation and integration with the current program. This elementary solid modeller uses one or more of the following approaches for modelling:

- Boundary surface representation
- Blending of surfces
- Other simple techniques like axisymmetric rotation, sweeping, translation of a surface over a curve etc.,

In general, shape optimization problems do not require very advanced solid modelling procedures. Generation of solids with the above techniques is usually sufficient for the initial purpose of modelling the object. Usually a redefinition of the solid as an assembleage of curved bricks is necessary for disceretization. Some of the simple solids modelled are shown in figures 7-9.

At the present stage of development of software this is complete except for the following:

FIG.7

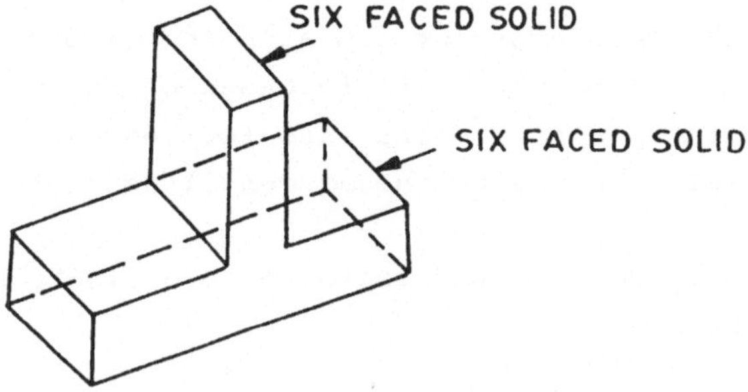
SIX FACED SOLID
SIX FACED SOLID

FIG.8

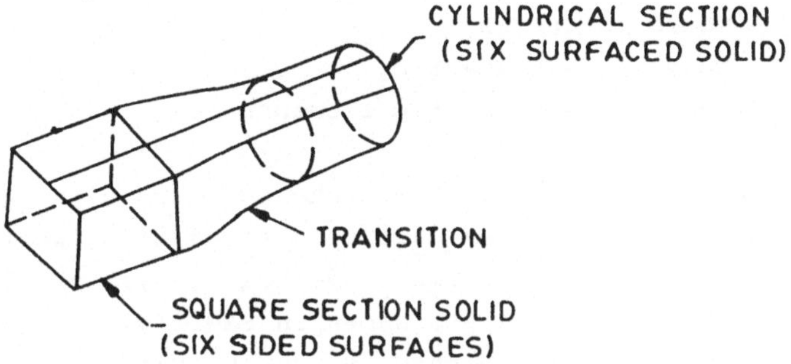
CYLINDRICAL SECTIION
(SIX SURFACED SOLID)
TRANSITION
SQUARE SECTION SOLID
(SIX SIDED SURFACES)

FIG.9

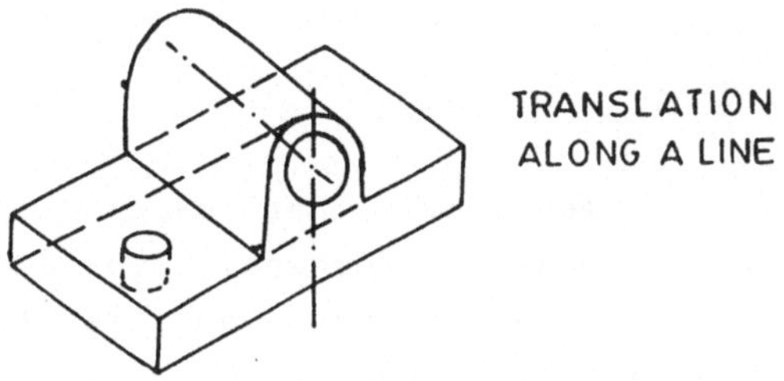
TRANSLATION
ALONG A LINE

- The determination of the surfaces of intersection.
- Carrying out certain boolean operations.

5. SEMI AUTOMATIC MESH GENERATION

The mesh generator incorporated in the program is a simple but very useful component. It is based on the concept of transfinite mapping of rectangular domains into four noded currilinear patches or cuboidal domains into eight noded curvilinear blocks.

The program divides the region into a finite number of user guided finite elements.

- 6-Noded Triangles.
- 8-Noded Iso-parametric.
- 9-Noded Iso-parametric elements.

These elements have been chosen because in the context of solid mechanics analysis they have been found to be well behaved. Some of the simple finite element meshes generated for selected shape optimization examples are shown in Fig. 10-12. It is to be noted that the element sizes within a patch can be uniformly varied along parametric directions.

5.1 Adaptive Gridding

It is frequently mentioned that mapped mesh generators are unsuitable for local mesh refinement. To a limited extent this difficulty may be overcome through an intelligent partition of patches or blocks, arbitrary mesh grading is difficult to achieve. However, it is quite possible to increase mesh refinement within a user specified patch without altering the neighbouring elements. This facility greatly helps in carrying out more accurate analyses.

In Fig. 13, the four patch fillet domain is discretized using the h-adaptive gridding. Such a facility is extremely useful in shape optimization if used in conjunction with error estimation procedures [11].

In the author's opinion, research should be directed towards the

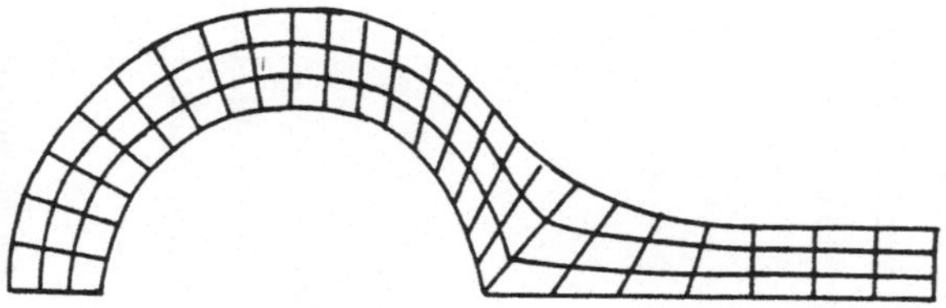

FIG.10 CONNECTING ROD PROBLEM

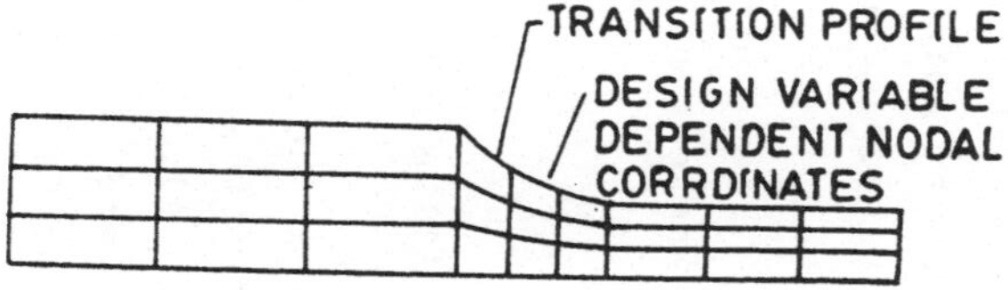

FIG.11 FILLET PROBLEM

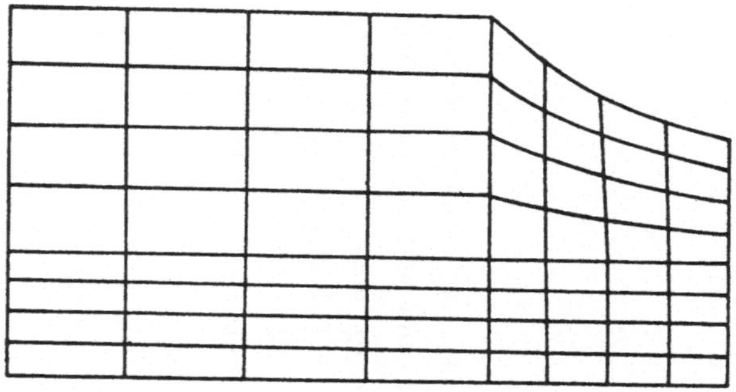

FIG. 12 FOUR PATCH FILLET DOMAIN

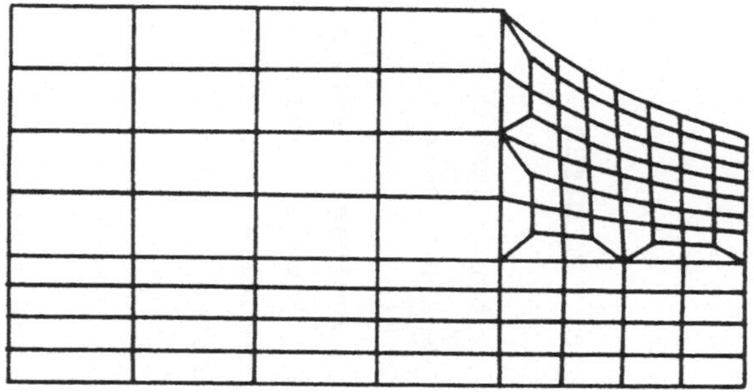

FIG.13 FOUR PATCH FILLET DOMAIN

estimation of error measures for design derivatives also in addition to those for stresses.

6. FRONT WIDTH OPTIMIZER

One of the immediate consequences of semiautomatic mesh generation is the generation of element and nodal nicknames in an arbitrary fashion which results in lack of control on frontwidth. Since the analyser employs Gaussian elimination with frontal housekeeping, the algorithm proposed by Sloan and Randoph [12] is utilised optionally for squashing of frontwidth. This is very essential since repeated solutions are carried out during optimization.

7. FINITE ELEMENT ANALYSER

The finite element analyser is based on conventional analysis procedures with an efficient frontal solver. Stress evaluations are done at the Gauss points and are extrapolated to the boundary regions. This has been generally found to be satisfactory.

The program has automatic facility to compute stresses only on those elements where constraints are present while this is skipped at other elements.

8. DESIGN DERIVATIVE COMPUTATIONS

The efficient computation of design derivatives with reasonable accuracy has been the subject of active research during the last two decades [1, 3, 4]. Since gradient based optimizers are certainly more efficient and accurate second derivative computations are impossible, accurate design sensitivities are important and the development of general purpose software for this purpose is essential. Here the alternatives available are critically analysed and some results of the author's computation are presented. These are

(i) Direct differencing procedure

(ii) Implicit Differentitation Algorithm with

a) Finite difference computation for $[B]_i'$ and $[J]_i'$ [15]

b) Finite difference computation for nodal coordinates only [1]

(iii) Variational Design Sensitivity Algorithm [13].

Broadly the characteristics of the different algorithms are presented in a tabular form below:

Algorithm	Accuracy	Ease of coding	Integrated with commercial codes	Computing time
DDA	Poor	Excellent	Excellent	High
IDA(a)	Good	Relatively Difficult	Difficult	Small
IDA(b)	Excellent	Difficult	Difficult	Small
VDSA (Domain)	Good	Okay	Possible	Small
VDSA(Boundary)	Poor	Okay	Possible	Small

It has been shown that the VDSA domain and boundary algorithms are mathematically identical to the IDA algorithm but only constitute different approaches. In the context of shape optimization, where shapes are altered in a small localized region as shown (Fig. 14) the auxiliary load vector in the nodes pertaining to these elements is complex and the auxiliary analysis for accurate stress derivatives requires better discretization.

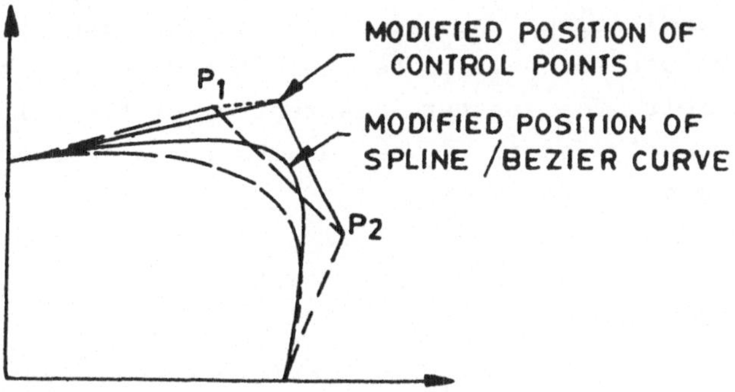

FIG.14 VARIATION OF SHAPE

This is impossible when a fixed mesh is utilized. Perhaps, the employment of an hybrid algorithm here may yield better results but further research is necessary.

In an program developed by the author the IDA algorithm with the use of finite difference for nodal coordinate derivatives is employed. To illustrate the point on the accuracy of derivatives, the following test example has been studied.

8.1 Test Example

Fillet in a Tension bar.

Fig. 15 below represents a simple flat tension bar having a transition fillet over a length of 10 cms. The semiwidths of the two ends are 10 cms and 5 cms. For the purpose of finite element discretization, lengths of 30 cms and 20 cms are considered along the wider and thinner portions of the bar respectively. The boundary conditions utilized are also shown. It is seen that the stress distribution at the ends represent a uniform σx condition justifying our choice of lengths for the wider and thinner regions. The transition profile has been modelled using cubic splines with three intervals giving rise to two control points whose 'x' and 'y'. Coordinates constitute the four design variables. The domain is modelled as a three patch assembly as described earlier and discretised using the semiautomatic mesh generator dividing each patch into a N x N mesh. The results of the stress derivative computations are ploted in Fig. 16. The corresponding results by the DDA are also shown indicating the accuracy of the algorithm. It is quite obvious that use of alternate meshes and increasing the number of design variabels for better descreption of the fillet etc., can be achieved extremely easily. The results for different discreliztions are also shown for comparison.

9. OPTIMIZATION ALGORITHM

While any powerful gradient based constrained optimization algorithm is acceptable, earlier investigations [16] have proved that

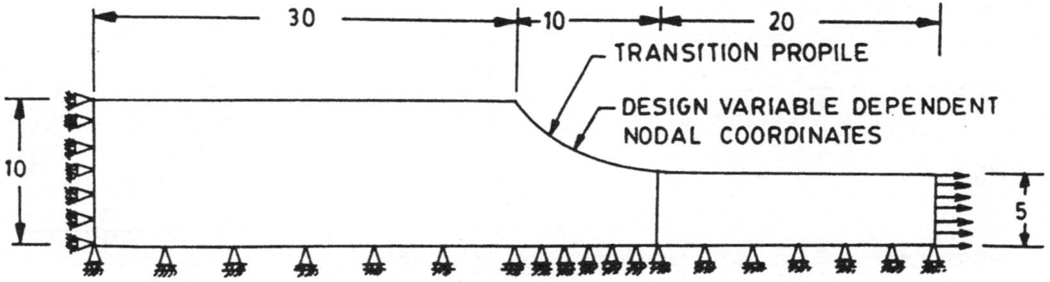

FIG. 15 FILLET IN A TENSION BAR

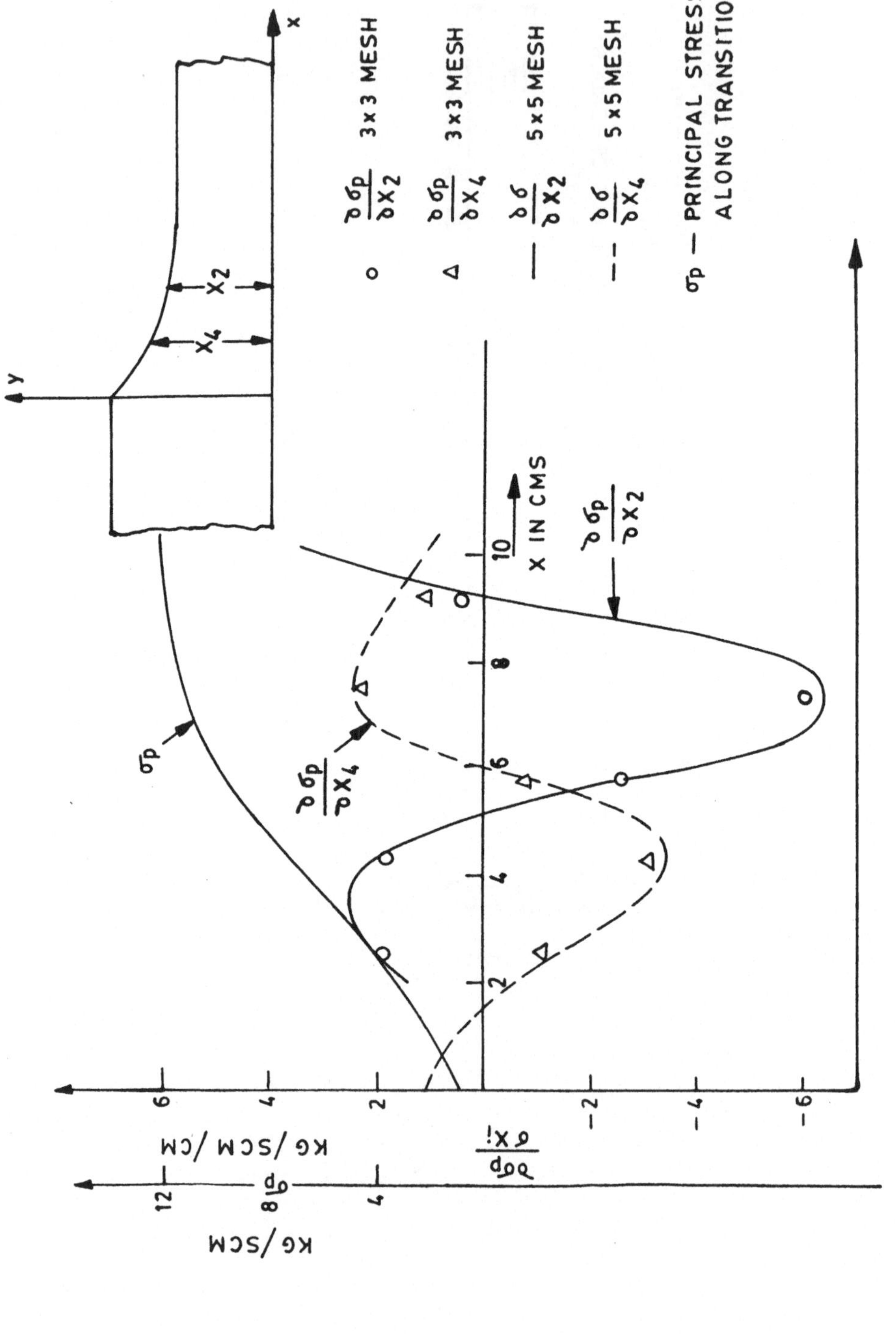

FIG. 16 STRESS DERIVATIVE COMPUTATIONS

the Improved Move Limit Method of Sequential Linear Programming is very general and the most suitble for shape optimization problems. This algorithm has been incorporated in the program. The algorithm has been tested on a series of bench mark problems for structural optimization and the method has been shown to be very competitive [17].

The entire program is being utilized to carryout detailed studies on various benchmark tests problems related to shape optimization. This will be published separately [18].

10. THREE DIMENSIONAL SHAPE OPTIMIZATION

While generalizing the program for 3-D shape optimization using the 3-D modeller described earlier, certain additional facilities have to be incorporated in the context of automated mesh generation. After developing the solid model on the screen, further model development as an assembleage of blocks is necessary. Each block is represented as a six surfaced solid. Determination of all these surfaces and their assembly without ambiguity is essential. After achieving this on the screen, the orthographic projections of the surface are obtained. By suitably inputting the positions of the corner points and control points through an interactive device all the necessary information can be supplied.

To illustrate the point, one segment of a cylinder-cylinder intersection is shown in Fig. 17. The solid has to be modelled as an assembly of three solid blocks. Two surfaces of intersection have to be clearly represented as shown and these surfaces have to be generated through suitable analytical procedures on the screen.

The procedure for inputting details for each block is described in Fig. 18 where one quarter of a nut is represented as an assemblege of three blocks. By first obtaining the orthographic projections and inputting details from these, the three solid blocks can be separately generated.

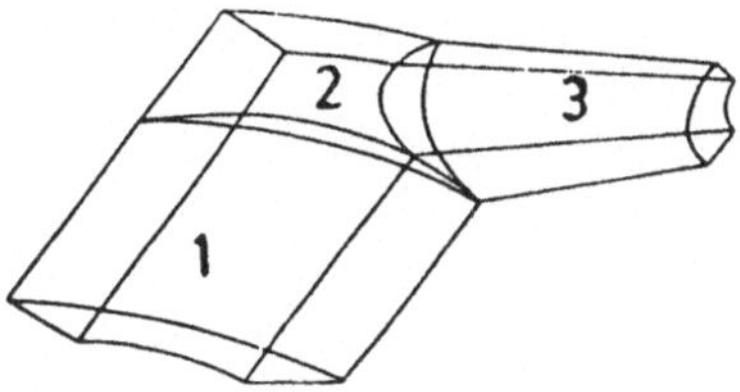

Fig. 17 Cylinder-nozzle intersection shown

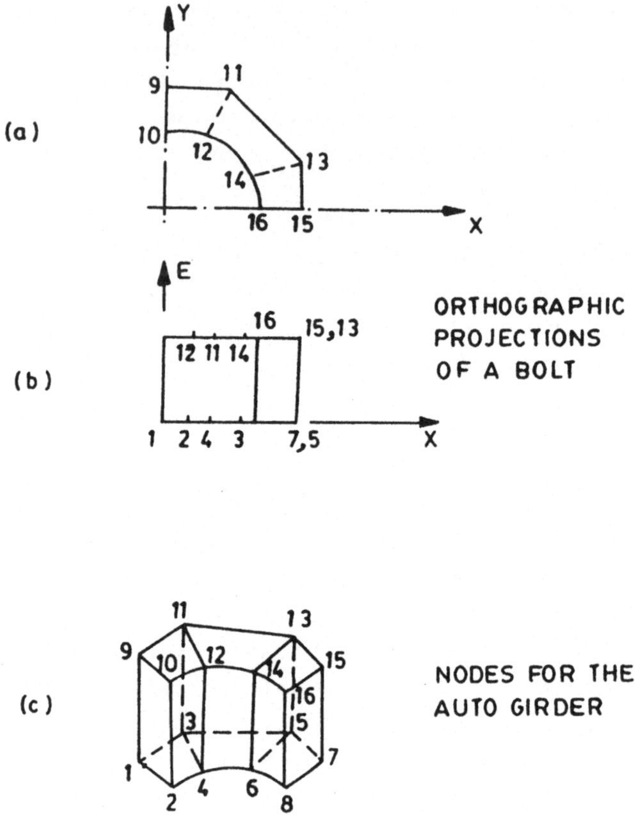

FIG.18 STEPS FOR INPUTTING INFORMATION TO THE MESH GENERATOR

ACKNOWLEDGEMENTS

Most of the software development and intergration of some of the existing software was carried out by the author during his stay at the Univ. College of Swansea, U.K. as a senior visitor under the Indo-British Collaborative Programme between IIT Delhi and UCS. The author gratefully acknowledges his useful discussions with K.V. John, of Delhi College of Engg. during the course of the implementation of this program at IIT Delhi.

REFERENCES

1. C.V. Ramakrishnan and A. Francavilla - "Structural Shape Optimization using Penalty Functions" J. Struct Mech. 3 (4), 403-422, 1975.
2. S.S. Bhavikatti and C.V. Ramakrishnan - "Optimum Shape Design of Shoulder Fillets in Tension Bars and T-Heads", Int. J. Mech. Sciences, Vol. 21, pp. 29-39, 1979.
3. S.S. Bhavikatti and C.V. Ramakrishnan - "Optimum Shape Design of Pressure Vessel and Nozzle Junction". Nuclear Engineering and Design, 1979.
4. E.J. Haug, K.K. Choi and V. Komkov - Design Sensitivity Analysis of Structural Systems. Academic Press, New York (1986).
5. O. Pironneau - Optimal Shape Design for Elliptical systems. Springer- Verlag (1984).
6. M.E. Botkin, R.J. Yang and J.A. Bennett - "Shape Optimization of Three Dimensional Stamped and Solid Automative Components"- in "The Optimum Shape" (Ed) J.A. Bennett and M.E. Botkin, Plenum Press, (1986).
7. K.K. Choi and H.G. Seong - "A Numerical Method for Shape Design Sensitivity Analysis and Optimization of Built-up Structures" - in "The Optimum Shape", Plenum Press, (1986).
8. M.S. Shephard and M.A. Yerry - Automatic Finite Element Modelling for use with Three Dimensional Shape Optimization". in "The Optimum Shape", Plenum Press (1986).
9. K. Chandran - 'Development of a Solid Modeller' - M.Tech Thesis, IIT Delhi (1987).
10. Sushil K. Gupta - "Development of Interactive Software for Automatic Mesh Generation", M.Tech. Thesis, IIT Delhi, Dec. 1987.

11. O.C. Zienkiewicz, A.W. Craig, J.Z. Zhu and R.H. Gallagher - "Adaptive Analysis Refinement and Shape Optimization-some New Possibilities" in "The Optimum Shape", Plenum Press, 1986.

12. S.W. Sloan and M.F. Randolph - Automatic Element Reordering for Finite Element Analysis with Frontal Solution Schemes" - Int. J. Num. Meth. for Engg., 19, pp.1153-1181 (1983).

13. E.J. Haug, and K.K. Choi - "Material Derivative Methods for Shape Design Sensitivity Analysis", in "The Optimum Shape", Plenum Press, 1986.

14. "Design Sensitivity Analysis" - NASA Symposium (1987).

15. G.A. Vanderplatts - in 'Design Sensitivity Analysis' NASA Symposium (1987).

16. C.V. Ramakrishnan and S.S. Bhavikatti - "Computational Efficiency of Improved Move Limit Method of Sequential Linear Programming for structural Optimization" - Computers and structures, 11, pp. 191-196, 1986.

17. K.V. John, C.V. Ramakrishnan and K.G. Sharma - Minimum Weight Design of Trusses using Improved Move Limit Method of Sequential Linear Programming' - Computers and Structures, Vol 27, 5, pp. 585-591 (1987).

.8. C.V. Ramakrishnan and K.V. John - Automated Optimum Shape Design of 2-D Elastic systems (Under Preparation).

Optimisation d'un système surfacique (équation de la chaleur)

B. ROUSSELET
Département de Mathématiques
Université de Nice
Parc Valrose - 06034 NICE Cedex

1. - INTRODUCTION

L'utilisation de coordonnées curvilignes est classique en mécanique des milieux continus ; l'utilisation de tenseurs permet d'écrire dans un tel système de coordonnées des formules explicites invariantes par changement de systèmes de coordonnées ; c'est-à-dire des formules pour les opérateurs différentiels et les intégrales qui ont un sens *intrinsèque*.

Après avoir rappelé brièvement les notations du calcul différentiel et intégral avec l'utilisation de coordonnées curvilignes, nous en déduisons très naturellement la dérivée d'intégrales sur Ω par rapport à des variations de Ω. Nous illustrons ceci en retrouvant la variation de l'équation de la chaleur par rapport à des variations de Ω (résultat maintenant classique : Céa [1986], Céa-Gioan-Michel [1974], Derms [1980], Masmoudi [1987], Murat-Simon [1976], Pironneau [1974], Rousselet [1982], Simon [1980]

Mais ce formalisme permet d'obtenir des formules de variation d'intégrales de surface par rapport à des variations de cette surface ; puis nous établissons une formule de *variation* de la solution de l'équation de la chaleur *posée sur cette surface* S .

Nous ne précisons pas les hypothèses de régularité pour lesquelles ces formules ont un sens ; cela pourrait être justifié à l'aide des techniques de Rousselet [1982].

Les présentations de théorie de coque par Valid [1977] et Destuynder [1985] sont à l'origine de ce travail.

2. - POSITION DES PROBLEMES

2.1 Le cas volumique

Soit ω un ouvert d'un espace vectoriel euclidien V_n de dimension n ; un difféomorphisme Φ avec un ouvert de référence Ω permet de définir un système de coordonnées curvilignes ; en tout point $x = \Phi(X)$ nous disposons d'une base locale :

(2.1) $$g_i = \frac{\partial \Phi}{\partial x} e_i$$ où e_i est une base orthonormée de l'espace euclidien V_n ;

nous désignons par g^{*i} la base duale : $< g^{*i}, g_j > = \delta^i_j$. Si v est un champ de vecteurs* :

---

* Nous utilisons systématiquement la convention des indices répétés : $v^i g_i = \sum_{i=1}^{3} v^i g_i$

$v = v^i g_i$

(2.2) on a $\frac{\partial v}{\partial x} = v^i{}_{|j}\, g_i \otimes g^{*j} \in E \otimes E^*$

(2.3) $v^i{}_{|j} = v^i{}_{|j} + \Gamma^i{}_{\ell j}\, v^l$ avec $\Gamma^i{}_{\ell j} = \langle g^{*i}, g_{\ell,j} \rangle$

Posons $g_{ij} = g_i \cdot g_j$ (première forme fondamentale) et soit g^{jk} le terme général de la matrice inverse; nous avons grad $v = g^{jk} v^i{}_{|j}\, g_i \otimes g_k \in E \otimes E$.

L'équation stationnaire de conduction de la *chaleur* dans le corps Ω s'écrit :

(2.4) $$\begin{cases} \operatorname{div} q = f \\ q = -c \operatorname{grad} u \end{cases} \quad \text{dans } \Omega$$

(2.5) $$\begin{cases} u = 0 & \text{sur } \Gamma_1 \\ q \cdot n = -g & \text{sur } \Gamma_2 \end{cases}$$

où f est la densité volumique de source de chaleur et q le flux de chaleur.

Classiquement ce système peut s'écrire variationnellement :

(2.6) $$\begin{cases} \forall v \in V \qquad a(u,v) = l(v) \\ \text{avec} \\ V = \{v \in H^1(\Omega) \mid v_{|\Gamma_1} = 0\} \\ a(u,v) = \int_\Omega c \frac{\partial v}{\partial x} \operatorname{grad} u \, d\Omega \\ l(v) = \int_\Omega f v \, d\Omega + \int_\Omega g v \, d\Omega . \end{cases}$$

2.2 Le cas surfacique.

Nous supposons maintenant que la surface S est localement paramétrée à l'aide d'un difféomorphisme Φ avec un ouvert de référence Ω ; nous disposons encore d'une base de l'espace tangent :

(2.7) $a_\alpha = \frac{\partial \Phi}{\partial \xi} e_\alpha$

Si $v_t = v^\alpha a_\alpha$ est un champ de vecteurs *tangents,* $\frac{\partial v_t}{\partial \xi_\alpha}$ n'est pas tangent ; toutefois

sa projection sur le plan tangent a un sens intrinsèque à la surface et l'on a :

(2.8) $$\Pi \frac{\partial v_t}{\partial m} = v^{\alpha}_{|\beta}\, a_{\alpha} \otimes a^{*\beta}$$

(2.9) où $$v^{\alpha}_{|\beta} = v^{\alpha}_{,\beta} + \Gamma^{\alpha}_{\lambda\beta}\, v^{\lambda} \quad \text{avec}$$

(2.10) $$\Gamma^{\lambda}_{\alpha\beta} = \langle a^{*\lambda}, a_{\alpha,\beta} \rangle$$

La *divergence* d'un champ de vecteurs tangents peut se définir par la formule :

(2.11) $$\int_S f \operatorname{div} v_t \, dS = - \int_S \frac{\partial f}{\partial m}\, v_t \, dS$$

pour tout f à *support compact* dans S ; cet opérateur différentiel peut s'exprimer en coordonnées curvilignes à l'aide de la formule ci-dessus en se rappelant que

(2.12) $$dS = \sqrt{a}\, d\xi \text{ avec } a = \det [a_{\alpha\beta}] \,,\, a_{\alpha\beta} = a_{\alpha} \cdot a_{\beta} \,.$$

On obtient :

(2.13) $$\operatorname{div} v_t = \frac{1}{\sqrt{a}} (v^{\alpha} \sqrt{a})_{,\alpha} = v^{\alpha}_{|\alpha} = g^{\alpha} . (\Pi \frac{\partial v_t}{\partial m} a_{\alpha}) \,.$$

En supposant que la conduction de la chaleur peut être bien approchée par une équation sur S, nous avons le système suivant :

(2.14) $$\begin{cases} \operatorname{div} q = f & \\ u = 0 & \text{sur } \Gamma_1 \\ q \,.\, n = -g & \text{sur } \Gamma_2 \\ q = -c \operatorname{grad} u & \end{cases} \qquad \text{en coordonnées} \qquad \begin{cases} q^{\alpha}_{|\alpha} = f \\ u = 0 \\ q^{\alpha} n_{\alpha} = -g \\ q^{\beta} = -c\, a^{\alpha\beta} u_{,\alpha} \end{cases}$$

n est normal à Γ_2 dans le plan *tangent* à S ; il est unitaire, sortant.

Comme dans le § 2.1, on montre encore que pour des données assez régulières (2.14) est équivalent au problème variationnel :

(2.15) $$\begin{cases} \forall v \in V \qquad a(u,v) = l(v) \quad \text{avec} \\ V = \{ v \in H^1(S) \mid \quad v|_{\Gamma_1} = 0 \} \\ a(u,v) = \int_S c \langle \frac{\partial v}{\partial m}, \operatorname{grad} u \rangle \, dS \\ l(v) = \int_S f\, v\, dS + \int_{\Gamma_2} g\, v\, d\sigma \end{cases}$$

avec les hypothèses habituelles sur a et l ; précisons que

(2.16) $$H^1(S) = \{ v \in L^2(S) \mid \operatorname{grad} v \in H_t \}$$

où

(2.17) $\quad H_t = \{v_t = v^\alpha a_\alpha \mid v^\alpha \in L^2(S)\}$.

3. - VARIATIONS D'INTEGRALES

Nous allons utiliser les notions rappelées au § 2 pour préciser la variation d'intégrales par rapport au domaine ω ou à la surface S. Ce genre de résultats est classique en mécanique des milieux continus (par exemple Germain [1979]) ; le cas volumique est très utilisé en optimisation de domaines (Céa [1986], Masmoudi [1987]); pour le cas surfacique et volumique, Simon [1980]. Le point de vue adopté ici permet une présentation *unifiée* et *naturelle*.

3.1 Le cas volumique

Comme dans le § 2.1 nous supposons disposer d'un difféomorphisme Φ avec un ouvert de référence Ω ; nous définissons les variations de ω à l'aide d'un champ de vecteurs ψ ; nous disposons aussi de $\tilde{\psi}(X) = \psi(\Phi(X))$; $\tilde{\psi}$ assez petit $(\Phi + \tilde{\psi})$ est un difféomorphisme de ω sur un nouveau domaine noté ω_ψ ; notons que ψ a un sens intrinsèque mais que $\tilde{\psi}$ dépend du paramétrage.

Définition 3.1

Soit f_ψ une fonction définie sur ω_ψ ; elle dépend explicitement de ψ et implicitement par la position du point où elle est évaluée ; nous considérons

$$\delta f = \lim_{t \to 0} \frac{f_{t\psi}(x + t\,\psi(x)) - f_0(x)}{t}$$

(dérivée particulaire de la mécanique des milieux continus).

Proposition 3.2

$$\delta \int_\omega f\, d\Omega = \int_\omega \delta f\, d\Omega + \int_\omega f \operatorname{div} \psi\, d\Omega$$

En effet $d\Omega = \sqrt{g}\, dX$ avec $g = \det[g_i \cdot g_j]$ et le résultat découle directement du lemme suivant.

Lemme 3.3

$$\delta \sqrt{g} = \sqrt{g} \operatorname{div} \psi$$

ce lemme découle, à l'aide de manipulations algébriques simples de $\delta g_i = \psi_i$.

3.2 Le cas surfacique.

La différence avec le § 3.1 réside dans l'utilisation d'un champ ψ *traverse* à la surface S. Nous rappelons d'abord quelques résultats sur le calcul différentiel avec les vecteurs transverses à une surface.

Le résultat de départ est la dérivée d'un vecteur normal unitaire ; on peut l'identifier à une forme linéaire sur l'espace tangent ; on note :

$\dfrac{\partial a_3}{\partial m} = - b^\alpha_\beta\, a_\alpha \otimes a^{*\beta}$ où (b^α_β) désigne le tenseur (mixte) de courbure ; ceci généralise le résultat analogue pour les courbes ; on peut alors montrer que

$$\frac{\partial a_\alpha}{\partial \xi^\beta} = \Gamma^\lambda_{\alpha\beta} a_\lambda + b_{\alpha\beta} a_3 \quad \text{et que}$$

$$\frac{\partial \psi}{\partial m} a_\beta = (\psi^\lambda_{|\beta} - b^\lambda_\beta \psi_3) a_\lambda + (\psi_{3,\lambda} + b_{\lambda\beta} \psi^\lambda) a_3 .$$

D'autre part on peut définir la *divergence tangentielle* d'un champ de vecteurs transverses par :

$$\mathrm{div}_S \psi = a^\beta . (\Pi \frac{\partial \psi}{\partial m} a_\beta) \equiv \psi^\beta_{|\beta} - b^\beta_\beta \psi_3 .$$

Remarquons que $\mathrm{div}_S a_3 = - b^\beta_\beta \equiv H$ la courbure moyenne et que si l'on décompose $\psi = \Pi\psi + \psi_3 a_3$ l'on a :

$$\mathrm{div}_S \psi = \mathrm{div}\, \Pi \psi + \psi_3 \mathrm{div}_S a_3 .$$

Enfin si ψ est définie au voisinage de S on a

$$\mathrm{div}_S \psi = \mathrm{div}\, \psi - a_3 . \frac{\partial \psi}{\partial n}$$

Pour la variation d'intégrales de surfaces, nous avons le résultat suivant.

Proposition 3.4

$$\delta \int_S f \, dS = \int_S \delta f \, dS + \int_S f \, \mathrm{div}_S \psi \, dS .$$

En effet, comme $dS = \sqrt{a}\, d\xi$ avec $a = \det[a_\alpha . a_\beta]$ la proposition découle directement du lemme suivant.

Lemme 3.5

$$\delta \sqrt{a} = \sqrt{a}\, \mathrm{div}_S \psi .$$

Démonstration du lemme.

Un calcul algébrique élémentaire montre que

$\frac{\partial a}{\partial a_{\alpha\beta}} = a\, a^{\alpha\beta}$ où $a_{\alpha\beta} = a_\alpha . a_\beta$ et $a^{\alpha\beta}$ est le terme général de la matrice inverse de $[a_{\alpha\beta}]$.

D'autre part, on a $\delta a_{\alpha\beta} = a_\alpha . \psi_{,\beta} + a_\beta . \psi_{,\alpha}$, comme

$\delta a = \frac{\partial a}{\partial a_{\alpha\beta}} \delta a_{\alpha\beta}$ nous obtenons $\delta a = a\, a^{\alpha\beta} (a_\alpha . \psi_{,\beta} + a_\beta . \psi_{,\alpha}) = 2a\, a^\alpha . \psi_{,\alpha}$ ou

$\delta a = 2a\, \mathrm{div}_S \psi$ d'où le lemme.

4. - VARIATION DE L'EQUATION DE LA CHALEUR

4.1 Le cas volumique

L'équation de la chaleur a été choisie pour sa simplicité mais la technique pourrait s'appliquer à d'autres situations comme par exemple les plaques étudiées avec une technique un peu différente dans Rousselet [1982]. Pour éviter toute difficulté liée à la régularité du problème mêlé, nous supposons que $\partial\Omega$ est constitué de deux composantes connexes Γ_1 et Γ_2 ; toutefois en utilisant la connaissance du comportement singulier (voir Grisvard [1985]), on peut traiter des exemples avec conditions aux limites mêlées (c.f. Delfour-Payre-Zolésio [1983]).

On suppose dans la suite $v_\psi(x + \psi(x)) = v_0(x)$ ou $\delta v = 0$, ce qui est compatible avec l'invariance de $H^1(\Omega)$ par changement de variables.

Proposition 4.1.

La dérivée δu de la solution u de l'équation (2.6) vérifie :

$$\forall\, v \in V \quad a(\delta u, v) = -\delta a(u,v) + \delta l(v)$$

avec

$$\delta a(u,v) = -\int_\omega c\left[< \frac{\partial u}{\partial x}, < \frac{d\psi}{\partial x}, \text{grad } v >> + < \frac{\partial v}{\partial x}, < \frac{\partial \psi}{\partial x}, \text{grad } u >> \right] d\omega + $$

$$+ \int_\omega c < \frac{\partial u}{\partial x}, \text{grad } v > \text{div } \psi \, d\omega$$

et

$$\delta l(v) = \int_{\Gamma_2} v\, \delta g \, d\sigma + \int_{\Gamma_2} v\, g \, \text{div}_{\Gamma_2} \psi \, d\sigma + \int_\omega v\, \delta f \, d\omega + \int_\omega f\, v \, \text{div } \psi \, d\omega .$$

La démonstration résulte immédiatement des deux lemmes suivants.

Lemme 4.2.

$$\delta g^{ij} = -(g^{ik} g^j + g^{ik} g^i) . \psi_{,k} .$$

Ce lemme résulte à l'aide de manipulations algébriques simples de l'identité $g^{ij} g_{jk} = \delta^i{}_k$.

Lemme 4.3.

Pour v qui satisfait $\delta v = 0$, nous avons :

$$\delta < \frac{\partial u}{\partial x}, \text{grad } v > = < \frac{\partial \delta u}{\partial x}, \text{grad } v > - < \frac{\partial u}{\partial x}, < \frac{\partial \psi}{\partial x}, \text{grad } v >> - < \frac{\partial v}{\partial x}, < \frac{\partial \psi}{\partial x}, \text{grad } u >>$$

Démonstration

$$< \frac{\partial u}{\partial x}, \text{grad } v > = g^{ij} u_{,j} v_{,i}$$

où

$$v_{,i} = \frac{\partial \tilde{v}}{\partial \xi_i} \qquad u_{,j} = \frac{\partial \tilde{u}}{\partial \xi_j}$$

avec

$$\tilde{u}(\xi) = u(\Phi(\xi) + \psi(\xi)) \qquad \tilde{v}(\xi) = v(\Phi(\xi) + \psi(\xi))$$

on a donc $\delta u_{,i} = (\delta u)_{,i}$ et $\delta v_{,i} = 0$.

D'autre part, le lemme 4.2 fournit :

$$(\delta g^{ij})\, u_{,j}\, v_{,i} = -(g^{ik} g^j + g^{jk} g^i) . \psi_{,k}\, u_{,j}\, v_{,i}$$

$$= -g^{ik}\, v_{,i}\, u_{,j}\, g^j . \psi_{,k} - g^{jk}\, u_{,j}\, v_{,i}\, g^i . \psi_{,k}$$

$$= -< \frac{\partial u}{\partial x}, < \frac{\partial \psi}{\partial x}, \text{grad } v >> - < \frac{\partial x}{\partial x}, < \frac{\partial \psi}{\partial x}, \text{grad } u >>$$

d'où ce lemme. ∎

Remarque : Dans la proposition ci-dessus, nous avons supposé l'existence de δu ; pour la justification de l'existence nous renvoyons à Rousselet [1982].

Dérivée d'une fonctionnelle

Pour fixer les idées, considérons

$$J(\omega, u) = \int_\omega \alpha(u)\, d\omega + \int_{\Gamma_2} \beta(u)\, d\sigma$$

Avec les lemmes 8.3 et 8.7, nous obtenons

$$\delta J = \int_\omega \alpha'(u)\, \delta u\, d\omega + \int_\omega \alpha(u) \operatorname{div} \psi\, d\omega + \int_{\Gamma_2} \beta'(u)\, \delta u\, d\sigma + \int_{\Gamma_2} \beta(u) \operatorname{div}_{\Gamma_2} \psi\, d\sigma .$$

Pour obtenir une formule plus utilisable, il est classique déliminer δu au profit d'une formule explicite en ψ .

Proposition 4.4

Soit

$$L(u, v) = J(u) + a(u, v) - l(v)$$

et soit p solution de l'équation d'état adjoint :

$$\forall\ w \in v \quad \frac{\partial L}{\partial u} w = 0$$

c'est-à-dire

$$\forall\ w \in v \quad \frac{\partial J}{\partial u} w + a(w, p) = 0$$

alors

$$\delta J = \delta L$$

où la dérivée de L est calculé avec u et p indépendants de la variation de ω ; c'est-à-dire :

$$\delta J = \delta a(u, p) - \delta l(p) + \int_\omega \alpha(u) \operatorname{div} \psi\, d\omega + \int_{\Gamma_2} \beta(u) \operatorname{div}_S \psi\, d\sigma$$

où δa et δl sont donnés dans la proposition 4.1.

Démonstration

Faisons $w = \delta u$ dans l'équation d'état adjoint :

$$a(\delta u, p) = -\int_{\omega} \alpha'(u)\, \delta u \, d\omega - \int_{\Gamma_2} \beta'(u)\, \delta u \, d\sigma$$

l'expression de δJ donnée ci-dessus peut donc s'écrire :

$$\delta J = -a(\delta u, p) + \int_{\omega} \alpha(u) \operatorname{div} \psi \, d\omega + \int_{\Gamma_2} \beta(u) \operatorname{div}_S \psi \, d\sigma$$

d'où avec l'équation satisfaite par δu dans la proposition 4.1 la formule indiquée dans la proposition ci-dessus. ∎

4.2. Le cas surfacique.

Comme dans le cas volumique on suppose que $v_\psi(m + \psi(m)) = v_0(m)$ ou $\delta v = 0$, ce qui est compatible avec l'invariance de $H^1(S)$ par changement de variable.

La proposition suivante est à rapprocher de la proposition ci-dessus relative au cas volumique.

Proposition 4.5.

La dérivée δu de la solution u de l'équation 2.15 par rapport à des variations de surface vérifie :

$$\forall \ v \in V \quad a(\delta u, v) = -\delta a(u, v) + \delta l(v)$$

avec

$$\delta a(u,v) = -\int_S c \left[< \frac{\partial u}{\partial m}, < \Pi \frac{d\psi}{\partial m}, \operatorname{grad} v >> + < \frac{\partial v}{\partial m}, < \Pi \frac{\partial \psi}{\partial m}, \operatorname{grad} u >> \right] dS$$

$$+ \int_S c < \frac{\partial v}{\partial m}, \operatorname{grad} u > \operatorname{div}_S \psi \, dS$$

$$\delta l(v) = \int_S v \, \delta f \, dS + \int_S f \, v \operatorname{div}_S \psi \, dS + \int_{\Gamma_2} v \, \delta g \, d\sigma + \int_{\Gamma_2} v \, g \operatorname{div}_{\Gamma_2} \psi \, d\sigma$$

La démonstration résulte comme dans le cas volumique de deux lemmes.

Lemme 4.6.

$$\delta a^{\alpha\beta} = -(a^{\alpha\nu} a^{\beta} + a^{\beta\nu} a^{\alpha}) \, . \, \psi_{,\nu}$$

Lemme 4.7.

Pour v qui satisfait $\delta v = 0$, nous avons

$$\delta < \frac{\partial u}{\partial m}, \operatorname{grad} v > = < \frac{\partial \delta u}{\partial m}, \operatorname{grad} v > - < \frac{\partial u}{\partial m}, \Pi \frac{\partial \psi}{\partial m} \operatorname{grad} v > - < \frac{\partial v}{\partial m}, \Pi \frac{\partial \psi}{\partial m} \operatorname{grad} u >$$

où $\Pi \frac{\partial \psi}{\partial m}$ est la dérivée *covariante* du champ de vecteurs ψ transverses à S.

Démonstration

Comme dans le cas volumique :

$$< \frac{\partial u}{\partial m}, \text{grad}\, v > \; = v_{,\alpha}\, u_{,\beta}\, a^{\alpha\beta}$$

et

$$\delta u_{,\alpha} = (\delta u)_{,\alpha} \qquad \delta v_{,\alpha} = 0 \,;$$

avec le lemme précédent, nous obtenons

$$v_{,\alpha}\, u_{,\beta}\, \delta a^{\alpha\beta} = -(a^{\alpha\nu} a^{\beta} + a^{\beta\nu} a^{\alpha}) \,.\, \psi_{,\nu}\, v_{,\alpha}\, u_{,\beta}$$

$$= -a^{\alpha\nu}\, v_{,\alpha}\, u_{,\beta}\, a^{\beta} \,.\, \psi_{,\nu} - a^{\beta\nu}\, u_{,\beta}\, v_{,\alpha}\, a^{\alpha} \,.\, \psi_{,\nu} \,.$$

Mais

$$\Pi \frac{\partial \psi}{\partial m} = \pi\, \psi_{,\nu}\, a^{*\nu}$$

$$\frac{\partial u}{\partial m} = u_{,\beta}\, a^{*\beta}$$

$$\text{grad}\, v = a^{\alpha\nu}\, v_{,\alpha}\, a_{\nu}$$

d'où le résultat annoncé.

Remarque.

Comme dans le cas volumique, la dérivabilité pourrait se démontrer à l'aide des résultats généraux de Rousselet [1982].

De même l'utilisation d'un état adjoint permet d'obtenir la dérivée d'une fonctionnelle exprimée de façon explicite par rapport à ψ .

CONCLUSION

Il ressort de cette étude que pour l'équation de la chaleur sur une surface la variation de la solution par rapport à cette surface peut s'exprimer à l'aide de formules intrinsèques très proches de celle du cas volumique ; toutefois leur expression dans un système de coordonnées curvilignes peut paraître bien plus compliquée, ce travail constitue un lien entre les techniques d'optimisation de domaines et les nouveaux développements pour l'optimisation d'arches et de coques (Chenais-Rousselet [1984], Chenais [1987], Chenais-Rousselet [1987], Bernadou-Palma-Rousselet [1988]).

BIBLIOGRAPHIE

BERNADOU, M.; PALMA F.; ROUSSELET, B. [1988] : Optimisation de forme d'une coque mince élastique sous différents critères, Rapport de recherche, INRIA.

CEA, J. [1986] : Conception optimale ou identification de formes : calcul rapide de la dérivée directionnelle de la fonction coût. Modélisations mathématiques et analyse numérique, Vol. 20, n° 3, pp. 371-402.

CEA, J.; GIOAN, A.; MICHEL, J. : [1974] Adaptation de la méthode du gradient à un problème d'identification de domaine. Computing methods in applied science and engineering. Springer-Verlag, Berlin.

CHENAIS, D. [1987] : Optimal design of midsurface of shells : differentiability proof and sensitivity computation. Appl. Math. Optim., vol. 16, pp. 93-133.

CHENAIS, D.; ROUSSELET, B. [1984] : Différentiation du champ de déplacements dans une arche par rapport à la forme de la surface moyenne en élasticité linéaire. C.R. Acad. Sci. Paris, 298, pp.533-536.

CHENAIS, D.; ROUSSELET, B. [1988] : Dependence of the buckling load of a shallow arch with respect to the shape of its midcurve. Soumis à Computer methods in applied Mech. and Eng.

DELFOUR, M.; PAYRE, G. ; ZOLESIO, J.P. [1983] : Optimal design of a minimum weight thermal diffuser with constraint on the output thermal power flux. Appl. Math. Optim., vol. 9, pp. 225-262.

DEMS, K. [1980] : Multiparameter shape optimization of elastic bars in torsion. Int. J. num. methods eng., vol. 15, pp. 1517-1539.

DESTUYNDER, P. ; [1985] : A classification of thin shell theories, Acta Applic. Mathem., vol. 4, pp. 15-63.

GERMAIN, P. [1973] : Cours de mécanique des milieux continus, Paris, Masson.

GRISVARD, P. [1985] : Elliptic problems in non smooth domains, London, Pitman.

MASMOUDI, M. [1987] : Outils pour la conception optimale de formes, thèse d'Etat, Université de Nice.

MURAT, F. ; SIMON, J. [1976] : Sur le contrôle par un domaine géométrique. Publication du L.A. 189, Université de Paris VI.

PIRONNEAU, O. [1984] : Optimal shape design for elliptic systems. Springer series in computational physics, Springer-Verlag.

ROUSSELET, B. [1982] : Quelques résultats en optimisation de domaines, thèse d'Etat, Université de Nice.

SIMON, J. [1980] : Variation par rapport au domaine dans des problèmes aux limites. Publication du L.A.189, Université de Paris VI.

VALID, R. [1981] : Mechanics of continuous media and analysis of structures, North-Holland.

OUTPUT LEAST SQUARE ESTIMATION OF POWER-LAWS IN DISTRIBUTED DAMPING MODELS: BEAMS

L.W. White
University of Oklahoma
Norman, Oklahoma 73019

1. PRELIMINARIES.

Let $\Omega = (0,1)$ and let $Q = \Omega \times (0,T)$. We consider the problem:

Find $u \in L^2(0,T;H_0^2)$ such that

$$u_{tt} + b|u_t|^\rho u_t + Au = f \text{ in } \Omega \tag{1.1)(i}$$

$$u(0) = 0 \tag{1.1)(ii}$$

$$u_t(0) = 0 \tag{1.1)(iii}$$

where we have suppressed the dependence on Ω so that $H^r = H^r(\Omega)$. The operator A is given formally as

$$A\varphi = (a\, \varphi_{xx})_{xx}$$

where we assume that

$$a \in L^\infty(\Omega) \text{ and } a(x) \geqq \nu > 0$$

for a constant ν.

Models with such damping terms are studied in the engineering literature in connection with flexible structures in which damping is small compared to contributions of inertial and stiffness elements [7]. Here we consider distributed problems where damping parameters b and ρ may depend on the spacial variable x and seek to apply the methods studied in [3,10] in this context.

The operator A is associated with a bilinear form on H_0^2

$$a(\varphi,\psi) = \int_\Omega a(x)\, \varphi_{xx}(x)\, \psi_{xx}(x)\, dx$$

for φ and ψ in H_0^2 such that

$$a(\varphi,\varphi) \geqq \nu_A \|\varphi\|_{H^2}^2 \tag{1.2)(i}$$

and

$$|a(\varphi,\psi)| \leqq C_A \|\varphi\|_{H^2} \|\psi\|_{H^2} \tag{1.2)(ii}$$

for positive constants ν_A and C_A.

The function f in (1.1)(i) is assumed to satisfy

$$f \in L^2(0,T;L^2) \text{ and } f_t \in L^2(0,T;L^2) \tag{1.3}$$

The parameters that we wish to identify are the coefficient b and the exponent ρ. We make the assumptions at the outset that

$$b \text{ and } \rho \in H^1 \text{ and } b \geqq 0 \text{ and } \rho \geqq 0. \tag{1.4}$$

Our aim is to estimate the functions b and ρ from observations of the displacement u. In this note we obtain regularity properties of optimal estimators obtained from the regularized output least squares method.

The existence of a unique solution to (1.1) for constant b and ρ is proved in [4] in the similar case of second order A. We provide a sketch of the proof since our case is somewhat different. Moreover, the estimates of the solutions of the Galerkin equations arising the proof are useful in the sequel.

Proposition 1.1. Let (1.2)-(1.4) hold. Then for each b and ρ satisfying (1.4) there exists a unique solution $u = u(\rho, b)$ to equation (1.1). Furthermore, there exists a function K of b and ρ that takes bounded sets in $H^1 \times H^1$ into bounded sets in $\mathbb{R}^+ \cup \{0\}$ such that $u(\rho, b)$ satisfies the estimates

$$\|u(\rho,b)\|_{L^\infty(0,T;H_0^2)} \leqq K$$
$$\|u_t(\rho,b)\|_{L^\infty(0,T;H_0^2)} \leqq K$$
$$\|u_{tt}(\rho,b)\|_{L^\infty(0,T;L^2)} \leqq K.$$

Proof. Let $\{\varphi_i : i = 1, \cdots, N\}$ be a linearly independent set of functions in H^2 that span a dense subspace of L^2 for example constructed from cubic B-splines cf. [8]. Denote by $S^N = span\{\varphi_i : i = 1, \cdots, N\}$ and we assume that $\bigcup_N S^N$ is dense in H^2. Set

$$u^N(x,t) = \sum_{k=1}^{N} c_k(t)\varphi_k(x)$$

Consider the system of equations for $j = 1, \cdots, N$

$$\begin{aligned} (u_{tt}^N, \varphi_j) + \int_\Omega b|u_t^N|^\rho u_t^N \varphi_j \, dx + a(u^N, \varphi_j) &= (f, \varphi_j) \\ u^N(0) &= 0 \\ u_t^N(0) &= 0 \end{aligned} \tag{1.5}$$

That there exists a solution $\underline{c}(t) = (c_1(t), \cdots, c_N(t))$ to the initial value problem (1.5) on $[0,T]$ follows from standard results from ordinary differential equations and the monotonicity of the damping term.

By arguments similar to those presented in [4], we have the following estimates.

$$\|u_t^N(t)\|_{L^2}^2 + \|u^N(t)\|_{H^2}^2 \leqq C_1 \tag{1.6)(i}$$
$$\|u_{tt}^N(t)\|_{L^2}^2 + \|u_t^N(t)\|_{H^2}^2 \leqq C_1 \tag{1.6)(ii}$$

for $t \in [0,T]$ where C_1 is a bounded function of b and ρ from $H^1 \times H^1$ into $\mathbb{R}^+ \cup \{0\}$. We consider only (1.6)(ii) since (1.6)(i) is straight forward. To obtain (1.6)(ii) we must observe the following concerning $u_{tt}(0)$:

$$\|u_{tt}^N(0)\|_{L^2}^2 \leqq (f(0), u_{tt}^N(0))$$

Since for any $\varphi \in H^1$ the inequality $\|\varphi\|_{L^\infty} \leqq C\|\varphi\|_{H^1}$ holds [1], where C depends on Ω, we obtain

$$\|u_{tt}^N(0)\|_{L^2} \leqq C_2\|f\|_{H^1(0,T;L^2)}$$

which is bounded independently of N. Using this inequality we may obtain inequality (1.6)(ii)

$$(u_{ttt}^N, u_{tt}^N) + \int_\Omega b(\rho+1)|u_t^N|^\rho (u_{tt}^N)^2 dx + a(u_t^N, u_{tt}^N) = \\ = (f_t, u_{tt}^N)$$

From this equation we obtain

$$\tfrac{1}{2}\tfrac{d}{dt}[\|u_{tt}^N\|_{L^2}^2 + a(u_t^N, u_t^N)] \leqq (f, u_{tt}^N)$$

and (1.6)(ii) follows.

From (1.6) it follows that there is a subsequence u^λ with the property that

$$u^\lambda \to u \ w^* \text{ in } L^\infty(0,T;H^2)$$
$$u_t^\lambda \to u_t \ w^* \text{ in } L^\infty(0,T;H^2)$$
$$u_{tt}^\lambda \to u_{tt} \ w^* \text{ in } L^\infty(0,T;H^2)$$

The subsequence may be chosen such that

$$u_t^\lambda \to u_t \text{ in } L^2(0,T;H^1) \text{ and almost everywhere in } Q. \tag{1.7}$$

Also, from (1.6) the following holds

$$|u_t^N(x,t)| \leqq \int_0^x |u_{tx}^N(\xi,t)|d\xi \\ \leqq C\|u_t^N\|_{L^\infty(0,T;H^1)}$$

Accordingly, from the Lebesgue dominated convergence theorem, it follows that

$$|u_t^\lambda|^\rho u_t^\lambda \to |u_t|^\rho u_t \text{ in } L^2(Q).$$

Hence, we see that

$$\int_\Omega |u_t^\lambda|^\rho u_t^\lambda \ \varphi_j \ dx \to \int_\Omega |u_t|^\rho u_t \ \varphi_j \ dx \text{ in } L^2(0,T)$$

for each φ_j. The remainder of the proof is analogous to that in [4].

From a study of the proof of Proposition 1.1 we may show the next result.

Corollary 1.1. Let (1.2)-(1.4). Suppose $b_N \to b$ and $\rho_N \to \rho$ weakly in H^1. Then

$$u^N(\rho_N, b_N) \to u(\rho,b)w^* \text{ in } L^\infty(0,T;H^2)$$
$$u_t^N(\rho_N, b_N) \to u_t(\rho,b)w^* \text{ in } L^\infty(0,T;H^2)$$
$$u_{tt}^N(\rho_N, b_N) \to u_{tt}(\rho,b)w^* \text{ in } L^\infty(0,T;L^2).$$

2. DIFFERENTIABILITY PROPERTIES.

We now consider the differentiability with respect to b and ρ. First we specify some assumptions that essentially deal with the identifiability of b and ρ. Define the subset $\mathcal{U}$ of Q by

$$\mathcal{U} = \{(x,t) : u_t(x,t) = 0\} \text{ with } \mathcal{V} = Q \backslash \mathcal{U},$$

(A) $m(\mathcal{U}) = 0$ where m is the Lebesgue product measure on Q.

Remark 2.1. As mentioned the assumption that $u_t \neq 0$ almost everywhere amounts to an assumption that we can observe the influence of b and ρ over the entire set Q. The set $\mathcal{U}$ is measurable. In general little can be said to relate a condition that $m(\mathcal{U}) > 0$ to a condition on f. However, if $\mathcal{V}$ is open, we may obtain the following.

Proposition 2.1. If $\mathcal{V}$ is open and not dense in Q, then there exist functions c_1 and c_2 of t such that

$$\int_\alpha^x (x-\xi) f_t(\xi,t)\, d\xi = c_1(t) + c_2(t)(x-\alpha) \tag{2.1}$$

almost everywhere in a rectangle R contained in $\mathcal{U}$.

Proof. If the open set $\mathcal{V}$ is not dense in Q, then there exists a rectangle $R = (\alpha,\beta) \times (t_1,t_2)$ contained in the closed set $\mathcal{U}$. Thus, $u_t = 0$ in an open rectangle R. Let $\varphi \in C_0^1(\Omega)$ with $\text{supp}(\varphi) \subset (\alpha,\beta)$ and let $\nu \in C_0^1(t_1,t_2)$. In this case we see that

$$-\int_{t_1}^{t_2} (u_t,\varphi)\nu_t\, dt + \int_{t_1}^{t_2} \int_\Omega b|u_t|^\rho u_t\, \varphi\, dx(\nu(t)dt +$$
$$+ \int_{t_1}^{t_2} a(u,\varphi)\nu(t)dt = \int_{t_1}^{t_2} (f,\varphi)\nu(t)dt.$$

From our assumption it follows that indeed

$$\int_{t_1}^{t_2} a(u,\varphi)\nu(t)dt = \int_{t_1}^{t_2} (f,\varphi)\nu(t)dt.$$

for all such ν. We conclude that in fact

$$a(u,\varphi) = (f,\varphi) \text{ almost everywhere in } (t_1,t_2), [9].$$

This is equivalent to (2.1).

Remark 2.2. Consider a forcing function f

$$f(x,t) = \sin(\omega t).$$

In this case for a rectangle R we see that

$$\omega \cos(\omega t) \int_\alpha^x (x-\alpha)dx = \omega \cos(\omega t)(x^2 - \alpha\, x)$$

Due to the x^2 term, this integral cannot satisfy (2.1)

We consider the Galerkin approximating systems of ordinary differential equations (1.6). In this case we may directly apply the results in [2] to obtain the derivatives of u^N with respect to $b > 0$ and $\rho > 0$. Accordingly, we find for $D_b u^N(h) = w^N$

$$(w_{tt}^N,\varphi_j) + \int_\Omega (\rho+1)b|u_t^N|^\rho \, w_t^N \, \varphi_j \, dx + a(w^N,\varphi_j) = \\ = -\int_\Omega h \, u_t^N |u_t^N|^\rho \, \varphi_j \, dx \\ w^N(0) = w_t^N(0) = 0 \tag{2.2}$$

for $D_\rho u^N(\eta) = v^N$ with $\rho > 0$

$$(v_{tt}^N,\varphi_j) + \int_\Omega (\rho+1)b|u_t^N|^\rho \, v_t^N \, \varphi_j \, dx + a(v^N,\varphi_j) = \\ = -\int_\Omega \eta \, b \, u_t^N |u_t^N|^\rho \, \ln|u_t^N| \varphi_j \, dx \\ v^N(0) = v_t^N(0) = 0 \tag{2.3}$$

Remark 2.3. From Remark 2.1 it follows that for a given $\varepsilon > 0$ then for sufficiently large N the set $\{(x,t) : u^N(x,t) = 0\}$ has measure less than ε. Hence $\ln|u_t^N|$ is infinite on a set of at most measure ε. Moreover, since u_t^N is a continuous function on Q and $\rho > 0$, we consider the product $u_t^N|u_t^N|^\rho \ln|u_t^N|$ to be zero on this set, and obtain a continuous function on Q. Under assumption (A) the product $u_t|u_t|^\rho \ln|u_t|$ is zero only on the set $\mathcal{U}$ of measure zero.

Lemma 2.1. Let $\psi \in H_0^1(\Omega)$ be such that $\psi(x) \geqq 0$ on Ω and $\{x \in \Omega : \psi(x) = 0\}$ is of measure zero. Then

$$\psi(x)|\ln(\psi(x))| \leqq 1 + \psi(x)^2$$

almost everywhere in Ω.

Theorem 2.1. Let (1.2)-(1.4) and (A) hold. Then

$$v^N \to v \ w^* \text{ in } L^\infty(0,T;H^2) \\ v_t^N \to v_t \ w^* \text{ in } L^\infty(0,T;H^2)$$

and

$$w^N \to w \ w^* \text{ in } L^\infty(0,T;H^2) \\ w_t^N \to w_t \ w^* \text{ in } L^\infty(0,T;H^2)$$

where

$$v_{tt} + (\rho+1)b|u_t|^\rho v_t + Av = -b\,\eta|u_t|^\rho u_t \ \ln|u_t| \\ v(0) = v_t(0) = 0 \tag{2.4}$$

and

$$w_{tt} + (\rho+1)b|u_t|^\rho w_t + Aw = -h|u_t|^\rho u_t \\ w(0) = w_t(0) = 0 \tag{2.5}$$

Proof. The result is based on estimate of the solutions of equations (2.2) and (2.3). We consider only (2.4) as (2.5) is a simpler case.

The first estimate for (2.4) arises from

$$\tfrac{1}{2}\tfrac{d}{dt}[\|v_t^N\|_{L^2}^2 + a(v^N, v^N)] + \int_\Omega b(\rho+1)|u_t^N|^\rho (v_t^N)^2\, dx = \\ = -\int_\Omega \eta\, b\, u_t^N |u_t^N|^\rho \ln|u_t^N|\; v_t^N\, dx$$

It follows that

$$\tfrac{1}{2}\tfrac{d}{dt}[v_t^N\|_{L^2}^2 + a(v^N, v^N)] + \int_\Omega b(\rho+1)|u_t^N|^\rho (v_t^N)^2\, dx \leqq \\ \leqq \|\eta\|_{H^1}\|b\|_{H^1}\int_\Omega |u_t^N|^\rho (1+|u_t^N|)|v_t^N|dx$$

and this implies

$$\|v_t^N(t)\|_{L^2}^2 + \|v^N(t)\|_{H^2}^2 \leqq C_4. \tag{2.6}$$

Now from the inequality (2.6) it follows that there is a subsequence λ such that

$$v^\lambda \to v\ w^* \text{ in } L^\infty(0,T;H^2)$$
$$v_t^\lambda \to v_t\ w^* \text{ in } L^\infty(0,T;L^2)$$

and from (1.7)

$$|u_t^\lambda|^\rho \to |u_t|^\rho \text{ in } L^2(Q).$$

where v is the solution of (2.4), cf. [5].

In an analogous manner the convergence of w^N to the solution w of (2.5) may be proved. Since the solutions of (2.4) and (2.5) are unique, it follows from subsequent arguments that the entire sequence converges.

The proof of the following is analogous to those discussed above.

Corollary 2.1. Let $\rho^N \to \rho$ and $b^N \to b$ weakly in $H^1(\Omega)$. Then

$$v^N(\rho^N, b^N) \to v(\rho,b)w^* \text{ in } L^\infty(0,T;H^2)$$
$$v_t^N(\rho^N, b^N) \to v_t(\rho,b)w^* \text{ in } L^\infty(0,T;L^2)$$

and

$$w^N(\rho^N, b^N) \to w(\rho,b)w^* \text{ in } L^\infty(0,T;H^2)$$
$$w_t^N(\rho^N, b^N) \to w_t(\rho,b)w^* \text{ in } L^\infty(0,T;L^2)$$

3. REGULARITY OF APPROACHABLE SOLUTIONS.

At this point we introduce the estimate problem. Suppose that data are given in the form of a function $z = z(x,t)$ in $L^2(Q)$. Here we view z as an interpolation of point measurements of displacements to obtain a function z. The output least square estimation problem is now formulated as an optimization problem.

($\mathcal{P}$) Find $a = (\rho, b) \in \mathcal{A}$ such that $J(a) = \{J(a) : a \in \mathcal{A}\}$ where the fit-to-data function J is given by

$$J(a) = \int_0^T \|u(t;a) - z(t)\|_{L^2}^2 dt + \beta_1\|\rho\|_{H^1}^2 + \beta_2\|b\|_{H^1}^2$$

and where the admissible set $\mathcal{A}$ is given by

$$\mathcal{A} = \{a \in H^1 \times H^1 : \rho \geqq \nu_\rho > 0 \text{ and } b \geqq \nu_b > 0\}$$

for constants ν_ρ and ν_b.

Proposition 3.1. There exists a solution $\underline{a} = (\underline{\rho}, \underline{b})$ to $\mathcal{P}$.

Proof. The proof follows from the continuity properties established in Section 1.

Also of interest are the approximating problems in which the systems (1.5) form the underlying equations.

$(\mathcal{P}^N)$ Find $a \in \mathcal{A}$ such that $J^N(a) = \{J^N(a) : a \in \mathcal{A}\}$

where the fit-to-data function J^N is given by

$$J^N(a) = \int_0^T \|u^N(t;a) - z(t)\|^2_{L^2} dt + \beta_1 \|\rho\|^2_{H^1} + \beta_2 \|b\|^2_{H^1}$$

Proposition 3.2. There exists a solution $\underline{a}^N = (\underline{\rho}^N, \underline{b}^N)$ to problem $(\mathcal{P}^N)$.

Remark 3.1. We observe that the set $\{\underline{a}^N\}_{N=1}^{\infty}$ is bounded by $H^1 \times H^1$. Hence, $\{a^N\}_{N=1}^{\infty}$ has weak cluster points in H^1. Since $\mathcal{A}$ is weakly closed, these cluster points belong to $\mathcal{A}$.

From Corollary 1.1 we have the following.

Proposition 3.3. Every weak H^1 limit point of the sequence $\{\underline{a}^N : N = 1, \cdots, \infty\}$ is a solution of $(\mathcal{P})$.

We make the following definition.

Definition 3.1. A solution $\underline{a}$ of $(\mathcal{P})$ is called approachable if there exists a sequence of solution $\underline{a}^N$ of $(\mathcal{P}^N)$ such that $\underline{a}^N \to \underline{a}$ weakly in $H^1 \times H^1$.

Remark 3.2. Denote the set of approachable elements by $\underline{\mathcal{A}}$ and note that Proposition 3.3 implies that $\underline{\mathcal{A}} \neq \phi$.

We now obtain regularity results of approachable solutions. Our approach is to apply the Kuhn-Tucker Theorem to the problems $(\mathcal{P}^N)$ to obtain a sequence of Lagrange multipliers λ^N with corresponding stationary point and complementarity conditions. We then consider the limit of these conditions as N approaches infinity. For simplicity we will prove our results for ρ and assume that b is fixed.

The first step in this program is to verify that the solutions $\underline{\rho}^N$ satisfy a regular point condition for the constraints imposed on the admissible parameters in $\mathcal{A}$, [6]. To this end define a function G that is a mapping of H^1 into H^α, for $\alpha \in (1/2, 1]$ by $G(\rho) = \nu_\rho - \rho$. Thus, the set of admissible parameters is given as $\mathcal{A} = \{\rho : G(\rho) \leqq 0\}$ where 0 is the zero element of H^α. Note that H^α embeds compactly into $C^0(\overline{\Omega})$ for α belonging to $(1/2, 1]$. Hence, H^α contains a positive cone with a nonempty interior.

Definition 3.2. We say an element a satisfies the regular point condition for the constraint $G(a)\leqq 0$ if there exist an element η in H^1 such that

$$G(\rho) + DG(\rho)(\eta) < 0.$$

That this condition holds for any admissible ρ is easy to verify. The Frechet derivative of G at a point ρ in the direction of η is given by

$$DG(\rho)(\eta) = -\eta.$$

Therefore, we have for $\rho \geqq \nu_\rho$

$$G(\rho) + DG(\rho)(\eta) = \nu_\rho - \rho - \eta$$

which is certainly strictly less than zero for any $\eta > 0$.

The Lagrangian functional is defined as follows for $\lambda \in (H^\alpha)^*$

$$L^N(\rho, \lambda) = J^N(\rho) + \langle \lambda, G(\rho) \rangle.$$

The Kuhn-Tucker Theorem [6] asserts that there is an element λ_N^* such that $\lambda_N^* \geqq 0$ in the sense that $\langle \lambda_N^*, \varphi \rangle \geqq 0$ for every nonnegative $\varphi \in H^\alpha$ such that

$$DL^N(\underline{\rho}^N, \lambda_N^*)(\eta) = DJ^N(\underline{\rho}^N)(\eta) + \langle \lambda_N^*, DG(\underline{\rho}^N)(\eta) \rangle = 0 \tag{3.1}$$

for all $\eta \in H^1$ and

$$\langle \lambda_N^*, G(\underline{\rho}^N) \rangle = 0. \tag{3.2}$$

Equations (3.1) and (3.2) may be rewritten as

$$D_\rho J^N(\underline{\rho}^N)(\eta) - \langle \lambda_N^*, \eta \rangle = 0 \tag{3.3}$$

and

$$\langle \lambda_N^*, \nu_\rho - \underline{\rho}^N \rangle = 0 \tag{3.4}$$

Let us now consider (3.3). Thus, we have

$$2(u^N(\underline{\rho}^N) - z,\ v^N(\underline{\rho}^N))_{L^2} + 2\beta(\underline{\rho}^N, \eta)_{H^1} - \langle \lambda_N^*, \eta \rangle = 0. \tag{3.5}$$

for all $\eta \in H^1$ where $v^N(\underline{\rho}^N)$ is the solution of equation (2.3). Let us now introduce an "adjoint" equation of (2.3)

$$\begin{aligned} (p_{tt}^N, \varphi_j) - \textstyle\int_\Omega (\underline{\rho}^N + 1) b[|u_t^N|^{\underline{\rho}^N} p^N]_t\, \varphi_j\, dx + a(p^N, \varphi_j) &= \\ = (u^N(\underline{\rho}^N) - z, \varphi_j)& \\ p^N(T) = p_t^N(T) = 0.& \end{aligned} \tag{3.6}$$

In equation (3.6) replacing φ_j by the solution $v^N(\underline{\rho}^N)$ of equation (1.12) and integrating by parts we obtain

$$\begin{aligned} &(u^N(\underline{\rho}^N) - z, v^N(\underline{\rho}^N))_{L^2(0,T;L^2(\Omega))} = \\ &= -\textstyle\int_\Omega \eta\ b(\int_0^T u_t^N |u_t^N|^{\underline{\rho}^N} \ln |u_t^N| p_t^N\ dt) dx. \end{aligned} \tag{3.7}$$

Now substituting equation (3.7) into (3.5) we obtain the following

$$-2\int_\Omega (b\int_0^T u_t^N|u_t^N|^{\underline{\rho}^N} \ln|u_t^N| p_t^N\, dt)\eta\, dx +$$
$$+2\beta(\underline{\rho}^N,\eta)_{H^1} - \langle\lambda_N^*,\eta\rangle = 0 \tag{3.8}$$

for all $\eta \in H^1$. From (3.8) we have the following result.

Lemma 3.1. The set of Lagrange multipliers $\{\lambda_N^*\}$ associated with the constraint $G(\rho)\leqq 0$ is bounded in $(H^1)^*$. That is,

$$\|\lambda^*\|_{(H^1)^*} \leqq C(\tilde{\rho}, f, z)$$

where $C(\tilde{\rho}, f, z)$ is a function depending only on the upper bound $\tilde{\rho}$ of the sequence $\underline{\rho}^N$. Since $(H^1)^*$ is reflexive, it follows that there exists a subsequence λ_μ^* such that $\lambda_\mu^* \to \lambda^*$ weakly in $(H^1)^*$.

By arguments similar to those above and those given in [5], we may determine the following.

Proposition 3.4. Let $\underline{\rho}$ be an approachable solution of $(\mathcal{P})$ with $\underline{\rho}^N \to \underline{\rho}$ weakly in $H^1(\Omega)$. There exists a unique solution to

$$(p_{tt},\varphi) - \int_\Omega (\underline{\rho}+1)\underline{b}[|u_t|^{\underline{\rho}} p]_t\, \varphi\, dx + a(p,\varphi) =$$
$$= (u(\underline{\rho},\underline{b}) - z, \varphi) \tag{3.8}$$
$$p(T) = p_t(T) = 0.$$

such that

$$p \in L^\infty(0,T;H^2(\Omega))$$
$$p_t \in L^\infty(0,T;L^2(\Omega))$$

and the solutions of (3.6) have the property that

$$p^N \to p\ w^* \text{ in } L^\infty(0,T;H_0^2)$$
$$p_t^N \to p_t\ w^* \text{ in } L^\infty(0,T;L^2).$$

In the limit equation (3.7) yields the identity

$$(u(\underline{\rho}) - z,\ v(\underline{\rho}))_{L^2(0,T;L^2(\Omega))} =$$
$$= -\int_\Omega \eta\, b(\int_0^T u_t|u_t|^{\underline{\rho}} \ln|u_t| p_t\, dt) dx \tag{3.9}$$

for all $\eta \in H^1(\Omega)$. Accordingly from (3.8) and (3.9), it follows that in the limit, we have

$$-2\int_\Omega (b\int_0^T u_t|u_t|^{\underline{\rho}} \ln|u_t| p_t\, dt)\eta\, dx +$$
$$+2\beta_1(\underline{\rho},\eta)_{H^1} - \langle\lambda^*,\eta\rangle = 0 \tag{3.10}$$

where $\underline{\rho}^N \to \underline{\rho}$ weakly in H^1.

Now recall that weak convergence in H^1 implies at least subsequential convergence in $C^0(\overline{\Omega})$, [1]. If an approachable solution $\underline{\rho}$ of $(\mathcal{P})$ is strictly greater than ν_ρ at a point $\tilde{x}$ in Ω, then there is an $\varepsilon > 0$ and an interval (x_1, x_2) in Ω containing x such that $\underline{\rho}(x) \geqq \nu_\rho + \varepsilon$ in (x_1, x_2). Since there is a subsequence $\underline{\rho}^\mu \to \underline{\rho}$ in $C^0(\overline{\Omega})$, it follows that for sufficiently large μ we have $\underline{\rho}^\mu \geqq \nu_\rho + \varepsilon/2$ in (x_1, x_2). Accordingly, we see that if $\varphi \in H^1$ with $\text{supp}(\varphi) \in (x_1, x_2)$, then

$$\langle \lambda_\mu^*, \varphi \rangle = 0.$$

Moreover, it is clear that the subsequence may be chosen such that $\lambda_\mu^* \to \lambda^*$ weakly in $(H^1)^*$. Hence, we have $\langle \lambda^*, \varphi \rangle = 0$ for all such $\varphi \in H^1$. From (3.10) we now have

$$-2 \int_\Omega (b \int_0^T u_t |u_t|^{\underline{\rho}} \ln |u_t| p_t \, dt) \eta \, dx + 2\beta(\underline{\rho}, \eta)_{H^1} = 0 \tag{3.11}$$

for all $\eta \in H^1$ with $\text{supp}(\eta) \in (x_1, x_2)$. From (3.11) it follows that $\underline{\rho} \in H^2(x_1, x_2)$. Globally we have the following.

Theorem 3.1. Let (A) hold and let $\underline{\rho} > \nu_\rho$ be an approachable solution of $(\mathcal{P})$. Then $\underline{\rho} \in H^2$.

We have the local result.

Theorem 3.2. Let (A) hold. An approachable solution $\underline{\rho}$ belongs to H^1. If $\underline{\rho} > \nu_\rho$ on a subinterval (x_1, x_2), then $\underline{\rho} \in H^2(x_1, x_2)$.

Acknowledgement: This work was supported in part by funding through AFOSR grant 87-0368.

REFERENCES

[1] Adams, R.A., *Sobolev Spaces*, Academic Press, New York, 1975.
[2] Dieudonne, J., *Foundations of Modern Analysis*, Academic Press, New York, 1960.
[3] Kunisch, K. and L.W. White, "Regularity properties in parameter estimation of diffusion coefficients in elliptic boundary value problems," *J. Appl. Anal.* 21(1986), pp. 71-88.
[4] Lions, J.L., *Quelques Methodes de Resolution des Problems aux Limites Nonlineaires*, Dunod, Paris, 1969.
[5] Lions, J.L., *Optimal Control of Systems Governed by Partial Differential Equations*, Springer-Verlag, 1971.
[6] Luenberger, D.G., *Optimization by Vector Space Methods*, Wiley, 1969.
[7] Mottershead J.E. and R. Stanway, "Identification of N-th power velocity damping," *J. of Sound and Vibration*, 105(1986), pp. 309-319.
[8] Schultz, M., *Spline Analysis*, Prentice Hall, Englewood Cliffs, N.J., 1973.
[9] Temam, R., *Navier-Stokes Equations and Numerical Analysis*, North Holland, New York, 1979.
[10] White, L.W., "Estimation of elastic parameters in beams and certain plates: H^1 regularization," *J. Opt. Thy. Appl.* 60 (1989), to appear.

SOME REMARKS ON THE BOUNDARY STABILIZABILITY OF THE WAVE EQUATION

Enrike Zuazua
Departamento de Matemáticas, Universidad del País Vasco
Apartado 644. 48080 Bilbao. Spain

Abstract

The boundary stabilizability of the wave equation is studied. Dirichlet-Neumann and Dirichlet boundary conditions are both considered. In the first case we generalize a recent result of V. Komornik and the author by modifying the feedback law which becomes more robust. In the second one we give a simpler proof for a result due to I. Lasiecka and R. Triggiani by using a " strengthening norms " argument which simplifies the functional setting of the problem.

1. Introduction

This paper is devoted to the study of the boundary stabilizability of the wave equation in a bounded domain $\Omega \subset \mathbf{R}^n$, $n \geq 1$, with smooth boundary $\Gamma = \partial\Omega$:

$$y'' - \Delta y = 0 \qquad \text{in} \quad \Omega \times (0, \infty). \tag{1.1}$$

(In (1.1) we have used the notation $' = \partial/\partial t$). We denote by $\nu(x)$ the unit normal vector oriented towards Ω at the point $x \in \Gamma$ and by $\partial\cdot/\partial\nu$ the derivative in this direction.

The paper is divided in two parts, each one corresponding to a different type of boundary conditions. In the first one we consider mixed boundary conditions of the type

$$\partial y/\partial \nu = - k_0(x)y - k_1(x)y' \qquad \text{on} \quad \Gamma(x^0) \times (0, \infty) \tag{1.2}$$

$$y = 0 \qquad \text{on} \quad \Gamma_*(x^0) \times (0, \infty) \tag{1.3}$$

where $k_i \in L^\infty(\Gamma)$, $k_i \geq 0$ for $i = 0, 1$ and $(\Gamma(x^0), \Gamma_*(x^0))$ is the usual partition of the boundary Γ given by

$$\Gamma(x^0) = \{ x^0 \in \mathbf{R}^n : m(x) \cdot \nu(x) > 0 \} \tag{1.4}$$

$$\Gamma_*(x^0) = \{ x^0 \in \mathbf{R}^n : m(x) \cdot \nu(x) \leq 0 \} \tag{1.5}$$

where $x^0 \in \mathbf{R}^n$, $m(x) = x - x^0$ and $\cdot$ denotes the scalar product in $\mathbf{R}^n$.

There is a large number of publications on this problem concerning the case $k_0=0$. J. P. Quinn and D. L. Russell in [14] proved that the energy

$$E(t) = \frac{1}{2} \int_\Omega \{ |\nabla y(x, t)|^2 + |y'(x, t)|^2 \} \, dx \tag{1.6}$$

goes to zero as $t \to +\infty$ for every finite energy solution, i. e. such that $E(0) < \infty$ and $y^0 = 0$ on $\Gamma_*(x^0)$ provided $k_1 \equiv 0$. Subsequently, G. Chen in [2], [3], [4] proved the exponential decay rate when $k_1(x) \geq k > 0$ for all $x \in \Gamma_0$ under the additional geometrical restriction

$$m(x) \cdot \nu(x) \geq \gamma > 0 \quad \text{for all } x \in \Gamma(x^0). \tag{1.7}$$

In fact, in [2], [3], [4] this result was proven for multipliers m more general than the radial one $m(x) = x - x^0$. This result was later generalized by J. Lagnese in [9], [10] for a larger class of multipliers but always under a geometrical restriction of type (1.7). Subsequently, J. L. Lions in [13] gave a simpler proof to the results of [2], [3], [4], [9]. We note that (1.7) excludes non star-shaped simply connected domains.

In a recent work of V. Komornik and the author [7], [8] the exponential energy decay was proven without hypothesis (1.7) by taking $k_1(x) = m(x)\cdot\nu(x)$ (this result is restricted, for technical reasons, to dimensions $n \leq 3$). The fact that $k_1(x)$ vanishes linearly at the interface points $\overline{\Gamma(x^0)} \cap \overline{\Gamma_*(x^0)}$ is essential in the proof of [7], [8]. The proof uses a recent result due to P. Grisvard [5] concerning the exact controllability when singularities occur that will be mentioned in section 2. Subsequently, by using weights of type $k_1(x) = m(x)\cdot\nu(x)$, J. Lagnese in [11] generalized the results of [7] to a class of vector fields larger than the radial one.

We observe that in the case where int $\{ \Gamma_*(x^0) \} = \emptyset$ (i.e. Ω is star shaped with respect to x^0), even if the energy decays exponentially as $t \to +\infty$, there exist infinitely many non-trivial stationary solutions of (1.1)-(1.3). In fact, any constant function is a solution to the problem. Therefore, *even if the energy decays exponentially, the* $H^1(\Omega) \times L^2(\Omega)$ - *uniform stabilization property* (i.e. the exponential decay of the $H^1(\Omega) \times L^2(\Omega)$ - norm of every solution) *is not satisfied.*

The aim of section 2 is to show that when the additional term $-k_0 y$ is added in the boundary condition (1.2) with k_0 a positive constant, the uniform stabilization in $H^1(\Omega) \times L^2(\Omega)$ holds even in the case where int $\{ \Gamma_*(x^0) \} = \emptyset$. The introduction of this additional term is motivated by the fact that the energy

$$E_{k_0}(t) = \frac{1}{2} \int_\Omega \{ |\nabla y(x, t)|^2 + |y'(x, t)|^2 \} \, dx + \frac{k_0}{2} \int_{\Gamma(x^0)} |y(x, t)|^2 \, d\Gamma \tag{1.8}$$

is coercive in $H^1(\Omega) \times L^2(\Omega)$ ($d\Gamma$ denotes the surface measure on Γ). This excludes the existence of non trivial rest points and, as we will prove bellow, implies the exponential decay in

$H^1(\Omega) \times L^2(\Omega)$. The proof is a slight generalization of the methods of [7], [8].

The second part of this work is devoted to the study of the stabilization of the wave equation under the following boundary feedback

$$y = \frac{\partial (G y')}{\partial \nu} \quad \text{on} \quad \Gamma \times (0, \infty). \tag{1.9}$$

where $G := (-\Delta)^{-1} : H^{-1}(\Omega) \to H^1_0(\Omega)$. The natural space for the exponential decay is then $L^2(\Omega) \times H^{-1}(\Omega)$. This problem , much more delicate than the precedent one, has been studied by I. Lasiecka and R. Triggiani in [12]. They have proven the exponential decay under very strict geometrical conditions (the result applies mainly for strictly convex domains) the proof being very involved. The object of section 3 is not to improve this result but to give a simpler proof of it in the particular case where the domain Ω is a ball.We use a technique of " strengthening norms " due to J. L. Lions which transforms the boundary conditions (1.7)-(1.8) in some boundary conditions of type (1.2)-(1.3) and allows us to work in the space $H^1(\Omega) \times L^2(\Omega)$ and to apply some of the techniques of section 2.

2. Dirichlet - Neumann boundary conditions

Let us consider the following wave equation

$$\left| \begin{array}{ll} y'' - \Delta y = 0 & \text{in } \Omega \times (0, \infty) \\ \partial y/\partial \nu = - \{m(x) \cdot \nu(x)\}\, y' - \alpha\, y & \text{on } \Gamma(x^0) \times (0, \infty) \\ y = 0 & \text{on } \Gamma_*(x^0) \times (0, \infty) \\ y(0) = y^0 \in V,\ y'(0) = y^1 \in L^2(\Omega) & \end{array} \right. \tag{2.1}$$

where $V = H^1_{\Gamma_*(x^0)}(\Omega) = \{u \in H^1(\Omega): u = 0 \text{ on } \Gamma_*(x^0)\}$ and $\alpha > 0$. We introduce the corresponding energy

$$E_\alpha(t) = \frac{1}{2} \int_\Omega \{|\nabla y(x, t)|^2 + |y'(x, t)|^2\}\, dx + \frac{\alpha}{2} \int_{\Gamma(x^0)} |y(x, t)|^2\, d\Gamma. \tag{2.2}$$

Our main result is as follows.

Theorem 2.1. *Let be* $n \leq 3$. *For any* $x^0 \in \mathbf{R}^n$ *and* $\alpha > 0$ *there exist some positive constants* $C > 1$, $\omega = \omega(x^0, \alpha)$ *such that*

$$E_\alpha(t) \leq C\, E_\alpha(0) \exp\{-\omega t\} \qquad \forall\, t \geq 0 \tag{2.3}$$

for every solution $y=y(x,t)$ *of* (2.1) *corresponding to initial data* $\{y^0, y^1\} \in V \times L^2(\Omega)$.

Remark 2.2. Note that $(E_\alpha(t))^{1/2}$ is a norm in $V \times L^2(\Omega)$ equivalent to the usual one. Therefore (2.3) is equivalent to

$$\| y(t) \|_{H^1(\Omega)} + \| y'(t) \|_{L^2(\Omega)} \leq C \{ \| y^0 \|_{H^1(\Omega)} + \| y^1 \|_{L^2(\Omega)} \} \exp\{-\frac{\omega}{2} t\} \tag{2.4}$$

for another constant $C>0$. Thus (2.3) implies the uniform stabilization in $H^1(\Omega) \times L^2(\Omega)$. •

Proof of Theorem 2.1. The following identity is easy to verify

$$\frac{dE_\alpha(t)}{dt} = E_\alpha'(t) = - \int_{\Gamma(x^0)} \{m(x) \cdot \nu(x)\} \mid y'(x,t) \mid^2 d\Gamma \leq 0, \quad \forall\, t \geq 0. \tag{2.5}$$

Therefore, it is sufficient to prove the existence of $T > 0$ and $C > 0$ such that

$$E_\alpha(T) \leq C \int_0^T \int_{\Gamma(x^0)} \{m(x) \cdot \nu(x)\} \mid y'(x,t) \mid^2 d\Gamma\, dt. \tag{2.6}$$

Indeed, from (2.5) (2.6) we easily deduce

$$E_\alpha(T) \leq \frac{C}{1+C} E_\alpha(0). \tag{2.7}$$

Taking into account the fact that $(E_\alpha)^{1/2}$ is a norm in $V \times L^2(\Omega)$ equivalent to the norm induced by $H^1(\Omega) \times L^2(\Omega)$ we deduce that the semigroup $S(t)$ associated to (2.1) satisfies

$$\| S(T) \| < 1. \tag{2.8}$$

We note that the fact that $\alpha > 0$ is crucial at this level. When $\alpha = 0$ and int $\{\Gamma_*(x^0)\} = \emptyset$, $(E_\alpha)^{1/2}$ does not define a norm in $V \times L^2(\Omega) = H^1(\Omega) \times L^2(\Omega)$.

The estimate (2.8), combined with (2.5) and the semigroup property implies

$$\| S(t) \| \leq C e^{-\omega t}, \forall\, t \geq 0 \text{ with } C = 1/\|S(T)\| \text{ and } \omega = -\log \|S(T)\|^{1/T}. \tag{2.9}$$

Therefore, the proof of the theorem reduces to the obtention of (2.6). In which follows, in order to simplify the notations, we omit the variables x, t of the functions under the integral sign, i. e. we shall write y, y', ∇y, m, ν instead of $y(x,t)$, $y'(x,t)$, $\nabla y(x,t)$, $m(x)$, $\nu(x)$, etc. On the other hand, we note by $(\cdot,\cdot)$ the scalar product in $L^2(\Omega)$.

We multiply the equation $(2.1)_1$ by $m \cdot \nabla y$. Integrating by parts we obtain

$$(y', m \cdot \nabla y)\Big|_0^T + \frac{n}{2} \int_0^T \int_\Omega | y' |^2 dx\, dt - \frac{1}{2} \int_0^T \int_{\Gamma(x^0)} \{m \cdot \nu\} \mid y' \mid^2 d\Gamma\, dt - \int_0^T \int_\Omega \Delta y\, m \cdot \nabla y\, dx\, dt = 0. \tag{2.10}$$

In order to treat the last term of the identity (2.10) we apply an inequality due to P. Grisvard [5] and adapted in [7] , [8] to this type of boundary conditions (note that when $\overline{\Gamma(x^0)} \cap \overline{\Gamma_*(x^0)} = \emptyset$ the solutions of (2.1) are not sufficiently smooth to apply Green's formula). We get

$$(y', m\cdot\nabla y)\Big|_0^T + \frac{1}{2}\int_0^T E_\alpha(t)\,dt + \frac{n-1}{2}\int_0^T\int_\Omega \{|y'|^2 - |\nabla y|^2\}\,dx\,dt \le$$

$$\le \frac{1}{2}\int_0^T\int_{\Gamma(x^0)} \{m\cdot\nu\}\,|y'|^2\,d\Gamma\,dt + \int_0^T\int_\Gamma \frac{\partial y}{\partial\nu}\,m\cdot\nabla y\,d\Gamma\,dt - \frac{1}{2}\int_0^T\int_\Gamma \{m\cdot\nu\}\,|\nabla y|^2\,d\Gamma\,dt \le$$

$$\le \frac{1}{2}\int_0^T\int_{\Gamma(x^0)} \{m\cdot\nu\}\,|y'|^2\,d\Gamma\,dt + \int_0^T\int_{\Gamma(x^0)} \frac{\partial y}{\partial\nu}\,m\cdot\nabla y\,d\Gamma\,dt - \frac{1}{2}\int_0^T\int_{\Gamma(x^0)} \{m\cdot\nu\}\,|\nabla y|^2\,d\Gamma\,dt \qquad (2.11)$$

since

$$\int_0^T\int_{\Gamma_*(x^0)} \frac{\partial y}{\partial\nu}\,m\cdot\nabla y\,dx\,dt - \frac{1}{2}\int_0^T\int_{\Gamma_*(x^0)} \{m\cdot\nu\}\,|\nabla y|^2\,d\Gamma\,dt = \frac{1}{2}\int_0^T\int_{\Gamma_*(x^0)} \{m\cdot\nu\}\,\Big|\frac{\partial y}{\partial\nu}\Big|^2\,d\Gamma\,dt \le 0.$$

On the other hand

$$\int_0^T\int_{\Gamma(x^0)} \frac{\partial y}{\partial\nu}\,m\cdot\nabla y\,d\Gamma\,dt = -\int_0^T\int_{\Gamma(x^0)} \{m\cdot\nu\}\,y'\,m\cdot\nabla y\,d\Gamma\,dt - \alpha\int_0^T\int_{\Gamma(x^0)} y\,m\cdot\nabla y\,d\Gamma\,dt =$$

$$= -\int_0^T\int_{\Gamma(x^0)} \{m\cdot\nu\}\,y'\,m\cdot\nabla y\,d\Gamma\,dt - \alpha\int_0^T\int_{\Gamma(x^0)} \{m\cdot\nu\}\,y\frac{\partial y}{\partial\nu}\,d\Gamma\,dt - \alpha\int_0^T\int_{\Gamma(x^0)} y\,m\cdot\nabla_\Gamma y\,d\Gamma\,dt =$$

$$= -\int_0^T\int_{\Gamma(x^0)} \{m\cdot\nu\}\,y'\,m\cdot\nabla y\,d\Gamma\,dt - \alpha\int_0^T\int_{\Gamma(x^0)} \{m\cdot\nu\}\,y\frac{\partial y}{\partial\nu}\,d\Gamma\,dt - \frac{\alpha}{2}\int_0^T\int_{\Gamma(x^0)} m\cdot\nabla_\Gamma(|y|^2)\,d\Gamma\,dt =$$

$$= -\int_0^T\int_{\Gamma(x^0)} \{m\cdot\nu\}\,y'\,m\cdot\nabla y\,d\Gamma\,dt - \alpha\int_0^T\int_{\Gamma(x^0)} \{m\cdot\nu\}\,y\frac{\partial y}{\partial\nu}\,d\Gamma\,dt + \frac{\alpha}{2}\int_0^T\int_{\Gamma(x^0)} (\mathrm{div}_\Gamma m)\,|y|^2\,d\Gamma\,dt \le$$

$$\le \frac{1}{2}\int_0^T\int_{\Gamma(x^0)} \{m\cdot\nu\}\,|\nabla y|^2\,d\Gamma\,dt + C\int_0^T\int_{\Gamma(x^0)} [\{m\cdot\nu\}\,|y'|^2 + |y|^2]\,d\Gamma\,dt \qquad (2.12)$$

for $C > 0$ large enough depending only on α and $R = \|m\|_{L^\infty(\Omega)}$. We have used here the divergence formula on the variety $\Gamma(x^0)$:

$$\int_0^T\int_{\Gamma(x^0)} m\cdot\nabla_\Gamma(|y|^2)\,d\Gamma\,dt = -\int_0^T\int_{\Gamma(x^0)} (\mathrm{div}_\Gamma m)\,|y|^2\,d\Gamma\,dt$$

where ∇_Γ (resp. div_Γ) is the surface gradient (resp. divergence) on Γ. This formula is valid since $y = 0$ on $\overline{\Gamma(x^0)} \cap \overline{\Gamma_*(x^0)} = \partial(\Gamma(x^0))$.

On the other hand, multiplying the equation $(2.1)_1$ by y and integrating by parts we obtain

$$\int_0^T \int_\Omega \{ |y'|^2 - |\nabla y|^2 \} \, dx\, dt = (y, y')\Big|_0^T - \int_0^T \int_{\Gamma(x^0)} \frac{\partial y}{\partial \nu} y \, d\Gamma\, dt \leq$$

$$\leq (y, y')\Big|_0^T + C \int_0^T \int_{\Gamma(x^0)} [\{m \cdot \nu\} |y'|^2 + |y|^2] \, d\Gamma\, dt. \tag{2.13}$$

for $C > 0$ large enough. From (2.11) (2.12) (2.13) we deduce

$$(y', m \cdot \nabla y + \frac{n-1}{2} y)\Big|_0^T + \int_0^T E_\alpha(t)\, dt \leq C \int_0^T \int_{\Gamma(x^0)} \{\{m \cdot \nu\} \; |y'|^2 + |y|^2 \} \, d\Gamma\, dt. \tag{2.14}$$

From (2.5) we have clearly for some $T_0 > 0$,

$$|(y', m \cdot \nabla y + \frac{n-1}{2} y)\Big|_0^T| \leq \frac{T_0}{2} \{E_\alpha(0) + E_\alpha(T) \} = T_0 E_\alpha(t) + \frac{T_0}{2} \int_0^T \int_{\Gamma(x^0)} \{m \cdot \nu\} |y'|^2 \, d\Gamma\, dt \tag{2.15}$$

and

$$\int_0^T E_\alpha(t)\, dt = T E_\alpha(T) + \int_0^T \int_s^T \int_{\Gamma(x^0)} \{m \cdot \nu\} |y'|^2 \, d\Gamma\, dt\, ds \geq T E_\alpha(T). \tag{2.16}$$

By combining (2.14) (2.15) and (2.16) we deduce

$$(T - T_0) E_\alpha(T) \leq C \int_0^T \int_{\Gamma(x^0)} [\{m \cdot \nu\} |y'|^2 + |y|^2] \, d\Gamma\, dt \tag{2.17}$$

for all $T > T_0$ with $C > 0$ large enough.

Therefore, it is enough to prove the following estimate

$$\int_0^T \int_{\Gamma(x^0)} |y|^2 \, d\Gamma\, dt \leq C \int_0^T \int_{\Gamma(x^0)} \{m \cdot \nu\} |y'|^2 \, d\Gamma\, dt \tag{2.18}$$

for $T > T_0$ and for some $C > 0$. We argue by contradiction. If (2.18) is not satisfied for some $C>0$ there exists a sequence (y_k) of solutions of (2.1) such that

$$\int_0^T \int_{\Gamma(x^0)} |y_k|^2 \, d\Gamma\, dt = 1, \ \forall\, k = 1, 2, \dots \text{ and } \int_0^T \int_{\Gamma(x^0)} \{m \cdot \nu\} |y_k'|^2 \, d\Gamma\, dt \to 0 \text{ as } k \to +\infty. \tag{2.19}$$

From (2.18) (2.19) we deduce that

(y_k) is bounded in $L^\infty(0, T; H^1(\Omega)) \cap W^{1,\infty}(0, T; L^2(\Omega))$

and then

$$(y_k|_{\Gamma \times (0,T)}) \text{ is relatively compact in } L^2((0,T) \times \Gamma).$$

For a subsequence of (y_k) (still denoted for simplicity by (y_k)) we have

$$y_k \to y \quad \text{in } L^\infty(0, T; H^1(\Omega)) \cap W^{1,\infty}(0, T; L^2(\Omega)) \text{ weak } *, \text{ as } k \to +\infty$$

where $y = y(x, t)$ is a solution of (2.1) satisfying in addition

$$\int_0^T \int_{\Gamma(x^0)} |y|^2 d\Gamma\, dt = 1 \tag{2.20}$$

and

$$y' = 0 \qquad \text{on } \Gamma(x^0) \times (0, T). \tag{2.21}$$

But (2.21) implies tha $z = y'$ satisfies

$$\left| \begin{array}{lll} z'' - \Delta z = 0 & \text{in} & \Omega \times (0,T) \\ \partial z/\partial \nu = 0 & \text{on} & \Gamma(x^0) \times (0,T) \\ z = 0 & \text{on} & \Gamma \times (0,T) \end{array} \right.$$

and from Holmgren's uniqueness theorem $z \equiv 0$. Then, y is a rest point for (2.1) and since $\alpha > 0$, $y \equiv 0$ which contradicts (2.20). The proof of Theorem 2.1 is now completed. •

Remark 2.3. In Theorem 2.1 we have assumed $n \leq 3$ in order to apply Grisvard's inequality in (2.11). The result is probably true for all n. The method of proof of Theorm 2.1 would apply with a generalization of this inequality to dimensions $n \geq 4$. Of cours, when $\partial(\Gamma(x^0)) = \varnothing$ Theorem 2.1 remains valid even if $n \geq 4$. In this case, Green's formula can be applied in (2.11) since the solutions are sufficiently smooth. •

Remark 2.4. When $\text{int}\{\Gamma_*(x^0)\} \neq \varnothing$ Theorem 2.1 applies also if $\alpha = 0$ or even if $\alpha < 0$ and $|\alpha|$ is small enough. Indeed, in those conditions $(E_\alpha(\cdot))^{1/2}$ still defines a norm in $V \times L^2(\Omega)$. •

Remark 2.5. The same result is valid for the boundary conditions

$$\frac{\partial y}{\partial \nu} = -k_1(x)\{m(x)\cdot\nu(x)\}y' - k_0(x)\, y \qquad \text{on } \Gamma(x^0) \times (0, \infty)$$

where $k_0(x) \in C^1(\Gamma(x^0))$ (some regularity assumption on k_0 is needed in order to apply the divergence formula on $\Gamma(x^0)$), $k_0(x) \geq 0$, $k_0 \not\equiv 0$ and $k_1(x) \in L^\infty(\Gamma(x^0))$, $k_1(x) \geq k_1 > 0$.

One may also take $k_0(x) = \alpha(x)\,\{m(x)\cdot v(x)\}^{1/2}$ with $\alpha(x) \in L^\infty(\Gamma_0)$, $\alpha(x) \geq 0$, $\alpha \not\equiv 0$. In this case the divergence formula is not needed in (2.12) since we can directly apply the following estimate

$$\left| \int_0^T \int_{\Gamma(x^0)} \alpha\{m\cdot v\}^{1/2} m\cdot \nabla y \, d\Gamma \, dt \right| \leq \varepsilon \int_0^T \int_{\Gamma(x^0)} \{m\cdot v\} |\nabla y|^2 d\Gamma \, dt + C_\varepsilon \int_0^T \int_{\Gamma(x^0)} |y|^2 d\Gamma \, dt. \quad \bullet$$

Remark 2.6. The same results apply to a wave equation perturbed by a lower order term

$$y'' - \Delta y + V(x)y = 0 \qquad \text{in } \Omega \times (0, T)$$

where the potential V is nonnegative and satisfies

$$V \in L^p(\Omega) \ \text{ with } \ \begin{array}{l} p = 2 \text{ if } n = 1 \\ p > n \text{ if } n \geq 2. \end{array} \quad \bullet$$

Remark 2.7. Semilinear problems

$$y'' - \Delta y + f(y) = 0 \qquad \text{in } \Omega \times (0, T)$$

can also be treated as in [8] under the natural sign and growth assumptions:

$$f(s)s \geq 0 \ \ \forall\, s > 0;\ |f'(s)| \leq C(1+|s|^p) \text{ with } (n-2)p \leq 2.$$

The exponential decay of the energy

$$E_\alpha(t) = \frac{1}{2}\int_\Omega \{|\nabla y(t)|^2 + |y'(t)|^2\}\, dx + \int_\Omega F(y(t))\, dx + \frac{\alpha}{2}\int_{\Gamma(x^0)} |y(t)|^2 d\Gamma$$

with $F(z) = \int_0^z f(s)\, ds$ can be proven when $|\alpha|$ is small enough.

Nonlinear boundary conditions may also be considered by obtaining decay rates other that the exponential one (see [17]). •

Remark 2.8. Microlocal analysis allows to establish stabilization results for partitions of the boundary (Γ_0, Γ_1) much more general than $(\Gamma(x^0), \Gamma_*(x^0))$ but always under the assumption $\overline{\Gamma_0} \cap \overline{\Gamma_1} = \emptyset$ ([1,], [15]). •

Remark 2.9. Let us assume that Ω is star-shaped with respect to x^0. Then the feedback law (1.2) (1.3) is more robust when $k_0 > 0$ than it is when $k_0 = 0$, since the exponential decay is not lost by passing to the limit as $x^0_\varepsilon \to x^0$, i. e. the stabilization property is conserved under continuous perturbations of the support of the feedback control (see [18] for the study of this question). •

3. Dirichlet boundary conditions

The aim of this section is to study the exponential decay in the space $L^2(\Omega) \times H^{-1}(\Omega)$ of the solutions of the following system

$$\left|\begin{array}{ll} y'' - \Delta y = 0 & \text{in } \Omega \times (0,\infty) \\ y = \dfrac{\partial(Gy')}{\partial\nu} & \text{on } \Gamma \times (0,\infty) \\ y(0) = y^0 \in L^2(\Omega);\ y'(0) = y^1 \in H^{-1}(\Omega) & \end{array}\right. \tag{3.1}$$

where $G = (-\Delta)^{-1}: H^{-1}(\Omega) \to H^1_0(\Omega)$.

This problem has been solved by I. Lasiecka and R. Triggiani [12] in the case where the domain $\Omega = \Omega_1 \setminus \Omega_2$ is the difference of two strictly convex domains Ω_i, i =1, 2 with smooth boundaries and $\overline{\Omega_2} \subset \Omega_1$. We shall consider here the particular case where the domain Ω is a ball of $\mathbf{R}^n$ and give a simpler proof for this stabilization result.

First of all let us define the new dependent variable

$$u(x,t) = u^0(x) + \int_0^t y(x,s)\,ds \tag{3.2}$$

$$u^0 = -Gy^1. \tag{3.3}$$

We have

$$u' = y,\ u'' = y'$$

and integrating the equation $(3.1)_1$ with respect to t it is easy to check that u satisfies

$$u'' - \Delta u = 0 \qquad \text{in } \Omega \times (0,\infty). \tag{3.4}$$

On the other hand

$$u(0) = u^0 = -Gy^1 \in H^1_0(\Omega);\ u'(0) = u^1 = y^0 \in L^2(\Omega) \tag{3.5}$$

and $(3.2)_2$ gives

$$y = u' = \frac{\partial Gu''}{\partial\nu} = \frac{\partial G\Delta u}{\partial\nu}. \tag{3.6}$$

We now calculate $G\Delta u$ for all $u \in H^1_0(\Omega)$. We define Bu, the " harmonic part " of u, by

$$-\Delta Bu = 0 \quad \text{in } \Omega; \ Bu = u \quad \text{on } \Gamma. \tag{3.7}$$

Then

$$-\Delta(u - Bu) = -\Delta u, \quad u - Bu \in H^1_0(\Omega)$$

or equivalently

$$u - Bu = G(-\Delta u) = - G\Delta u \tag{3.8}$$

and (3.6) becomes

$$u' = \frac{\partial(Bu - u)}{\partial \nu}. \tag{3.9}$$

Therefore u is solution of the following system

$$\left| \begin{array}{ll} u'' - \Delta u = 0 & \text{in } \Omega \times (0, \infty) \\ u(0) = u^0 \in H^1_0(\Omega); \ u'(0) = u^1 \in L^2(\Omega) & \\ \dfrac{\partial(u - Bu)}{\partial \nu} = - u' & \text{on } \Gamma \times (0, \infty) \end{array} \right. \tag{3.10}$$

where B is defined by (3.7).

Taking into account that the solution y of (3.1) is in the class

$$y \in L^\infty(0, T; L^2(\Omega)) \cap W^{1,\infty}((0, \infty; H^{-1}(\Omega)); \ y = \frac{\partial Gy'}{\partial \nu} \in L^2(\Gamma \times (0, \infty))$$

we deduce

$$\left| \begin{array}{l} u' = y \in L^\infty(0, T; L^2(\Omega)) \\ u'' = \Delta u = y' \in L^\infty(0, \infty; H^{-1}(\Omega)). \end{array} \right. \tag{3.11}$$

On the other hand, integrating (3.6) we deduce

$$u - \frac{\partial Gu'}{\partial \nu} = u^0 - \frac{\partial Gu^1}{\partial \nu} \qquad \text{on } \Gamma \times (0, \infty)$$

and since $u^0 \in H^1_0(\Omega)$ we have

$$u - \frac{\partial G u'}{\partial \nu} = - \frac{\partial G u^1}{\partial \nu} \qquad \text{on } \Gamma \times (0, \infty)$$

but

$$\frac{\partial G u'}{\partial \nu} \in L^\infty(0, \infty; H^{1/2}(\Gamma)); \quad \frac{\partial G u^1}{\partial \nu} \in H^{1/2}(\Gamma)$$

and then

$$u|_\Gamma \in L^\infty(0, \infty; H^{1/2}(\Gamma))$$

that, combined with (3.11), implies

$$u \in L^\infty(0, \infty; H^1(\Omega)) \cap W^{1,\infty}(0, \infty; L^2(\Omega)). \tag{3.12}$$

It is easy to see, by standard techniques, that for any $\{u^0, u^1\} \in H^1{}_0(\Omega) \times L^2(\Omega)$ there exists a unique solution u of (3.10) in the class (3.12).

We note that $\{y(t), y'(t)\}$ decays exponentially in $L^2(\Omega) \times H^{-1}(\Omega)$ if and only if $\{u'(t), u''(t)\} = \{u'(t), \Delta(u(t) - Bu(t))\}$ does and this is equivalent to the exponential decay of $\{u(t) - Bu(t), u'(t)\}$ in the espace $H^1{}_0(\Omega) \times L^2(\Omega)$.

Therefore, let us consider the problem of the exponential decay of the energy

$$E(t) = \frac{1}{2} \int_\Omega \{|\nabla(u(t) - Bu(t))|^2 + |u'(t)|^2\} \, dx. \tag{3.13}$$

We have the following result.

Theorem 3.1. *Let* Ω *be a ball of* $\mathbf{R}^n$. *Then, there exist some constants* $C > 1$ *and* $\omega > 0$ *such that for every solution* $u = u(x, t)$ *of* (3.10) *the following estimate holds*

$$E(t) \le C\, E(0) \exp\{-\omega t\} \qquad \forall\, t \ge 0. \tag{3.14}$$

Proof: Let be $\Omega = B(x^0, r)$, the ball centered at x^0 with radious r in $\mathbf{R}^n$.

First of all we note that

$$\int_\Omega |\nabla(u - Bu)|^2 = \int_\Omega \{|\nabla u|^2 - |\nabla Bu|^2\} \, dx$$

since

$$\int_\Omega \nabla u \cdot \nabla Bu \;\, dx = \int_\Omega |\nabla Bu|^2 \, dx$$

and then

$$E(t) = \frac{1}{2} \int_\Omega \{ |u'(t)|^2 + |\nabla u(t)|^2 - |\nabla Bu(t)|^2 \} \, dx.$$

On the other hand, it is easy to check that

$$\frac{d\,E(t)}{dt} = E'(t) = \int_\Omega \{ u''(t)\, u'(t) + \nabla u(t)\cdot\nabla u'(t) - \nabla Bu(t)\cdot\nabla Bu'(t) \}\, dx =$$

$$= \int_\Gamma \frac{\partial u(t)}{\partial \nu}\, u'(t)\, d\Gamma - \int_\Gamma \frac{\partial Bu(t)}{\partial \nu}\, Bu'(t)\, d\Gamma =$$

$$= \int_\Gamma \frac{\partial (u(t)-Bu(t))}{\partial \nu}\, u'(t)\, d\Gamma = - \int_\Gamma | u'(t) |^2\, d\Gamma \le 0 \quad \forall t \ge 0. \tag{3.15}$$

From (3.15) we deduce, in particular, that the energy $E(\cdot)$ is nonincreasing.

As in the proof of Theorem 2.1, the proof of this theorem will be completed if we prove the existence of some positive constants $T, C > 0$ such that

$$E(T) \le C \int_0^T \int_\Gamma | u'(t) |^2\, d\Gamma\, dt. \tag{3.16}$$

In order to prove (3.15) we use the multiplier technique. By multiplying $(3.10)_1$ by $m\cdot\nabla(u - Bu)$, with $m(x) = x - x^0$, x^0 being the center of Ω, we obtain

$$(u'(t),\ m\cdot\nabla(u(t) - Bu(t)))\Big|_0^T + \int_0^T E(t)\, dt + \frac{n-1}{2}\int_0^T\int_\Omega \{ | u' |^2 - | \nabla(u - Bu) |^2 \}\, dx\, dt +$$

$$+ \int_0^T\int_\Omega u'\ m\cdot \nabla Bu'\, dx\, dt = \frac{1}{2}\int_0^T\int_\Gamma \{m\cdot \nu\}\, | u' |^2\, d\Gamma\, dt. \tag{3.17}$$

Next, by multiplying $(3.10)_1$ by $u-Bu$ we deduce:

$$\int_0^T\int_\Omega \{ | u' |^2 - | \nabla(u - Bu) |^2 \}\, dx\, dt = (u'(t), u(t) - Bu(t))\Big|_0^T + \int_0^T\int_\Omega u'\ Bu'\, dx\, dt. \tag{3.18}$$

Combining (3.16) (3.17) we obtain

$$X_1 + X_2 + \int_0^T E(t)\, dt = \frac{1}{2}\int_0^T\int_\Gamma \{m\cdot\nu\}\, | u' |^2\, d\Gamma dt \tag{3.19}$$

with

$$\left| \begin{aligned} X_1 &= (u'(t),\ m\cdot\nabla(u(t) - Bu(t)) + \frac{n-1}{2}(u(t) - Bu(t)))\Big|_0^T \\ X_2 &= \int_0^T\int_\Omega u'(m\cdot \nabla Bu' + \frac{n-1}{2} Bu')\, dx\, dt. \end{aligned} \right. \tag{3.20}$$

The quantity X_1 may be easily estimated by following V. Komornik [6]. We have:

$$| X_1 | \le r\, (E(0) + E(T)) = 2r\, E(T) + r\int_0^T\int_\Gamma | u' |^2\, d\Gamma dt. \tag{3.21}$$

On the other hand, since E(t) is nonincreasing,

$$T\,E(T) \le \int_0^T E(t)\,dt. \tag{3.22}$$

From (3.19) (3.21) and (3.22) we deduce

$$(T - 2r)\,E(T) + X_2 \le C \int_0^T \int_\Gamma |u'|^2\,d\Gamma\,dt. \tag{3.23}$$

We need now an estimate for X_2. We write $X_2 = Z_1 + \frac{n-1}{2} Z_2$ with

$$Z_1 = \int_0^T \int_\Omega u'\ m\cdot\nabla B u'\ dx\,dt\,; \quad Z_2 = \int_0^T \int_\Omega u'\,Bu'\ dx\,dt. \tag{3.24}$$

Concerning Z_2 we have

$$Z_2 = -\int_0^T \int_\Omega \Delta(Gu')Bu'\ dx\,dt = -\int_0^T \int_\Gamma \frac{\partial(Gu')}{\partial\nu} Bu'\ d\Gamma\,dt + \int_0^T \int_\Omega \nabla(Gu')\cdot\nabla(Bu')\,dx\,dt =$$

$$= \int_0^T \int_\Gamma \frac{\partial(Gu')}{\partial\nu}\frac{\partial(u - Bu)}{\partial\nu}\,d\Gamma\,dt = -\int_0^T \int_\Gamma \frac{\partial(Gu')}{\partial\nu}\frac{\partial(G\Delta u)}{\partial\nu}\,d\Gamma\,dt = -\int_0^T \int_\Gamma \frac{\partial(Gu')}{\partial\nu}\frac{\partial(Gu'')}{\partial\nu}\,d\Gamma\,dt =$$

$$= -\frac{1}{2}\int_0^T \frac{d}{dt}\int_\Gamma \left|\frac{\partial(Gu')}{\partial\nu}\right|^2 d\Gamma\,dt = -\frac{1}{2}\int_\Gamma \left|\frac{\partial(Gu'(t))}{\partial\nu}\right|^2 d\Gamma\,\Big|_0^T$$

and then

$$|Z_2| \le C\,(\,E(0) + E(T)\,) \le C\,(\,E(T) + \int_0^T\int_\Gamma |u'|^2\,d\Gamma\,dt\,) \tag{3.25}$$

for $C > 0$ which does not depend on T.

By applying Green's formula we have the following identity for Z_1:

$$Z_1 = -\int_0^T \int_\Omega \Delta(Gu')\ m\cdot\nabla(Bu')\,dx\,dt = -\int_0^T \int_\Gamma \frac{\partial(Gu')}{\partial\nu} m\cdot\ \nabla(Bu')\ d\Gamma\,dt\ - \int_0^T\int_\Omega Gu'\ \Delta(m\cdot\nabla(Bu'))\ dx\,dt\ +$$

$$+\int_0^T\int_\Gamma Gu'\frac{\partial(m\cdot\ \nabla(Bu'))}{\partial\nu}d\Gamma\ dt = -\int_0^T\int_\Gamma\frac{\partial(Gu')}{\partial\nu} m\cdot\nabla(Bu')\ d\Gamma\,dt = -\,r\int_0^T\int_\Gamma\frac{\partial(Gu')}{\partial\nu}\frac{\partial(Bu')}{\partial\nu}\,d\Gamma\,dt. \tag{3.26}$$

since $Gu' = 0$ on Γ, $m = r\,\nu$ on Γ and $\Delta(m\cdot\nabla(Bu')) = 0$ in Ω. The last term of the identity (3.26) must be understood in the sense of the duality between $L^2(0, T; H^{1/2}(\Gamma))$ and $L^2(0, T; H^{-1/2}(\Gamma))$.

By applying Lemma 3.3 of I. Lasiecka and R. Triggiani [12] to this particular case we deduce the following identity

$$Z_1 = -\frac{r}{2}\int_\Gamma \frac{\partial(Gu'(t))}{\partial\nu}\,\frac{\partial\big(B(\frac{\partial(Gu'(t))}{\partial\nu})\big)}{\partial\nu}\,d\Gamma\,\Big|_0^T. \tag{3.27}$$

Let us prove it for the sake of completness. From (3.26) it is enough to prove

$$\frac{1}{2}\frac{d}{dt}\int_\Gamma \frac{\partial(Gu'(t))}{\partial\nu}\,\frac{\partial\big(B(\frac{\partial(Gu'(t))}{\partial\nu})\big)}{\partial\nu}\,d\Gamma = \int_\Gamma \frac{\partial(Gu'(t))}{\partial\nu}\,\frac{\partial(Bu'(t))}{\partial\nu}\,d\Gamma. \tag{3.28}$$

Dropping the explicit dependence on t, we compute

$$\frac{d}{dt}\int_\Gamma \frac{\partial(Gu')}{\partial\nu}\,\frac{\partial\big(B(\frac{\partial(Gu')}{\partial\nu})\big)}{\partial\nu}\,d\Gamma = \int_\Gamma \frac{\partial(Gu')}{\partial\nu}\,\frac{\partial\big(B(\frac{\partial(Gu'')}{\partial\nu})\big)}{\partial\nu}\,d\Gamma + \int_\Gamma \frac{\partial(Gu'')}{\partial\nu}\,\frac{\partial\big(B(\frac{\partial(Gu')}{\partial\nu})\big)}{\partial\nu}\,d\Gamma$$

and by applying the general formula

$$\int_\Gamma \frac{\partial(Gv)}{\partial\nu}\,\frac{\partial\big(B(\frac{\partial(Gw)}{\partial\nu})\big)}{\partial\nu}\,d\Gamma = \int_\Gamma \frac{\partial\big(B(\frac{\partial(Gv)}{\partial\nu})\big)}{\partial\nu}\,\frac{\partial(Gw)}{\partial\nu}\,d\Gamma$$

we deduce

$$\frac{d}{dt}\int_\Gamma \frac{\partial(Gu')}{\partial\nu}\,\frac{\partial\big(B(\frac{\partial(Gu')}{\partial\nu})\big)}{\partial\nu}\,d\Gamma = 2\int_\Gamma \frac{\partial(Gu')}{\partial\nu}\,\frac{\partial\big(B(\frac{\partial(Gu'')}{\partial\nu})\big)}{\partial\nu}\,d\Gamma =$$

$$= 2\int_\Gamma \frac{\partial(Gu')}{\partial\nu}\,\frac{\partial\big(B(\frac{\partial(G\Delta u)}{\partial\nu})\big)}{\partial\nu}\,d\Gamma = 2\int_\Gamma \frac{\partial(Gu')}{\partial\nu}\,\frac{\partial\big(B(\frac{\partial(Bu-u)}{\partial\nu})\big)}{\partial\nu} = 2\int_\Gamma \frac{\partial(Gu')}{\partial\nu}\,\frac{\partial(Bu')}{\partial\nu}\,d\Gamma. \tag{3.29}$$

On the other hand,

$$\Big|\int_\Gamma \frac{\partial(Gu'(t))}{\partial\nu}\,\frac{\partial\big(B(\frac{\partial(Gu'(t))}{\partial\nu})\big)}{\partial\nu}\,d\Gamma\Big| \le C\,\Big\|\frac{\partial(Gu'(t))}{\partial\nu}\Big\|_{H^{1/2}(\Gamma)}\,\Big\|\frac{\partial\big(B(\frac{\partial(Gu'(t))}{\partial\nu})\big)}{\partial\nu}\Big\|_{H^{-1/2}(\Gamma)} \le$$

$$\le C\,\|Gu'(t)\|_{H^2(\Omega)}\,\Big\|\frac{\partial(Gu'(t))}{\partial\nu}\Big\|_{H^{1/2}(\Omega)} \le C\,\|Gu'(t)\|^2_{H^2(\Omega)} \le C\,\|u'(t)\|^2_{L^2(\Omega)}, \quad \forall t\ge 0$$

and then, from (3.27),

$$|Z_1| \le C\{\|u'(0)\|^2_{L^2(\Omega)} + \|u'(T)\|^2_{L^2(\Omega)}\} \le C\{E(0)+E(T)\} \le C\{E(T) + \int_0^T\int_\Gamma |u'|^2\,d\Gamma\,dt\} \tag{3.30}$$

for $C > 0$ large enough which does not depend on T.

Combining (3.23) (3.25) and (3.30) we obtain (3.16). The proof of the theorem is now completed. •

Remark 3.2. In [12] the exponential decay of the $L^2(\Omega)\times H^{-1}(\Omega)$ - norm solutions to (3.1) was proven for every domain of the form $\Omega = \Omega_1\setminus\overline{\Omega_2}$ where Ω_i, i =1, 2 are two strictly convex domains with smooth boundaries and such that $\overline{\Omega_2} \subset \Omega_1$. In this case the radial vector field must be replaced by a more general one and the estimates above become more complicated. We have omited here this general case for the sake of simplicity. We refer to [12] for the general case. •

Remark 3.3. The system (3.10) is well posed in $H^1(\Omega) \times L^2(\Omega)$, i. e. for every initial data $\{u^0, u^1\} \in H^1(\Omega) \times L^2(\Omega)$ there exists a unique solution of (3.10) in the class (3.12). The exponential decay rate (3.14) of the energy E(t) can be proven for all these solutions. However, we note that every constant function is a solution of (3.10) and then, in particular, there is not exponential decay of the $H^1(\Omega) \times L^2(\Omega)$ - norm. •

Remark 3.4. Theorem 3.1 may be generalized to equations of the form

$$\left|\begin{array}{ll} u'' - \Delta u + V(x)\, u = 0 & \text{in } \Omega \times (0, \infty) \\ u(0) = u^0 \in H^1_0(\Omega);\ u'(0) = u^1 \in L^2(\Omega) & \\ \dfrac{\partial(u - Bu)}{\partial\nu} = -u' & \text{on } \Gamma \times (0, \infty) \end{array}\right. \tag{3.31}$$

where V = V(x) is a nonnegative potential on the conditions of Remark 2.6. The operator B is now defined by

$$-\Delta(Bu) + V\,Bu = 0 \quad \text{in } \Omega; \quad u = 0 \quad \text{on } \Gamma.$$

This implies the exponential decay in the espace $L^2(\Omega) \times H^{-1}(\Omega)$ for every solution of the system

$$\left|\begin{array}{ll} y'' - \Delta y + V(x)\, y = 0 & \text{in } \Omega \times (0, \infty) \\ y = \dfrac{\partial(Gy')}{\partial\nu} & \text{on } \Gamma \times (0, \infty) \\ y(0) = y^0 \in L^2(\Omega);\ y'(0) = y^1 \in H^{-1}(\Omega) & \end{array}\right. \tag{3.32}$$

where $G = (-\Delta + V(x)\, I)^{-1}$ with I = identity. •

Aknowledgements. The author wishes to thank Prof. J. L. Lions for fruitful discussions and particularly for suggesting the study of the problem of section 3 on the framework of the system (3.10). This work was partially supported by grant G. V. 127. 310 - 1/87 of the " Eusko Jaurlaritza " (Basque Government).

References

[1] C. Bardos, G. Lebeau & J. Rauch, Contr ôle et stabilisation dans les problèmes hyperboliques, Appendix 2 in the Lecture Notes of J. L. Lions: *Contrôlabilité exacte de systèmes distribués. Méthode HUM.* Masson. Collection RMA. 1988.

[2] G. Chen, Energy decay estimates and exact boundary value controllability for the wave equation in a bounded domain, J. Math. Pures et Appl. (9) **58** (1979) 249-274.

[3] G. Chen, Control and stabilization for the wave equation in a bounded domain I-II, SIAM J. Control and Opt. **17** (1979), 66-81, **19** (1981) 114-122.

[4] G. Chen, A note on the boundary stabilization of the wave equation, SIAM J. Control and Opt. **19** (1981) 106-113.

[5] P. Grisvard, Contrôlabilité exacte avec des conditions mêlées, C. R. Acad. Sci. Paris, Sér. I, Math. **305** (1987) 363-366.

[6] V. Komornik, Contrôlabilité exacte en un temps minimal, C. R. Acad. Sci. Paris, Sér. I Math. **304** (1987) 223-225.

[7] V. Komornik & E. Zuazua, Stabilisation frontière de l'équation des ondes: Une méthode directe, C. R. Acad. Sci. Paris, Sér. I Math. **305** (1987) 605-608.

[8] V. Komornik & E. Zuazua, A direct method for the boundary stabilization of the wave equation, J. Math. Pures et Appl., to appear.

[9] J. Lagnese, Decay of solutions of wave equations in a bounded region with boundary dissipation, J. Diff. Equations **50** (1983), 163-182.

[10] J. Lagnese, Boundary stabilization of linear elastodynamic systems, SIAM J. Control and Opt. **21** (1983) 968-984.

[11] J. Lagnese, Note on the boundary stabilization of wave equations, SIAM J. Control and Opt., to appear.

[12] I. Lasiecka & R. Triggiani, Uniform exponential decay in a bounded region with $L^2(0,T;L^2(\Sigma))$ - feedback control in the Dirichlet boundary conditions, J. Diff. Equations **66** (1987) 340-390.

[13] J. L. Lions, Exact controllability, stabilization and perturbations for distributed systems, SIAM Review **30** (1988) 1-68.

[14] J. P. Quinn & D. L. Russell, Asymptotic stability and energy decay rates for solutions of hyperbolic equations with boundary damping, Proc. Roy. Soc. Edinburgh, Sect. A **77** (1977) 97-127.

[15] J. Rauch & M. E. Taylor, Exponential decay of solutions to hyperbolic equations in bounded domains, Indiana Univ. Math J. **24** (1974) 79-86.

[16] D. L. Russell, Controllability and stabilizability for linear partial differential equations. Recent progress and open questions, SIAM Rev. **20**(1978) 639-739.

[17] E.Zuazua, Uniform stabilization of the wave equation by nonlinear boundary feedback, in preparation.

[18] E. Zuazua, Robustesse du feedback de stabilisation par contrôle frontière, C. R. Acad. Sc. Paris, to appear.

Lecture Notes in Control and Information Sciences

Edited by M. Thoma and A. Wyner

Vol. 81: Stochastic Optimization
Proceedings of the International Conference,
Kiew, 1984
Edited by I. Arkin, A. Shiraev, R. Wets
X, 754 pages, 1986.

Vol. 82: Analysis and Algorithms
of Optimization Problems
Edited by K. Malanowski, K. Mizukami
VIII, 240 pages, 1986.

Vol. 83: Analysis and Optimization
of Systems
Proceedings of the Seventh International
Conference of Analysis and Optimization
of Systems
Antiba, June 26-27, 1986
Edited by A. Bensoussan, J. L. Lions
XVI, 901 pages, 1986.

Vol. 84: System Modelling
and Optimization
Proceedings of the 12th IFIP Conference
Budapest, Hungary, September 2–6, 1985
Edited by A. Prékopa, J. Szelezsán, B. Strazicky
XII, 1046 pages, 1986.

Vol. 85: Stochastic Processes
in Underwater Acoustics
Edited by Charles R. Baker
V, 205 pages, 1986.

Vol. 86: Time Series and
Linear Systems
Edited by Sergio Bittanti
XVII, 243 pages, 1986.

Vol. 87: Recent Advances in
System Modelling and
Optimization
Proceedings of the IFIP-WG 7/1
Working Conference
Santiago, Chile, August 27-31, 1984
Edited by L. Contesse, R. Correa, A. Weintraub
IV, 199 pages, 1987.

Vol. 88: Bruce A. Francis
A Course in H_∞ Control Theory
XI, 156 pages, 1987.

Vol. 88: Bruce A. Francis
A Course in H_∞ Control Theory
X, 150 pages, 1987.
Corrected - 1st printing 1987

Vol. 89: G. K. H. Pang/A. G. J. McFarlane
An Expert System Approach to
Computer-Aided Design
of Multivariable Systems
XII, 223 pages, 1987.

Vol. 90: Singular Perturbations
and Asymptotic Analysis
in Control Systems
Edited by P. Kokotovic,
A. Bensoussan, G. Blankenship
VI, 419 pages, 1987.

Vol. 91 Stochastic Modelling
and Filtering
Proceedings of the IFIP-WG 7/1
Working Conference
Rome, Italy, Decembre 10-14, 1984
Edited by A. Germani
IV, 209 pages, 1987.

Vol. 92: L. T. Grujić, A. A. Martynyuk,
M. Ribbens-Pavella
Large-Scale Systems Stability Under
Structural and Singular Perturbations
XV, 366 pages, 1987.

Vol. 93: K. Malanowski
Stability of Solutions to Convex
Problems of Optimization
IX, 137 pages, 1987.

Vol. 94: H. Krishna
Computational Complexity
of Bilinear Forms
Algebraic Coding Theory and
Applications to Digital
Communication Systems
XVIII, 166 pages, 1987.

Vol. 95: Optimal Control
Proceedings of the Conference on
Optimal Control and Variational Calculus
Oberwolfach, West-Germany, June 15-21, 1986
Edited by R. Bulirsch, A. Miele, J. Stoer
and K. H. Well
XII, 321 pages, 1987.

Vol. 96: H. J. Engelbert/W. Schmidt
Stochastic Differential Systems
Proceedings of the IFIP-WG 7/1
Working Conference
Eisenach, GDR, April 6-13, 1986
XII, 381 pages, 1987.

Lecture Notes in Control and Information Sciences

Edited by M. Thoma and A. Wyner

Vol. 97: I. Lasiecka/R. Triggiani (Eds.)
Control Problems for Systems
Described by Partial Differential Equations
and Applications
Proceedings of the IFIP-WG 7.2
Working Conference
Gainesville, Florida, February 3-6, 1986
VIII, 400 pages, 1987.

Vol. 98: A. Aloneftis
Stochastic Adaptive Control
Results and Simulation
XII, 120 pages, 1987.

Vol. 99: S. P. Bhattacharyya
Robust Stabilization Against
Structured Perturbations
IX, 172 pages, 1987.

Vol. 100: J. P. Zolésio (Editor)
Boundary Control and Boundary Variations
Proceedings of the IFIP WG 7.2 Conference
Nice, France, June 10-13, 1987
IV, 398 pages, 1988.

Vol. 101: P. E. Crouch,
A. J. van der Schaft
Variational and Hamiltonian
Control Systems
IV, 121 pages, 1987.

Vol. 102: F. Kappel, K. Kunisch,
W. Schappacher (Eds.)
Distributed Parameter Systems
Proceedings of the 3rd International Conference
Vorau, Styria, July 6–12, 1986
VII, 343 pages, 1987.

Vol. 103: P. Varaiya, A. B. Kurzhanski (Eds.)
Discrete Event Systems:
Models and Applications
IIASA Conference
Sopron, Hungary, August 3-7, 1987
IX, 282 pages, 1988.

Vol. 104: J. S. Freudenberg/D. P. Looze
Frequency Domain Properties of Scalar
and Multivariable Feedback Systems
VIII, 281 pages, 1988.

Vol. 105: Ch. I. Byrnes/A. Kurzhanski (Eds.)
Modelling and Adaptive Control
Proceedings of the IIASA Conference
Sopron, Hungary, July 1986
V, 379 pages, 1988.

Vol. 106: R. R. Mohler (Editor)
Nonlinear Time Series and
Signal Processing
V, 143 pages. 1988.

Vol. 107: Y. T. Tsay, L.-S. Shieh, St. Barnett
Structural Analysis and Design
of Multivariable Systems
An Algebraic Approach
VIII, 208 pages, 1988.

Vol. 108: K. J. Reinschke
Multivariable Control
A Graph-theoretic Approach
274 pages, 1988.

Vol. 109: M. Vukobratović/R. Stojić
Modern Aircraft Flight Control
VI, 288 pages, 1988.

Vol. 110: In preparation

Vol. 111: A. Bensoussan, J. L. Lions (Eds.)
Analysis and Optimization
of Systems
XIV, 1175 pages, 1988.

Vol. 112: Vojislav Kecman
State-Space Models of Lumped
and Distributed Systems
IX, 280 pages, 1988

Vol. 113: M. Iri, K. Yajima (Eds.)
System Modelling and Optimization
Proceedings of the 13th IFIP Conference
Tokyo, Japan, Aug. 31 – Sept. 4, 1987
IX, 787 pages, 1988.

Vol. 114: A. Bermúdez (Editor)
Control of Partial Differential Equations
Proceedings of the IFIP WG 7.2
Working Conference
Santiago de Compostela, Spain, July 6–9, 1987
IX, 318 pages, 1989

Vol. 115: H.J. Zwart
Geometric Theory for Infinite
Dimensional Systems
VIII, 156 pages, 1989.

Vol. 116: M.D. Mesarovic, Y. Takahara
Abstract Systems Theory
VIII, 439 pages, 1989

Lecture Notes in Control and Information Sciences

Edited by M. Thoma and A. Wyner

Vol. 117: K.J. Hunt
Stochastic Optimal Control Theory
with Application in Self-Tuning Control
X, 308 pages, 1989.

Vol. 118: L. Dai
Singular Control Systems
IX, 332 pages, 1989

Vol. 119: T. Başar, P. Bernhard
Differential Games and Applications
VII, 201 pages, 1989

Vol. 120: L. Trave, A. Titli, A. M. Tarras
Large Scale Systems:
Decentralization, Structure Constraints
and Fixed Modes
XIV, 384 pages, 1989

Vol. 121: A. Blaquière (Editor)
Modeling and Control of Systems
in Engineering, Quantum Mechanics,
Economics and Biosciences
Proceedings of the Bellman Continuum
Workshop 1988, June 13–14, Sophia Antipolis, France
XXVI, 519 pages, 1989

Vol. 122: J. Descusse, M. Fliess, A. Isidori,
D. Leborgne (Eds.)
New Trends in Nonlinear Control Theory
Proceedings of an International
Conference on Nonlinear Systems,
Nantes, France, June 13–17, 1988
VIII, 528 pages, 1989

Vol. 123: C. W. de Silva, A. G. J. MacFarlane
Knowledge-Based Control with
Application to Robots
X, 196 pages, 1989

Vol. 124: A. A. Bahnasawi, M. S. Mahmoud
Control of Partially-Known
Dynamical Systems
XI, 228 pages, 1989

Vol. 125: J. Simon (Ed.)
Control of Boundaries and Stabilization
Proceedings of the IFIP WG 7.2 Conference
Clermont Ferrand, France, June 20–23, 1988
IX, 266 pages, 1989